Quorum Sensing and its Biotechnological Applications

Vipin Chandra Kalia
Editor

Quorum Sensing and its Biotechnological Applications

Springer

Editor
Vipin Chandra Kalia
Department of Chemical Engineering
Konkuk University
Seoul, Republic of Korea

ISBN 978-981-13-4516-6 ISBN 978-981-13-0848-2 (eBook)
https://doi.org/10.1007/978-981-13-0848-2

Softcover re-print of the Hardcover 1st edition 2018

Printed on acid-free paper

This Springer imprint is published by the registered company Springer Nature Singapore Pte Ltd.
The registered company address is: 152 Beach Road, #21-01/04 Gateway East, Singapore 189721, Singapore

Dedicated to my lovely wife Amita

Preface

Microorganisms have long been used in various areas of biotechnology. Microbial diversity and community dynamics provide insights into their potential to degrade organic matter, which otherwise is regarded as environmental pollutants. Associating bioremediation process with energy generation is an economical proposal for reducing pollution and managing biowastes. Biological processes driven by a single bacterial strain are always at the risk of getting contaminated and destabilized leading to lower efficiency. This major limitation can be circumvented through the use of defined cultures, which can withstand adverse conditions, outcompete contaminating microorganisms, and drive the process successfully. In the recent times what has gained importance is the communication among microbes, known as quorum sensing (QS). QS allows a large bacterial population to work together in a coordinated manner to carry out metabolic activities, which individual bacterium cannot. Fascinating information has been generated on understanding the significance of QS. So far, the major objective of studying QS was to understand their role in causing infectious diseases and identifying drug targets to inhibit the process of QS especially the virulence factors. This area of QS in medical processes is still in its incipient stage and it may take some more time before we can exploit it to fight bacterial atrocities. It is advisable that we shift the focus on exploiting QS for other biotechnological applications such as in generating bioproducts, bioenergy, bioremediation, biosensors, health and agricultural activities. It is becoming an integral part of synthetic biology for genetic circuits for producing: (i) novel products, (ii) biosensors, (iii) bioactive molecules, etc. Here, we are covering a few biotechnological applications of QS in Environment, Bioremediation, Energy, Agriculture, and Health sectors. This piece of scientific documentation is intended to explore these diverse possibilities, present scholarly views and opinions, and to serve mankind with novel, innovative, and long-lasting strategies, in the book entitled: *Quorum Sensing and its Biotechnological Applications*. It is the contributions of passionate scientists, who are always working hard to gain insights into the unknown world and have volunteered to share their wonderful knowledge with the curious minds of young researchers and develop strategies for the economic benefits of human beings. This book is a vivid reflection of the sincerity with which scientific minds are dedicated to the welfare of the community. These contributions will help to steer the researchers around the globe into an interesting and healthy future. I have been truly inspired by my author colleagues and associates as this piece of work has taken this shape only because of their faith in me and the constant and untiring support of – my parents – Mr. R.B. Kalia and Late Mrs. Kanta Kalia (parents); Daksh and Bhrigu (sons); Nivedita, Sunita Bhardwaj, Ravi Bhardwaj, Anuradha Dube and Nikhil Dube, Dr Amit Ghosh (mentor), friends – Rup Lal, Hemant J. Purohit, Ramesh C. Kuhad, Tapan K. Adhya, Yogendra Singh, Prince Sharma, Amulya K. Panda, Appa Rao Podile, Sunil Khanna, J.S. Virdi, Pratyoosh Shukla, Sunil Pabbi, Sunil K. Khare. Shashi K. Das, Ramchandra, Gopal Reddy of AMI clan, Jyoti, Neeru, Ritusree, Malabika, Chinoo, Mamta Kapila, Madhurima Kahali, my young friends – Sr. Pastor Hyoungmin Kim, Pastor Elliot Lee, Young Choi, Jasmine Park, Juyeon Oh, Jinny Jeon, Dr. Seulji Lee, Prof. Doo Hwan Kim, Ms Hyesoon Cho, Isabella, Judy Gopal, Manikandan Muthu, Iyyakkannu Sivanesan, Diby Paul. I must also acknowledge the support of my student friends – Sadhana, Mamtesh, Sanjay Patel, Sanjay Yadav, Rahul, Virender, Subhasree, Shikha, Jyotsana, and Priyanka.

Seoul, Republic of Korea Vipin Chandra Kalia

Contents

About the Editor

Dr. Vipin Chandra Kalia is presently working as Professor, National Research Foundation (Korea), Department of Chemical Engineering, Konkuk University, Seoul, Republic of Korea. Previously, he was working as CSIR-Emeritus Scientist. He has been the Chief Scientist and Deputy Director at Microbial Biotechnology and Genomics, CSIR-Institute of Genomics and Integrative Biology, Delhi. He is a Professor, AcSIR, who obtained his M.Sc. and Ph.D. in Genetics from Indian Agricultural Research Institute, New Delhi. He has been elected as Fellow of the National Academy of Sciences (FNASc), Fellow of the National Academy of Agricultural Sciences (FNAAS), and Fellow of the Association of Microbiologists of India (FAMSc). His main areas of research are microbial biodiversity, bioenergy, biopolymers, genomics, microbial evolution, quorum sensing, quorum quenching, drug discovery, and antimicrobials. He has published more than 100 papers in scientific journals such as *Nature Biotechnology*, *Biotechnology Advances*, *Trends in Biotechnology*, *Annual Review of Microbiology*, *Critical Reviews in Microbiology*, *Bioresource Technology*, *PLoS ONE*, *BMC Genomics*, *International Journal of Hydrogen Energy*, and *Gene*. He has authored 28 book chapters. His works have been cited 5276 times with an h index of 39 and an i10 index of 87 (http://scholar.google.co.in/citations?hl=en&user=XaUw-VIAAAAJ). He has edited 12 books: (i) *Quorum Sensing vs Quorum Quenching: A Battle with No End in Sight* (2015), http://link.springer.com/book/10.1007/978-81-322-1982-8; (ii) *Microbial Factories – Biofuels, Waste Treatment: Volume 1* (2015), http://link.springer.com/book/10.1007%2F978-81-322-2598-0; (iii) *Microbial Factories – Biodiversity, Biopolymers, Bioactive Molecules: Volume 2* (2015), http://link.springer.com/book/10.1007%2F978-81-322-2595-9; *Waste Biomass Management – A Holistic Approach* (2017), http://www.springer.com/in/book/9783319495941; (v) *Drug Resistance in Bacteria, Fungi, Malaria, and Cancer* – editors: Arora, Gunjan, Sajid, Andaleeb, Kalia, Vipin Chandra, http://www.springer.com/in/book/9783319486826; (vi) *Microbial Applications Volume 1* Kalia, V. (Ed), Kumar, P. (Ed) (2017), http://www.springer.com/in/book/9783319526652; *Microbial Applications Volume 2* Kalia, V. (Ed) (2017), http://www.springer.com/in/book/9783319526683; (viii) Saini AK, Kalia VC (Eds.) *Metabolic Engineering for Bioactive Compounds: Strategies and Processes* (2017), http://www.springer.com/in/book/9789811055102 (Springer Nature); (ix) Kalia VC, Shouche Y, Purohit HJ, Rahi P (Eds.) *Mining of Microbial Wealth and MetaGenomic* (2017), http://www.springer.com/in/book/9789811057076 (Springer Nature); (x) Purohit HJ, Kalia VC, Vaidya A, Khardenavis AA P (Eds.) *Optimization of Applicability of Bioprocesses* (2018), http://www.springer.com/in/book/9789811068621 (Springer Nature); (xi) Purohit HJ, Kalia VC, More R (Eds.) *Soft Computing for Biological Systems* (2018) (Springer Nature) (In press); and (xii) Kalia VC (Ed.) *Biotechnological Applications of Quorum Sensing Inhibitors* (2018) (Springer Nature). He is presently the editor-in-chief of the *Indian Journal of Microbiology* (2013 through 2021) and editor of *Journal of Microbiology and Biotechnology* (Korea), *International Scholarly Res. Network ISRN Renewable Energy*, *Dataset Papers in Microbiology*, and *PLoS ONE*. He is a life member of the following scientific societies: Society of Biological Chemists of India; Society for Plant Biochemistry and Biotechnology, India; Association of Microbiologists of India; Indian Science Congress Association; BioEnergy Society of India; and the Biotech Research Society of India (BRSI). He has been a member of the American Society for Microbiology (2010–2015). He has been conferred the following awards: (i) Prof. S.R. Vyas Memorial Award, Association of Microbiologists of India (2016); (ii) ASM-IUSSF Indo-US Professorship Program, American Society of Microbiologists, USA (2014); (iii) INSA Bilateral Exchange Programme, Indian National Science Academy, India (2006); (iv) DBT Overseas Associateship, Department of Biotechnology, Government of India (2005–2006); (v) Dr. J.V Bhat Award for Best Paper in Indian J of Microbiology, Association of Microbiologists of India (2012, 2015, 2016, 2017) and Faculty Research Award in Microbiology, 2018. He can be contacted at: vckaliaku@gmail.com

Part I

Environment

Quorum Sensing and Its Inhibition: Biotechnological Applications

1

Vipin Chandra Kalia, Jyotsana Prakash, Shikha Koul, and Subhasree Ray

Abstract

Microorganisms have long been used in various areas of biotechnology. In the recent times what has gained fascination is the communication among microbes, known as Quorum sensing (QS). Fascinating information has been generated on understanding the significance of QS, and its inhibition (QSI), especially in plant, animal and human pathogenesis. Focus has now shifted on exploiting QS and QSIs for biotechnological applications in designing: (i) genetic circuits for producing novel products, (ii) biosensors, (iii) molecules for cancer therapy, etc. Here, we cover a few applications in Health, Agriculture, Aquaculture, Energy and Bioremediation sectors.

Keywords

Aquaculture · Bioenergy · Biofouling · Bioremediation · Cancer · Food · Health · Plants · Pathogens · Quorum sensing

V. C. Kalia (✉)
Molecular Biotechnology Lab, Department of Chemical Engineering, Konkuk University, Seoul, Republic of Korea
e-mail: vckaliaku@gmail.com

J. Prakash · S. Koul · S. Ray
Microbial Biotechnology and Genomics, CSIR – Institute of Genomics and Integrative Biology (IGIB), Delhi University Campus, Delhi, India

Academy of Scientific & Innovative Research (AcSIR), New Delhi, India

V. C. Kalia (ed.), *Quorum Sensing and its Biotechnological Applications*,
https://doi.org/10.1007/978-981-13-0848-2_1

Abbreviations

3OC12HSL	N-(3-oxododecanoyl)-L-HSL
3OC6HSL	N-(3-oxohexanoyl)-L-HSL
3OC8HSL	N-(3-oxooctanoyl)-L-HSL
3OHC6HSL	N-(3-hydroxyhexanoyl)-L-HSL
AHL	Acylhomoserine lactone
AI	Autoinducer
AIP	Autoinducing peptides
C10-HSL	N-decanoyl-L-HSL
C12HSL	N-dodecanoyl-L-HSL
C12HSL	N-dodecanoyl-L-HSL
C4HSL	N-butanoyl-L-homoserine lactone
C6HSL	N-hexanoyl-L-HSL
HSL	Homoserine lactone

1.1 Introduction

Microbial associations exist with plants (rhizosphere and phyllosphere), animals and human beings (skin surface and gut) (Doebeli and Ispolatov 2010; Ryall et al. 2012). A few of these associations especially the pathogenic, lead to economic losses. Efforts are being made to inhibit microbial growth and pathogenicity and restrict the damage. The discovery of antibiotics was hailed as a boon and perceived as a solution to fight off all bacterial attacks. However, bacteria were quick to respond to this threat to render the novel drug – antibiotic – ineffective (Davies and Davies 2010). The pharmaceutical industries are getting apprehensive and are not keen on investing in R&D for developing novel antibiotics. On the other hand, scientists are looking for novel and innovative ways to deal with pathogenic microbes.

Infectious diseases are generally caused by biofilm forming microbes, through a cell density dependent phenomenon – quorum sensing (QS). Biofilm shield bacteria, which can now resist antibiotic concentrations, up to 1000 times more than those which are enough to kill their free living counterparts (Rasmussen and Givskov 2006). QS allows bacteria to sense their neighbouring cell density through the release of signal molecules, leading to the expression of virulent behaviour. Inhibiting the synthesis or interaction of signal molecules with receptors and their transcription, also known as QS inhibition (QSI) can repress virulence and the bioactive molecules so employed can act as drugs to fight diseases (Kalia 2013). On the other hand, this biofilm formation can also be exploited in scenarios like bioremediation, where it can act as "immobilization" support and ensure high cell density.

1.2 Biotechnological Applications of Quorum Sensing Systems

A few areas, where QS has been found to have the potential to generate bioproducts of high values include bioenergy, waste treatment, food preservatives, biosensors, health, and agricultural activities, etc.

1.2.1 Bio-energy

Bio-hydrogen (H_2) has been recognised as the cleanest fuel of the future. Microbes have an ability to produce H_2 from different substrates including bio-wastes (Patel and Kalia 2013; Yasin et al. 2013). An innovative strategy can be to combine H_2-production with biofilm formation abilities of potential H_2-producers: *Bacillus, Clostridium, Streptococcus, Sinorhizobium, Enterobacter, Klebsiella, Caldicellulosiruptor* and *Escherichia* (Kalia and Purohit 2008; Pawar et al. 2015). QS has also been reported to be of use in bioethanol and biodiesel production. Generation of bioelectricity by Microbial Fuel Cells and H_2 and biomethane by Microbial Electrolysis Cells require strong biofilms (Zhou et al. 2013; Hu et al. 2015).

1.2.2 Bioremediation

Bioremediation process is limited by the insolubility and hence availability of the pollutants to the bacteria. Bio surfactants – Rhamnolipids and Surfactins produced through QS, find use in removal of oil and toxic metals from contaminated sites and soils (Chakraborty and Das 2014; Oslizlo et al. 2014). QS mediated processes like denitrification, ammonium oxidation and exoenzyme production are reported to enhance biodegradation (Shukla et al. 2014; Yong et al. 2015).

1.2.3 Food and Health

QS signalling molecules – AIPs -Nisin, Bacteriocins or lantibiotics, produced by *Lactococcus lactis, Bacillus,* etc. have antimicrobial properties useful for pharma industries and as preservatives in food industry (Nishie et al. 2012; Dobson et al. 2012; Camargo et al. 2016). Various other lantibiotics including cinnamycin, plantaricin C, pep5, epidermin, MU1140, hold a potential as novel antibiotics with some under clinical trials (Dischinger et al. 2014; Field et al. 2015; Li and Tian 2015).

1.2.4 Detecting Metals and Pathogens

QS has also been used to design plasmid biosensors by integrating AHL receptors in RP4 vector, for detecting microbes in the environment (Choudhary and Schmidt-Dannert 2012; Hsu et al. 2016).

1.2.5 Cancer Therapy

Certain oligopeptides of gut microbiota are reported to promote the angiogenesis having a potential to influence metastasis. The toxins, cytokines, tumor antigens can be exploited as a novel and effective treatment of cancer (Hong et al. 2014; Kwon et al. 2014; De Spiegeleer et al. 2015; Wynendaele et al. 2015).

1.2.6 Industrial Products

Microbial QSS produces various extracellular products such as enzymes, rhamnolipids, isobutanol and 1,3-propanediol, 2,3-butanediol with commercial applications (Bernstein and Carlson 2012; Liu and Lu 2015; Chang et al. 2015).

1.2.7 Genetic Devices

Engineering based techniques have long been used in biology to construct synthetic gene networks (Davis et al. 2015). Engineered LuxI/LuxR system fused with antigen proteins have been used to produce vaccines (Choudhary and Schmidt-Dannert 2012; Sturbelle et al. 2013; Chu et al. 2015).

1.3 Biotechnological Applications of QSI

The topic has been reviewed during the last few years (Kalia and Purohit 2011; Kalia 2013). The information described below emphasizes on the developments on the biotechnological applications of QSI, during the last 3 years.

1.3.1 Food Industry

Fruits and vegetables processing industries face economical as well as safety issues due to bacteria which are responsible for problems like. Plant and fruit extracts known as phytochemicals including limonoids, flavanoids, polyphenols, furocoumarins, phenolics, etc. have been tested and reported to act as QSIs to prevent food poisoning and spoilage (Kerekes et al. 2015; Zhu et al. 2015; Oliveira et al. 2016; Venkadesaperumal et al. 2016). Essential oils such as from ginger, eucalyptus, rose and tea tree are reported to have anti-QS effects and may find use as sanitizers and as food preservatives (Kerekes et al. 2015).

1.3.2 Aquaculture

Aquaculture being an important food producing industry world-wide suffers heavy losses due to aquatic pathogens like *Vibrio* spp.and *Aeromonas* spp. killing fishes, prawns, shrimps and molluscs (Niu et al. 2014; Zhao et al. 2014). Halogenated furanones have been holding great promise in protecting fishes (Defoirdt et al. 2007; Benneche et al. 2011). However, thiophenones are proving to be less toxic (Defoirdt et al. 2012; Yang et al. 2015). Supplementation of fish feed with variant of lactonases and peptides have been reported to inhibit QS mediated pathogenicity (Zhang et al. 2015; Sun and Zhang 2016).

1.3.3 Health Care

In infectious diseases, such as cystic fibrosis, bacterial endocarditis, chronic prostatitis, oral cavities, etc., bacteria express their virulent behaviour through QS mediated biofilm formation. Recent efforts have been focussed on developing a strategy to effectively disarm a pathogen through the use of QSIs. Bioactive molecules and nanoparticles are being searched aggressively for they have the potential usage as antimicrobials and QSIs (Gui et al. 2014; Arasu et al. 2015; Balakrishnan et al. 2015; Bandyopadhyay et al. 2015; Bose and Chatterjee 2015; Dobrucka and Długaszewska 2015; Go et al. 2015; Szweda et al. 2015; Begum et al. 2016; Wadhwani et al. 2016; Ahiwale et al. 2017; Azman et al. 2017; Saini and Keum 2017).

1.3.3.1 QSIs from Microbes

Prokaryotes being easy to culture and handle, become a choice of many when it comes to the need of producing novel health care products (Karumuri et al. 2015; Shiva Krishna et al. 2015; Jeyanthi and Velusamy 2016; Varsha et al. 2016; Sanchart et al. 2017; Thakur et al. 2017). Norspermidine, Maniwamycins, Solonamides play role in QSI – inhibit biofilm formation in *S. aureus*, *S. epidermis* and *E. coli* (Nesse et al. 2015; Baldry et al. 2016; Qu et al. 2016). Efforts have also been made to develop synthetic compounds – Thiazolidinedione (TZD) that could inhibit QS in gram negative as well as in gram positive bacteria (Lidor et al. 2015).

1.3.3.2 QSIs of Plant Origin

Epigallocatechin-3-gallate (EGCG), a major catechin in *Camellia sinensis,* green tea leaves is known for its antioxidative, anticancerous and antimicrobial properties (Yin et al. 2015; Fournier-Larente et al. 2016). Secondary metabolites produced by plants – dietary phytochemicals – ajoene, iberin, limonoids, furocoumarins, through their antimicrobial and QSI activities are known to provide health benefits (Kazemian et al. 2015; Sarkar et al. 2015; Brackman et al. 2016). Plant extracts such as *Piper betle* extract, vanillin, flavanoids, etc. also inhibit biofilm formation and thus hold the potential to be used as antifouling agents (Igarashi et al. 2015; Siddiqui et al. 2015). Quercetin, a flavanol is found in various fruits and vegetables such as apples, grapes, onion and tomatoes and finds use as anticancerous, antiapoptotic and anti-oxidative agent (Ouyang et al. 2016).

1.3.3.3 QSIs of Animal Origin

Meat extracts such as those from turkey and beef patties, chicken breast and beef steak have shown AI-2 signalling inhibition (Lu et al. 2004; Soni et al. 2008). Cattle's milk has anti-QSIs compounds (Abolghait et al. 2016; Hernández-Saldaña et al. 2016). Mare colostrum inhibits biofilm forming and virulence factors producing characteristic of *S. aureus* (Srivastava et al. 2015).

1.3.4 Anti-biofouling

Quenching the bacterial communication to target bio-fouling might be a boon to a number of sectors including aqua industries, naval departments and water treatment plants (Ponnusamy et al. 2013; Cheong et al. 2014; Lee et al. 2014; Kim et al. 2015; Wu et al. 2015).

1.3.5 Agriculture

Rhizospheric bacteria, certain epiphytes, essential oils, facilitates easy clearance of pathogen (Corral-Lugo et al. 2016; Des Essarts et al. 2016). Biocontrol agents *Bacillus* spp. A24 as well as *Pseudomonas fluorescens* expressing *aiiA* gene showed reduced rot and gall symptoms by the phytopathogens (Helman and Chernin 2015; Sánchez-Elordi et al. 2015; Kang et al. 2016).

1.4 Synergism Between Antibiotics and QSIs

Another encouraging feature in this battle against pathogens, is the potential synergism between QSI and antibiotics (Zhang et al. 2011; Ma et al. 2012).

1.5 Fais Attention à QSI

A few studies have shown evidence of emergence of microbial resistance to QSI (Kalia et al. 2014; García-Contreras et al. 2016; Koul et al. 2016). Hence, we may need to be more cautious and look for QSI which are QS signal independent (Lee et al. 2015).

1.6 Field Trials

In spite of the fact that mechanisms of QS and QSI have been widely studied, their testing at field and clinical level has been limited (Reuter et al. 2016). Application of QSI under field conditions is necessary before the same can be extended for commercialization (Kim et al. 2014, 2017).

1.6.1 Protecting Plants

Chromobacterium sp. has been successfully employed to control ginseng and pepper from *Alternaria* and *Phytophthora* infections (Kim et al. 2008, 2010, 2014). The pre-treated plants of rice, tomato and wheat were protected from their respective pathogens in a large scale trial (Kim et al. 2017).

1.6.2 Drinking Water

In order to tackle the trouble caused by biofouling encountered during drinking water treatment, QSI enzyme-acylase coated nanofiltration membranes and by using encapsulated bacteria were shown to be stable and enzymatic activity (Kim et al. 2011; Jahangir et al. 2012; Maqbool et al. 2015; Lee et al. 2016).

1.6.3 Health Sector

Short term clinical trial on guinea pig and 24 h trial on human volunteers were tried with QSIs – fimbrolides or furanones. Clinical trial on employing QSI as therapeutics for treating cystic fibrosis in human patients was reported by Prof. Givskov and his team (Smyth et al. 2010). A QSIs formulation patented by Colgate-palmolive for oral care to inhibit biofilm formations is an encouraging sign (Grandclément et al. 2016).

1.7 Opinion

Human trials on the effects of QSIs on infectious diseases have been conducted on a limited scale. We may have to wait for some more time till the confidence level goes up. In this scenario, a stop-gap arrangement has to be made to provide QSIs through: (i) the use of dietary sources rich in phyto-nutrients, (ii) use as inducers to cause pseudo-induction of QS and make bacteria susceptible to immune system, and eradicate pathogens with low antibiotic doses, and (iii) attack those bacteria which promote QSS in pathogens.

Acknowledgements This work was supported by Brain Pool grant (NRF-2018H1D3A2001746) by National Research Foundation of Korea (NRF) to work at Konkuk University (VCK).

References

Abolghait SK, Garbaj AM, Moawad AA (2016) Raw cow's milk relatively inhibits quorum sensing activity of Chromobacterium violaceum in comparison to raw she-camel's milk. Open Vet J 1:35–38

Ahiwale SS, Bankar AV, Tagunde S, Kapadnis BP (2017) A bacteriophage mediated gold nanoparticle synthesis and their anti-biofilm activity. Indian J Microbiol 57:188–194. https://doi.org/10.1007/s12088-017-0640-x

Arasu MV, Al-Dhabi NA, Rejiniemon TS, Lee KD, Huxley VAJ, Kim DH, Duraipandiyan V, Karuppiah P, Choi KC (2015) Identification and characterization of *Lactobacillus brevis* P68 with antifungal, antioxidant and probiotic functional properties. Indian J Microbiol 55:19–28. https://doi.org/10.1007/s12088-014-0495-3

Azman CA-S, Othman I, Fang C-M, Chan K-G, Goh B-H, Lee L-H (2017) Antibacterial, anticancer and neuroprotective activities of rare actinobacteria from mangrove forest soils. Indian J Microbiol 57:177–187. https://doi.org/10.1007/s12088-016-0627-z

Balakrishnan D, Bibiana AS, Vijayakumar A, Santhosh RS, Dhevendaran K, Nithyanand P (2015) Antioxidant activity of bacteria associated with the marine sponge *Tedania anhelans*. Indian J Microbiol 55:13–18. https://doi.org/10.1007/s12088-014-0490-8

Baldry M, Kitir B, Frøkiær H, Christensen SB, Taverne N, Meijerink M, Franzyk H, Olsen CA, Wells JM, Ingmer H (2016) The *agr* inhibitors Solonamide B and analogues alter immune responses to *Staphylococccus aureus* but do not exhibit adverse effects on immune cell functions. PLoS One 11:e0145618. https://doi.org/10.1371/journal.pone.0145618

Bandyopadhyay P, Mishra S, Sarkar B, Swain SK, Pal A, Tripathy PP, Ojha SK (2015) Dietary *Saccharomyces cerevisiae* boosts growth and immunity of IMC *Labeo rohita* (Ham.) juveniles. Indian J Microbiol 55:81–87. https://doi.org/10.1007/s12088-014-0500-x

Begum IF, Mohankumar R, Jeevan M, Ramani K (2016) GC–MS analysis of bioactive molecules derived from *Paracoccus pantotrophus* FMR19 and the antimicrobial activity against bacterial pathogens and MDROs. Indian J Microbiol 56:426–432. https://doi.org/10.1007/s12088-016-0609-1

Benneche T, Herstad G, Rosenberg ML, Assev S, Scheie AA (2011) Facile synthesis of 5-(alkylidene)thiophen-2(5H)-ones. A new class of antimicrobial agents. RSC Adv 1:323–332. https://doi.org/10.1039/C1RA00254F

Bernstein HC, Carlson RP (2012) Microbial consortia engineering for cellular factories: in vitro to in silico systems. Comput Struct Biotechnol J 3:e201210017. https://doi.org/10.5936/csbj.201210017

Bose D, Chatterjee S (2015) Antibacterial activity of green synthesized silver nanoparticles using Vasaka (*Justicia adhatoda* L.) leaf extract. Indian J Microbiol 55:163–167. https://doi.org/10.1007/s12088-015-0512-1

Brackman G, Breyne K, De Rycke R, Vermote A, Van Nieuwerburgh F, Meyer E, Van Calenbergh S, Coenye T (2016) The quorum sensing inhibitor hamamelitannin increases antibiotic susceptibility of *Staphylococcus aureus* Biofilms by affecting peptidoglycan biosynthesis and eDNA release. Sci Rep 6:20321. https://doi.org/10.1038/srep20321

Camargo AC, de Paula OA, Todorov SD, Nero LA (2016) In vitro evaluation of bacteriocins activity against *Listeria monocytogenes* biofilm formation. Appl Biochem Biotechnol 178:1239–1251. https://doi.org/10.1007/s12010-015-1941-3

Chakraborty J, Das S (2014) Biosurfactant-based bioremediation of toxic metals. In: Das S (ed) Microbial biodegradation and bioremediation. Elsevier, Philadelphia, pp 167–202. https://doi.org/10.1016/B978-0-12-800021-2.00007-8

Chang MW, Lo TMS, Poh CL (2015) Expression construct for sensing cell density and substrate availability and its use in conversion of hydroxycinnamic acids. Patent: US 20150176039 A1

Cheong WS, Kim SR, Oh HS, Lee SH, Yeon KM, Lee CH, Lee JK (2014) Design of quorum quenching microbial vessel to enhance cell viability for biofouling control in membrane bioreactor. J Microbiol Biotechnol 24:97–105. https://doi.org/10.4014/jmb.1311.11008

Choudhary S, Schmidt-Dannert C (2012) Quorum sensing in microbial biosensors. In: Patel VB (ed) Biosensors and environmental health. CRC Press, Boca Raton, pp 191–208. https://doi.org/10.1201/b12775-12

Chu T, Ni C, Zhang L, Wang Q, Xiao J, Zhang Y, Liu Q (2015) A quorum sensing-based in vivo expression system and its application in multivalent bacterial vaccine. Microb Cell Factories 14:37. https://doi.org/10.1186/s12934-015-0213-9

Corral-Lugo A, Daddaoua A, Ortega A, Espinosa-Urgel M, Krell T (2016) Rosmarinic acid is a homoserine lactone mimic produced by plants that activates a bacterial quorum-sensing regulator. Sci Signal 9:ra1. https://doi.org/10.1126/scisignal.aaa8271

Davies J, Davies D (2010) Origins and evolution of antibiotic resistance. Microbiol Mol Biol Rev 74:417–433. https://doi.org/10.1128/MMBR.00016-1

Davis RM, Muller RY, Haynes KA (2015) Can the natural diversity of quorum-sensing advance synthetic biology? Front Bioeng Biotechnol 3:30. https://doi.org/10.3389/fbioe.2015.00030

De Spiegeleer B, Verbeke F, D'Hondt M, Hendrix A, Van De Wiele C, Burvenich C, Peremans K, De Wever O, Bracke M, Wynendaele E (2015) The quorum sensing peptides PhrG, CSP and EDF promote angiogenesis and invasion of breast cancer cells in vitro. PLoS One 10:e0119471. https://doi.org/10.1371/journal.pone.0119471

Defoirdt T, Miyamoto CM, Wood TK, Meighen EA, Sorgeloos P, Verstraete W, Bossier P (2007) The natural furanone (5Z)4-bromo-5-(bromomethylene)-3-butyl-2(5H)-furanone disrupts quorum sensing-regulated gene expression in *Vibrio harveyi* by decreasing the DNA-binding activity of the transcriptional regulator protein LuxR. Environ Microbiol 9:2486–2495. https://doi.org/10.1111/j.1462-2920.2007.01367.x

Defoirdt T, Benneche T, Brackman G, Coenye T, Sorgeloos P, Aamdal Scheie A (2012) A quorum sensing-disrupting brominated thiophenone with a promising therapeutic potential to treat luminescent vibriosis. PLoS One 7:e41788. https://doi.org/10.1371/journal.pone.0041788

Des Essarts YR, Cigna J, Quêtu-Laurent A, Caron A, Munier E, Beury-Cirou A, Hélias V, Faure D (2016) Biocontrol of the potato blackleg and soft rot diseases caused by *Dickeya dianthicola*. Appl Environ Microbiol 82:268–278. https://doi.org/10.1128/AEM.02525-15

Dischinger J, Chipalu SB, Bierbaum G (2014) Lantibiotics: promising candidates for future applications in health care. Int J Med Microbiol 304:51–62. https://doi.org/10.1016/j.ijmm.2013.09.003

Dobrucka R, Długaszewska J (2015) Antimicrobial activities of silver nanoparticles synthesized by using water extract of *Arnicae anthodium*. Indian J Microbiol 55:168–174. https://doi.org/10.1007/s12088-015-0516-x

Dobson A, Cotter PD, Ross RP, Hill C (2012) Bacteriocin production: a probiotic trait? Appl Environ Microbiol 78:1–6. https://doi.org/10.1128/AEM.05576-11

Doebeli M, Ispolatov I (2010) Complexity and diversity. Science 328:494–497. https://doi.org/10.1126/science.118746

Field D, Cotter PD, Ross RP, Hill C (2015) Bioengineering of the model lantibiotic nisin. Bioengineered 6:187–192. https://doi.org/10.1080/21655979.2015.1049781

Fournier-Larente J, Morin MP, Grenier D (2016) Green tea catechins potentiate the effect of antibiotics and modulate adherence and gene expression in *Porphyromonas gingivalis*. Arch Oral Biol 65:35–43. https://doi.org/10.1016/j.archoralbio.2016.01.014

García-Contreras R, Maeda T, Wood TK (2016) Can resistance against quorum-sensing interference be selected? ISME J 10:4–10. https://doi.org/10.1038/ismej.2015.84

Go T-H, Cho K-S, Lee S-M, Lee O-M, Son H-J (2015) Simultaneous production of antifungal and keratinolytic activities by feather-degrading *Bacillus subtilis* S8. Indian J Microbiol 55:66–73. https://doi.org/10.1007/s12088-014-0502-8

Grandclément C, Tannières M, Moréra S, Dessaux Y, Faure D (2016) Quorum quenching: role in nature and applied developments. FEMS Microbiol Rev 40:86–116. https://doi.org/10.1093/femsre/fuv038

Gui Z, Wang H, Ding T, Zhu W, Zhuang X, Chu W (2014) Azithromycin reduces the production of α-hemolysin and biofilm formation in *Staphylococcus aureus*. Indian J Microbiol 54:114–117. https://doi.org/10.1007/s12088-013-0438-4

Helman Y, Chernin L (2015) Silencing the mob: disrupting quorum sensing as a means to fight plant disease. Mol Plant Pathol 16:316–329. https://doi.org/10.1111/mpp.12180

Hernández-Saldaña OF, Valencia-Posadas M, de la Fuente-Salcido NM, Bideshi DK, Barboza-Corona JE (2016) Bacteriocinogenic bacteria isolated from raw goat milk and goat cheese produced in the Center of México. Indian J Microbiol 56:301–308. https://doi.org/10.1007/s12088-016-0587-3

Hong H, Lim D, Kim GJ, Park SH, Sik Kim H, Hong Y, Choy HE, Min JJ (2014) Targeted deletion of the *ara* operon of *Salmonella typhimurium* enhances L-arabinose accumulation and drives PBAD promoted expression of anti-cancer toxins and imaging agents. Cell Cycle 13:3112–3120. https://doi.org/10.4161/15384101.2014.949527

Hsu Y, Chen BK, Hu RH, Chen BS (2016) Systematic design of a quorum sensing-based biosensor for enhanced detection of metal ion in *Escherichia coli*. IEEE Trans Biomed Circuits Syst 10:593–601. https://doi.org/10.1109/TBCAS.2015.2495151

Hu Y, Yang Y, Katz E, Song H (2015) Programming the quorum sensing-based AND gate in *Shewanella oneidensis* for logic gated-microbial fuel cells. Chem Commun 51:4184–4187. https://doi.org/10.1039/C5CC00026B

Igarashi Y, Gohda F, Kadoshima T, Fukuda T, Hanafusa T, Shojima A, Nakayama J, Bills GF, Peterson S (2015) Avellanin C, an inhibitor of quorum-sensing signaling in *Staphylococcus aureus*, from *Hamigera ingelheimensis*. J Antibiot (Tokyo) 68:707–710. https://doi.org/10.1038/ja.2015.50

Jahangir D, Oh SH, Kim SR, Park PK, Lee CH, Lee JK (2012) Specific location of encapsulated quorum quenching bacteria for biofouling control in an external submerged membrane bioreactor. J Membr Sci 411–412:130–136. https://doi.org/10.1016/j.memsci.2012.04.022

Jeyanthi V, Velusamy P (2016) Anti-methicillin resistant *Staphylococcus aureus* compound isolation from halophilic *Bacillus amyloliquefaciens* MHB1 and determination of its mode of action using electron microscope and flow cytometry analysis. Indian J Microbiol 56:148–157. https://doi.org/10.1007/s12088-016-0566-8

Kalia VC (2013) Quorum sensing inhibitors: an overview. Biotechnol Adv 31:224–245. https://doi.org/10.1016/j.biotechadv.2012.10.004

Kalia VC, Purohit HJ (2008) Microbial diversity and genomics in aid of bioenergy. J Ind Microbiol Biot 35:403–419. https://doi.org/10.1007/s10295-007-0300-y

Kalia VC, Purohit HJ (2011) Quenching the quorum sensing system: potential antibacterial drug targets. Crit Rev Microbiol 37:121–140. https://doi.org/10.3109/1040841X.2010.532479

Kalia VC, Wood TK, Kumar P (2014) Evolution of resistance to quorum-sensing inhibitors. Microb Ecol 68:13–23. https://doi.org/10.1007/s00248-013-0316-y

Kang JE, Han JW, Jeon BJ, Kim BS (2016) Efficacies of quorum sensing inhibitors, piericidin A and glucopiericidin A, produced by *Streptomyces xanthocidicus* KPP01532 for the control of potato soft rot caused by *Erwinia carotovora* subsp. atroseptica. Microbiol Res 184:32–41. https://doi.org/10.1016/j.micres.2015.12.005

Karumuri S, Singh PK, Shukla P (2015) In silico analog design for terbinafine against *Trichophyton rubrum*: a preliminary study. Indian J Microbiol 55:333–340. https://doi.org/10.1007/s12088-015-0524-x

Kazemian H, Ghafourian S, Heidari H, Amiri P, Yamchi JK, Shavalipour A, Houri H, Maleki A, Sadeghifard N (2015) Antibacterial, anti-swarming and anti-biofilm formation activities of *Chamaemelum nobile* against *Pseudomonas aeruginosa*. Rev Soc Bras Med Trop 48:432–436. https://doi.org/10.1590/0037-8682-0065-2015

Kerekes EB, Vidács A, Jenei JT, Gömöri C, Takó M, Chandrasekaran M, Kadaikunnan S, Alharbi NS, Krisch J, Vágvölgyi C (2015) Essential oils against bacterial biofilm formation and quorum sensing of food-borne pathogens and spoilage microorganisms. In: Méndez-Vilas A (ed) The battle against microbial pathogens: basic science, technological advances and educational programs, pp 429–437

Kim YC, Jung H, Kim KY, Park SK (2008) An effective biocontrol bioformulation against *Phytophthora* blight of pepper using growth mixtures of combined chitinolytic bacteria under different field conditions. Eur J Plant Pathol 120:373–382. https://doi.org/10.1007/s10658-007-9227-4

Kim YC, Lee JH, Bae YS, Sohn BK, Park SK (2010) Development of effective environmentally-friendly approaches to control *Alternaria* blight and anthracnose diseases of Korean ginseng. Eur J Plant Pathol 127:443–450. https://doi.org/10.1007/s10658-010-9610-4

Kim JH, Choi DC, Yeon KM, Kim SR, Lee CH (2011) Enzyme-immobilized nanofiltration membrane to mitigate biofouling based on quorum quenching. Environ Sci Technol 45:1601–1607. https://doi.org/10.1021/es103483j

Kim HJ, Choi HS, Yang SY, Kim IS, Yamaguchi T, Sohng JK, Park SK, Kim JC, Lee CH, McSpadden Gardener BB, Kim YC (2014) Both extracellular chitinase and a new cyclic lipopeptide, chromobactomycin, contribute to the biocontrol activity of *Chromobacterium* sp. C61. Mol Plant Pathol 15:122–132. https://doi.org/10.1111/mpp.12070

Kim SR, Lee KB, Kim JE, Won YJ, Yeon KM, Lee CH, Lim DJ (2015) Macroencapsulation of quorum quenching bacteria by polymeric membrane layer and its application to MBR for biofouling control. J Membr Sci 473:109–117. https://doi.org/10.1016/j.memsci.2014.09.009

Kim IS, Yang SY, Park SK, Kim YC (2017) Quorum sensing is a key regulator for the antifungal and biocontrol activity of chitinase-producing *Chromobacterium* sp. C61. Mol Plant Pathol 18:134–140. https://doi.org/10.1111/mpp.12379

Koul S, Prakash J, Mishra A, Kalia VC (2016) Potential emergence of multi-quorum sensing inhibitor resistant (MQSIR) bacteria. Indian J Microbiol 56:1–18. https://doi.org/10.1007/s12088-015-0558-0

Kwon SY, Jiang SN, Zheng JH, Choy HE, Min JJ (2014) *Rhodobacter sphaeroides*, a novel tumor-targeting bacteria that emits natural near-infrared fluorescence. Microbiol Immunol 58:172–179. https://doi.org/10.1111/1348-0421.12134

Lee B, Yeon KM, Shim J, Kim SR, Lee CH, Lee J, Kim J (2014) Effective antifouling using quorum-quenching acylase stabilized in magnetically-separable mesoporous silica. Biomacromolecules 15:1153–1159. https://doi.org/10.1021/bm401595q

Lee JH, Wood TK, Lee J (2015) Roles of indole as an interspecies and interkingdom signaling molecule. Trends Microbiol 23:707–718. https://doi.org/10.1016/j.tim.2015.08.001

Lee S, Park SK, Kwon H, Lee SH, Lee K, Nahm CH, Jo SJ, Oh HS, Park PK, Choo KH, Lee CH, Yi T (2016) Crossing the border between laboratory and field: bacterial quorum quenching for anti-biofouling strategy in an MBR. Environ Sci Technol 50:1788–1795. https://doi.org/10.1021/acs.est.5b04795

Li YH, Tian XL (2015) An alternative strategy as quorum-sensing inhibitor: pheromone-guided antimicrobial peptides. In: Kalia VC (ed) Quorum sensing vs quorum quenching: a battle with no end in sight. Springer, New Delhi, pp 327–334. https://doi.org/10.1007/978-81-322-1982-8_26

Lidor O, Al-Quntar A, Pesci EC, Steinberg D (2015) Mechanistic analysis of a synthetic inhibitor of the *Pseudomonas aeruginosa* LasI quorum-sensing signal synthase. Sci Rep 23:16569. https://doi.org/10.1038/srep16569

Liu H, Lu T (2015) Autonomous production of 1,4- butanediol via a de novo biosynthesis pathway in engineered *Escherichia coli*. Metab Eng 29:135–141. https://doi.org/10.1016/j.ymben.2015.03.009

Lu LG, Hume MF, Pillai SD (2004) Autoinducer-2-like activity associated with foods and its interaction with food additives. J Food Prot 67:1457–1462

Ma L, Liu X, Liang H, Che Y, Chen C, Dai H, Yu K, Liu M, Ma L, Yang CH, Song F (2012) Effects of 14-alpha-lipoyl andrographolide on quorum sensing in *Pseudomonas aeruginosa*. Antimicrob Agents Chemother 56:6088–6094. https://doi.org/10.1128/AAC.01119-12

Maqbool T, Khan SJ, Waheed H, Lee CH, Hashmi I, Iqbal H (2015) Membrane biofouling retardation and improved sludge characteristics using quorum quenching bacteria in submerged membrane bioreactor. J Membrane Sci 483:75–83. https://doi.org/10.1016/j.memsci.2015.02.011

Nesse LL, Berg K, Vestby LK (2015) Effects of norspermidine and spermidine on biofilm formation by potentially pathogenic *Escherichia coli* and *Salmonella enterica* wild-type strains. Appl Environ Microbiol 81:2226–2232. https://doi.org/10.1128/AEM.03518-14

Nishie M, Nagao J, Sonomoto K (2012) Antibacterial peptides "bacteriocins": an overview of their diverse characteristics and applications. Biocontrol Sci 17:1–16. https://doi.org/10.4265/bio.17.1

Niu Y, Defoirdt T, Baruah K, Van de Wiele T, Dong S, Bossier P (2014) *Bacillus* sp. LT3 improves the survival of gnotobiotic brine shrimp (*Artemia franciscana*) larvae challenged with *Vibrio campbellii* by enhancing the innate immune response and by decreasing the activity of shrimp-associated vibrios. Vet Microbiol 173:279–288. https://doi.org/10.1016/j.vetmic.2014.08.007

Oliveira BD, Rodrigues AC, Cardoso BM, Ramos AL, Bertoldi MC, Taylor JG, da Cunha LR, Pinto UM (2016) Antioxidant, antimicrobial and anti-quorum sensing activities of *Rubus rosaefolius* phenolic extract. Ind Crop Prod 84:59–66. https://doi.org/10.1016/j.indcrop.2016.01.037

Oslizlo A, Stefanic P, Dogsa I, Mandic-Mulec I (2014) Private link between signal and response in *Bacillus subtilis* quorum sensing. Proc Natl Acad Sci U S A 111:1586–1591. https://doi.org/10.1073/pnas.1316283111

Ouyang J, Sun F, Feng W, Sun Y, Qiu X, Xiong L, Liu Y, Chen Y (2016) Quercetin is an effective inhibitor of quorum sensing, biofilm formation and virulence factors in *Pseudomonas aeruginosa*. J Appl Microbiol 120:966–974. https://doi.org/10.1111/jam.13073

Patel SKS, Kalia VC (2013) Integrative biological hydrogen production: an overview. Indian J Microbiol 53:3–10. https://doi.org/10.1007/s12088-012-0287-6

Pawar SS, Vongkumpeang T, Grey C, van Niel Ed WJ (2015) Biofilm formation by designed co-cultures of *Caldicellulosiruptor* species as a means to improve hydrogen productivity. Biotechnol Biofuels 8:19. https://doi.org/10.1186/s13068-015-0201-7

Ponnusamy K, Kappachery S, Thekeettle M, Song JH, Kweon JH (2013) Anti-biofouling property of vanillin on *Aeromonas hydrophila* initial biofilm on various membrane surfaces. World J Microbiol Biotechnol 29:1695–1703. https://doi.org/10.1007/s11274-013-1332-2

Qu L, She P, Wang Y, Liu F, Zhang D, Chen L, Luo Z, Xu H, Qi Y, Wu Y (2016) Effects of norspermidine on *Pseudomonas aeruginosa* biofilm formation and eradication. Microbiol Open 5:402–412. https://doi.org/10.1002/mbo3.338

Rasmussen TB, Givskov M (2006) Quorum-sensing inhibitors as anti-pathogenic drugs. Int J Med Microbiol 296:149–161. https://doi.org/10.1016/j.ijmm.2006.02.005

Reuter K, Steinbach A, Helms V (2016) Interfering with bacterial quorum sensing. Perspect Med Chem 8:1–15. https://doi.org/10.4137/PMC.S13209

Ryall B, Eydallin G, Ferenci T (2012) Culture history and population heterogeneity as determinants of bacterial adaptation: the adaptomics of a single environmental transition. Microbiol Mol Biol Rev 76:597–625. https://doi.org/10.1128/MMBR.05028-11

Saini RK, Keum Y-S (2017) Progress in microbial carotenoids production. Indian J Microbiol 57:129–130. https://doi.org/10.1007/s12088-016-0637-x

Sanchart C, Rattanaporn O, Haltrich D, Phukpattaranont P, Maneerat S (2017) *Lactobacillus futsaii* CS3, a new GABA-producing strain isolated from Thai fermented shrimp (*Kung-Som*). Indian J Microbiol 57:211–217. https://doi.org/10.1007/s12088-016-0632-2

Sánchez-Elordi E, De los Ríos LM, Vicente C, Legaz ME (2015) Sugar cane arginase competes with the same fungal enzyme as a false quorum signal against smut teliospores. Phytochem Lett 14:115–122. https://doi.org/10.1016/j.phytol.2015.09.013

Sarkar R, Mondal C, Bera R, Chakraborty S, Barik R, Roy P, Kumar A, Yadav KK, Choudhury J, Chaudhary SK, Samanta SK, Karmakar S, Das S, Mukherjee PK, Mukherjee J, Sen T (2015) Antimicrobial properties of *Kalanchoe blossfeldiana*: a focus on drug resistance with particular reference to quorum sensing-mediated bacterial biofilm formation. J Pharm Pharmacol 67:951–962. https://doi.org/10.1111/jphp.12397

Shiva Krishna P, Sudheer Kumar B, Raju P, Murty MSR, Prabhakar Rao T, Singara Charya MA, Prakasham RS (2015) Fermentative production of pyranone derivate from marine *Vibrio* sp. SKMARSP9: isolation, characterization and bioactivity evaluation. Indian J Microbiol 55:292–301. https://doi.org/10.1007/s12088-015-0521-0

Shukla SK, Mangwani N, Rao TS, Das S (2014) Biofilm-mediated bioremediation of polycyclic aromatic hydrocarbons. In: Das S (ed) Microbial biodegradation and bioremediation. Elsevier, Oxford, pp 203–232 ISBN 978-0-12-800021-2

Siddiqui MF, Rzechowicz M, Harvey W, Zularisam AW, Anthony GF (2015) Quorum sensing based membrane biofouling control for water treatment: a review. J Water Proc Eng 30:112–122. https://doi.org/10.1016/j.jwpe.2015.06.003

Smyth AR, Cifelli PM, Ortori CA, Righetti K, Lewis S, Erskine P, Holland ED, Givskov M, Williams P, Cámara M, Barrett DA, Knox A (2010) Garlic as an inhibitor of *Pseudomonas aeruginosa* quorum sensing in cystic fibrosis–a pilot randomized controlled trial. Pediatr Pulmonol 45:356–362. https://doi.org/10.1002/ppul.21193

Soni KA, Lu L, Jesudhasan PR, Hume ME, Pillai SD (2008) Influence of autoinducer-2 (AI-2) and 2261 beef sample extracts on *E. coli* 0157:H7 survival and gene expression of virulence 2262 genes *yadK* and *hhA*. J Food Sci 73:M135–M139. https://doi.org/10.1111/j.1750-3841.2007.00654.x

Srivastava A, Singh BN, Deepak D, Rawat AK, Singh BR (2015) Colostrum hexasaccharide, a novel *Staphylococcus aureus* quorum sensing inhibitor. Antimicrob Agents Chemother 59:2169–2178. https://doi.org/10.1128/AAC.03722-14

Sturbelle RT, Conceição RC, Da Rosa MC, Roos TB, Dummer L, Leite FP (2013) The use of quorum sensing to improve vaccine immune response. Vaccine 32:90–95. https://doi.org/10.1016/j.vaccine.2013.10.065

Sun B, Zhang M (2016) Analysis of the antibacterial effect of an *Edwardsiella tarda* LuxS inhibitor. Springerplus 5:92. https://doi.org/10.1186/s40064-016-1733-4

Szweda P, Gucwa K, Kurzyk E, Romanowska E, Dzierżanowska-Fangrat K, Jurek AZ, Kuś PM, Milewski S (2015) Essential oils, silver nanoparticles and propolis as alternative agents against fluconazole resistant *Candida albicans, Candida glabrata* and *Candida krusei* clinical isolates. Indian J Microbiol 55:175–183. https://doi.org/10.1007/s12088-014-0508-2

Thakur R, Sharma KC, Gulati A, Sud RK, Gulati A (2017) Stress-tolerant *Viridibacillus arenosi* strain IHB B 7171 from tea rhizosphere as a potential broad-spectrum microbial inoculant. Indian J Microbiol 57:195–200. https://doi.org/10.1007/s12088-017-0642-8

Varsha KK, Nishant G, Sneha SM, Shilpa G, Devendra L, Priya S, Nampoothiri KM (2016) Antifungal, anticancer and aminopeptidase inhibitory potential of a phenazine compound produced by *Lactococcus* BSN307. Indian J Microbiol 56:411–416. https://doi.org/10.1007/s12088-016-097-1

Venkadesaperumal G, Rucha S, Sundar K, Shetty PK (2016) Anti-quorum sensing activity of spice oil nanoemulsions against food borne pathogens. LWT-Food Sci Technol 66:225–231. https://doi.org/10.1016/j.lwt.2015.10.044

Wadhwani SA, Shedbalkar UU, Singh R, Vashisth P, Pruthi V, Chopade BA (2016) Kinetics of synthesis of gold nanoparticles by *Acinetobacter* sp. SW30 isolated from environment. Indian J Microbiol 56:439–444. https://doi.org/10.1007/s12088-016-0598-0

Wu Y, Quan X, Si X (2015) Incorporation of brominated furanone into Nafion polymer enhanced anti-biofilm efficacy. Int Biodeterior Biodegradation 99:39–44. https://doi.org/10.1016/j.ibiod.2014.11.013

Wynendaele E, Verbekea F, D'Hondt M, Hendrix A, Van De Wiele C, Burvenichd C, Peremans K, De Wever O, Bracke M, De Spiegeleer B (2015) Crosstalk between the microbiome and cancer cells by quorum sensing peptides. Peptides 64:40–48. https://doi.org/10.1016/j.peptides.2014.12.009

Yang Q, Scheie AA, Benneche T, Defoirdt T (2015) Specific quorum sensing-disrupting activity (AQSI) of thiophenones and their therapeutic potential. Sci Rep 5:18033. https://doi.org/10.1038/srep18033

Yasin NH, Mumtaz T, Hassan MA, Abd Rahman N (2013) Food waste and food processing waste for biohydrogen production: a review. J Environ Manag 130:375–385. https://doi.org/10.1016/j.jenvman.2013.09.00

Yin H, Deng Y, Wang H, LiuW ZX, Chu W (2015) Tea polyphenols as an antivirulence compound disrupt quorum-sensing regulated pathogenicity of *Pseudomonas aeruginosa*. Sci Rep 5:16158. https://doi.org/10.1038/srep16158

Yong YC, Wu XY, Sun JZ, Cao YX, Song H (2015) Engineering quorum sensing signaling of *Pseudomonas* for enhanced wastewater treatment and electricity harvest: a review. Chemosphere 140:18–25. https://doi.org/10.1016/j.chemosphere.2014.10.020

Zhang SH, Yu X, Guo F, Wu ZY (2011) Effect of interspecies quorum sensing on the formation of aerobic granular sludge. Water Sci Technol 64:1284–1290. https://doi.org/10.2166/wst.2011.723

Zhang Y, Liu J, Tang K, Yu M, Coenye T, Zhang XH (2015) Genome analysis of *Flaviramulus ichthyoenteri* Th78(T) in the family Flavobacteriaceae: insights into its quorum quenching property and potential roles in fish intestine. BMC Genomics 16:38. https://doi.org/10.1186/s12864-015-1275-0

Zhao J, Chen M, Quan CS, Fan SD (2014) Mechanisms of quorum sensing and strategies for quorum sensing disruption in aquaculture pathogens. J Fish Dis 38:771–786. https://doi.org/10.1111/jfd.12299

Zhou M, Wang H, Hassett DJ, Gu T (2013) Recent advances in microbial fuel cells (MFCs) and microbial electrolysis cells (MECs) for wastewater treatment, bioenergy and bioproducts. J Chem Technol Biotechnol 88:508–518. https://doi.org/10.1002/jctb.4004

Zhu S, Wu H, Zeng M, Zunying L, Zhao Y, Dong S (2015) Regulation of spoilage-related activities of *Shewanella putrefaciens* and *Shewanella baltica* by an autoinducer-2 analogue, (Z)-5-(Bromomethylene)furan-2(5H)-one. J Food Process Preserv 39:719–728. https://doi.org/10.1111/jfpp.12281

Talking Through Chemical Languages: Quorum Sensing and Bacterial Communication

2

Mohini Mohan Konai, Geetika Dhanda, and Jayanta Haldar

Abstract

Bacteria constitute a large domain of prokaryotic microorganisms which have been cohabiting with us for a very long time. Nevertheless, understanding them is a magnificent task. Communication among bacteria, both inter-species and intra-species constitutes a highly specific but complicated process known as 'Quorum sensing'. Many essential group behaviours (such as bioluminescence, virulence, swarming, nodulation, biofilm formation and many more) in bacterial population are guided by quorum sensing which involves production of molecules, acting as signals. Recognition of the signals results in gene expression, which ultimately regulates the collective behaviour beneficial for bacterial survival. The signalling molecules are different for Gram-positive and Gram-negative bacteria. In this chapter, we have discussed various classes of signalling molecules, their production, recognition and signal transduction.

Keywords

Acyl homoserine lactone · Autoinducing peptide · Signal production · Signal recognition · Signal transduction · Quorum sensing

M. M. Konai · G. Dhanda · J. Haldar (✉)
Antimicrobial Research Laboratory, New Chemistry Unit, Jawaharlal Nehru Centre for Advanced Scientific Research, Bengaluru, Karnataka, India
e-mail: jayanta@jncasr.ac.in; http://www.jncasr.ac.in/jayanta/

V. C. Kalia (ed.), *Quorum Sensing and its Biotechnological Applications*,
https://doi.org/10.1007/978-981-13-0848-2_2

Abbreviations

3-hydroxy-7-cis-C14-HSL	*N*-(3-hydroxy- 7-cis-tetradecenoyl)-L-homoserine lactone,
3-oxo-C10-HSL	*N*-3-oxo-decanoyl-L-homoserine lactone,
3-oxo-C12-HSL	*N*-3-oxo-dodecanoyl-L-homoserine lactone,
3-oxo-C6-HSL	*N*-3-oxo-hexanoyl-L- homoserine lactone,
3-oxo-C8-HSL	*N*-3-oxo-octanoyl-L-homoserine lactone,
7-cis-C14-HSL	*N*-(7-cis-tetradecenoyl)-L-homoserine lactone,
ACP	Acyl carrier protein,
agr	Accessory gene regulator,
AHL	Acyl homoserine lactone,
AI-2	Autoinducer 2,
AIP	Autoinducing peptide,
AMR	Antimicrobial resistance,
C10-HSL	*N*-decanoyl-L-homoserine lactone,
C12-HSL	*N*-dodecanoyl-L-homoserine lactone,
C14-HSL	*N*-tetradecanoyl-L-homoserine lactone,
C4-HSL	*N*-butanoyl-L-homoserine lactone,
C6-HSL	*N*-hexanoyl-L-homoserine lactone,
C8-HSL	*N*-octanoyl-L-homoserine lactone,
CAI-I	Cholera autoinducer 1,
DPD	4,5-dihydroxy-2,3-pentanedione,
DPO	3,5-dimethylpyrazin-2-ol,
DSF	Diffusible signal factor,
HPK	Histidine protein kinase,
HSL	Homoserine lactone,
PQS	Pseudomonas quinolone signal
qrr	Quorum regulatory,
QS	Quorum sensing,
RR	response regulator,
SAM	*S*- adenosylmethionine

2.1 Introduction

The collective behaviour in bacteria is regulated by a cell-to-cell communication pathway known as Quorum sensing (QS). Through this QS process, bacteria control essential group behaviours in their community such as production of bioluminescence, virulence factor secretion, biofilm formation and so on (Koul et al. 2016; Papenfort and Bassler 2016). QS process in bacteria is regulated by various chemical signalling molecules which act like the 'specific language' for communication (Bassler 2002). These signalling molecules are referred to as autoinducers. Bacteria are known to produce signalling molecules for both intra- and inter-species communication. For intra-species communication, the signalling molecules produced act like a 'secret language' for a particular species of bacteria. On the contrary, for inter-species communication, a dedicated signalling pathway is conserved in the bacterial world and a 'universal language' is used for this purpose. Most of the Gram-negative bacteria use acyl homoserine lactones (AHLs) as their secret language of communication. Each species of Gram-negative bacteria produces AHLs specific to them. Similarly, Gram-positive bacteria mainly use autoinducing peptide (AIP) as their secret language. The universal language used for inter-species communication is the same for all bacteria which is called as autoinducer-2 (AI-2). The concentration of signalling molecules increases in the external environment as bacteria grow in their population. Through recognition of different signalling molecules, bacteria are capable to track the changes in the population of their own species as well as the population of other species. Once bacteria reach the threshold population density, necessary genes are expressed which regulate the corresponding group behaviour. Herein, we will discuss about various signalling molecules, mainly focusing on AHL and AIP. The details of signal production, recognition and transduction will be described for these two major classes of signalling molecules. In the end, other classes of signalling molecules will also be discussed briefly.

2.2 Acyl Homoserine Lactone (AHL) Based Quorum Sensing

AHLs are the signalling molecules used by Gram-negative bacteria for communication (Fig. 2.1 and Table 2.1). They are synthesised from *S*-adenosylmethionine (SAM) and fatty acid biosynthesis intermediates which contain a fatty acyl group. However, it has been discovered recently that non-fatty acyl groups can also be incorporated into the homoserine lactone-based signalling molecules (Schaefer et al. 2008). AHLs have two major properties. They are readily diffusible into the cell membrane and are bound by specific receptors either inside the cytoplasm or the inner membrane of the bacteria. AHLs, on binding to the receptors lead to a process called autoinduction which results in increased synthesis of the signalling molecule. This in turn stimulates quorum sensing and establishes a feed-forward loop which promotes expression of a particular gene in the bacterial population. The first AHL based system was discovered in the marine bacterium *Vibrio fischeri*.

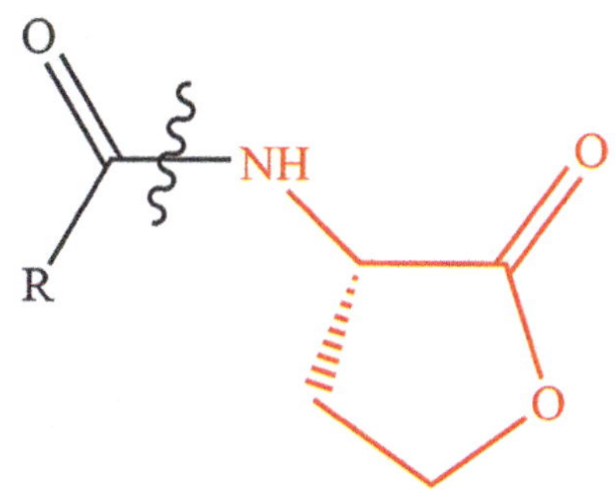

Fig. 2.1 General chemical structure of AHLs

Table 2.1 AHL signalling molecules produced by bacteria and their corresponding LuxI/LuxR homologous proteins (This table is generated from the information provided by the University of Nottingham, England- http://www.nottingham.ac.uk/quorum/table.htm)

Bacterium	Acyl group	AHL	LuxI/LuxR homologues	References
Aeromonas hydrophila	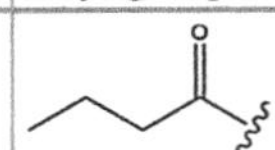	C4-HSL	AhyI/AhyR	Swift et al. (1997)
Aeromonas salmonicida	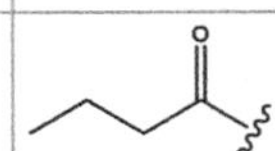	C4-HSL	AsaI/AsaR	Swift et al. (1997)
Agrobacterium tumefaciens	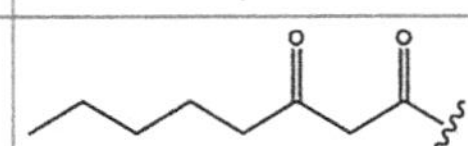	3-oxo-C8-HSL	TraI/TraR	Piper et al. (1999)
Burkholderia cepacia	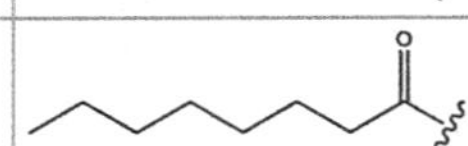	C8-HSL	CepI/CepR	Lewenza et al. (1999)
Chromobacterium violaceum	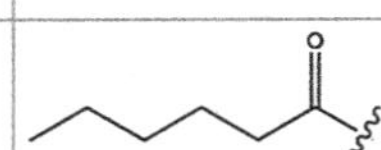	C6-HSL	CviI/CviR	McClean et al. (1997) and Chernin et al. (1998)
Enterobacter agglomerans	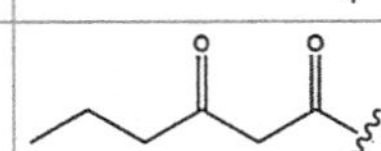	3-oxo-C6-HSL	EagI/EagR	Swift et al. (1993)

Erwinia carotovora	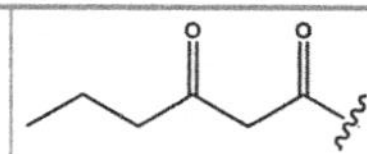	3-oxo-C6-HSL	CarI/ExpR	Bainton et al. (1992), Pirhonen et al. (1993), and Swift et al. (1993)
Erwinia chrysanthemi	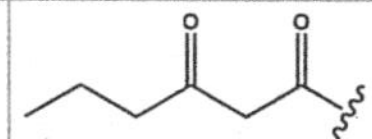	3-oxo-C6-HSL	ExpI/ExpR (EchI/EchR)	Nasser et al. (1998)
Pantoea stewartii	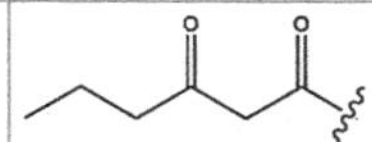	3-oxo-C6-HSL	EsaI/EsaR	von Bodman et al. (1998)
Pseudomonas aeruginosa	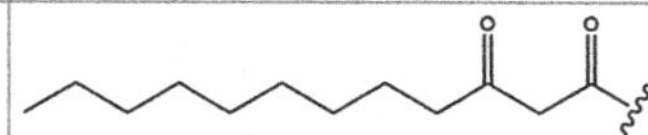	3-oxo-C12-HSL	LasI/LasR	Passador et al. (1993), Glessner et al. (1999), and Lee and Zhang (2015)
	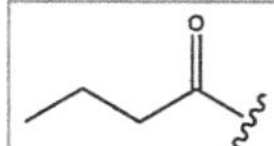	C4-HSL	RhlI/RhlR	
Pseudomonas aureofaciens	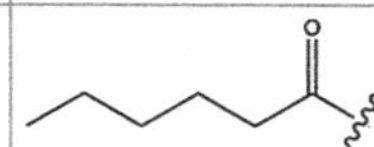	C6-HSL	PhzI/PhzR	Pierson et al. (1994) and Wood et al. (1997)
Rhizobium leguminosarum	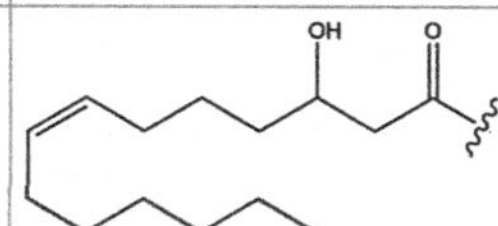	3-hydroxy-7-cis-C14-HSL	RhiR	Sanchez-Contreras et al. (2007)

(continued)

Table 2.1 (continued)

Bacterium	Acyl group	AHL	LuxI/LuxR homologues	References
Rhodobacter sphaeroides	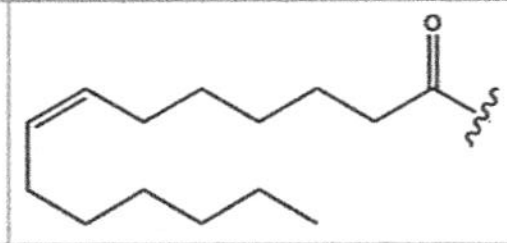	7-cis-C14-HSL	CerI/CerR	Puskas et al. (1997)
Serratia liquefaciens	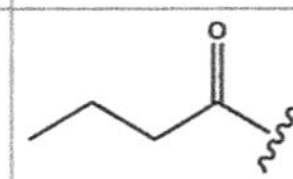	C4-HSL	SwrI/SwrR	Eberl et al. (1996) and Givskov et al. (1997)
Vibrio anguillarum	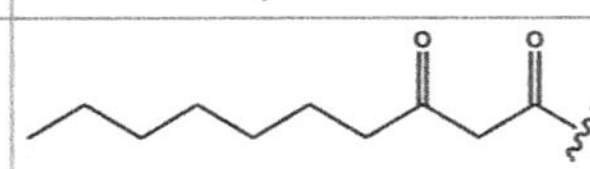	3-oxo-C10-HSL	VanI/VanR	Milton et al. (1997)
Vibrio fischeri	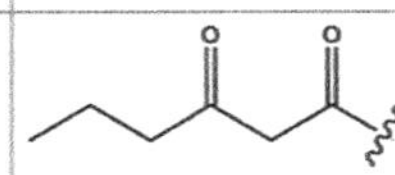	3-oxo-C6-HSL	LuxI/LuxR	Lupp and Ruby (2005)
Yersinia enterocolitica	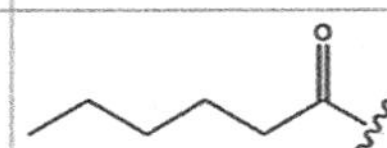	C6-HSL	YenI/YenR	Throup et al. (1995)
Yersinia pseudotuberculosis	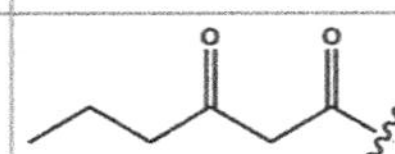	3-oxo-C6-HSL	YpsI/YpsR	Atkinson et al. (1999)

Some bacteria use a more complicated system for quorum sensing which involves more than one component either of the same kind or of different kind. For example, *Pseudomonas aeruginosa* uses four types of quorum sensing signalling systems, two of which are AHL-based systems and the other two are quinolone signal based (PQS) and carbaldehyde signal based (IQS) systems (Williams et al. 2007). All the four systems are inter-linked and are together responsible for quorum sensing in *P. aeruginosa.* Despite many different signalling systems found in Gram-negative bacteria, AHL-based quorum sensing is the most explored and commonly found. In this section, we will briefly discuss about AHL-based signalling molecules; their biosynthesis, signal recognition and transduction.

2.2.1 AHL-Based Signalling Molecules

Natural AHLs consist of L-homoserine lactone ring as the core which is *N*-acylated at the α-position and unsubstituted at the β and γ positions (Fig. 2.1 and Table 2.1). The acyl chain varies from bacterium to bacterium. The acyl chain may have modifications and consists of 4–18 carbon atoms (Galloway et al. 2011; Koul and Kalia 2017). Therefore, all AHLs produced by bacteria have a conserved homoserine lactone moiety with diversity in the acyl chain. Various AHL signals used by different bacteria are listed in the table below (Table 2.1).

The first AHL based system was discovered in the marine Gram-negative bacterium *Vibrio fischeri* which uses the signalling molecule, *N*-3-oxohexanoylhomoserine lactone (3-oxo-C6 HSL) to regulate the property of bioluminescence. *P. aeruginosa* uses two AHL-based signalling molecules, namely 3-oxo-C12-HSL which activates the LasI/LasR signalling system and C4-HSL which activates the RhlI/RhlR signalling system. LasI/LasR and RhlI/RhlR systems are LuxI/LuxR homologous systems. LuxI and LuxR type proteins are involved in signal production and signal recognition of AHL-based signalling molecules in Gram-negative bacteria respectively. The details about signal production and detection are discussed in the following sections.

2.2.2 Signal Production

The proteins responsible for synthesis of AHLs are usually LuxI synthases or they can be LuxI homologues as well. For example, two LuxI-type synthases have been found in *P. aeruginosa,* namely LasI and RhlI (Lee et al. 2013). LuxI/LuxI homologues are present in combination with their corresponding LuxR/LuxR homologous receptors. LuxI homologous proteins consist of about 200 amino acid residues. Certain conserved residues are present in the amino-terminal domain of LuxI homologues which are important for their enzymatic activity. Conserved amino acid residues present in the carboxy terminal domain are necessary for Acyl Carrier Protein (ACP) selection (Whitehead et al. 2001). The stepwise mechanism of AHL biosynthesis is outlined in Fig. 2.2. In the AHL structure, the homoserine lactone moiety is

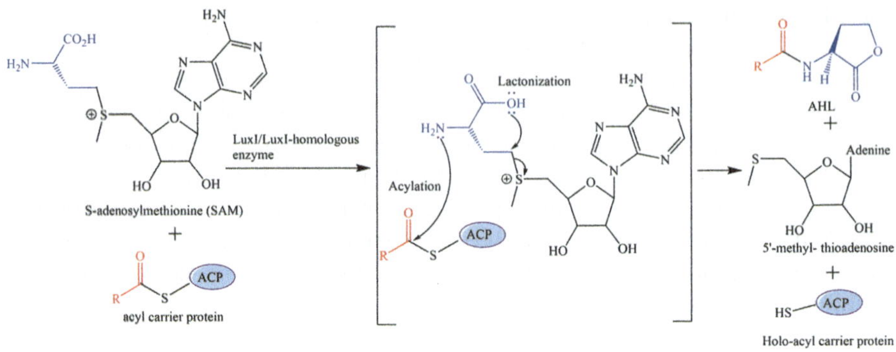

Fig. 2.2 Schematic diagram showing the synthesis of AHL

contributed by the *S*-adenosylmethionine (SAM). SAM binds to the active site of the LuxI/LuxI-homologous enzyme and the acyl group is transferred to this complex from ACP substrates. Therefore, ACP also plays a crucial role by integrating the acyl group to homoserine lactone moiety, thus leading to production of the final AHL signalling molecule. The homoserine lactone ring formation takes place through a basic lactonization mechanism. The amino donor in SAM attacks an intermediate of fatty acid biosynthesis and finally leads to the formation of *N*-acyl homoserine lactone through an addition-elimination mechanism (Hentzer and Givskov 2003; Rasmussen and Givskov 2006). There are some AHL-based signalling molecules which are not synthesized by LuxI or LuxI homologues. For example, the synthesis of 3-oxo-C4-HSL requires a LuxLM protein which does not show any structural similarity with LuxI homologues (Bassler et al. 1993; Schauder and Bassler 2001).

2.2.3 Signal Recognition and Transduction

Bacteria encounter complex mixtures of signalling molecules in heterogeneous environments. To ensure that they extract information from the right signalling molecule, they have receptors; either cytoplasmic or membrane-bound, which are highly specific for a particular signalling molecule. In case of AHLs, the LuxI or LuxI homologous synthases have complimentary LuxR-type receptors. These receptors recognize their respective signalling molecules and thus enable communication through quorum sensing. The LuxR receptors are one- component cytoplasmic receptors. AHL ligands passively diffuse out of the cell and are sensed by receptors inside the neighbouring cells when the 'specific quorum' of bacterial population is reached (Moore et al. 2015). As the bacterial population increases, the local concentration of the AHL also increases. On achieving a threshold population, the AHL binds with the LuxR or LuxR homologous receptor. This binding activates the quorum sensing target genes (Moore et al. 2015).

The LuxR proteins consist of two functional domains- (1) Amino (*N*)-terminal ligand binding domain and (2) Carboxy (*C*)-terminal DNA binding domain. The signalling molecule binds to the amino-terminal of the receptor protein (Smith et al. 2006). The *N*-terminal has three tryptophan moieties which interact with the HSL backbone of the AHL-based signalling molecule and define the orientation of binding. Residues present in the *N*-terminal, providing hydrophobic and van der Waal's interaction are less conserved and decide specificity. The LuxR receptor protein binds to the DNA through the *C*-terminal. The binding with DNA takes place when the LuxR-type receptor forms a complex with the signalling molecule (Smith et al. 2006). In the absence of signalling molecules, the LuxR or LuxR homologue proteins do not fold and degrade. LuxR-proteins bound to their respective AHL molecule dimerize and bind to DNA (Swem et al. 2009). AHL-LuxR complexes associate with DNA sequences which are known as 'lux boxes'. Lux boxes are present upstream of the target genes. On binding of the AHL-LuxR complex to these 'lux boxes', genes promoting the production of signalling molecules and genes coding for processes like virulence, biofilm formation, luminescence get activated (Fig. 2.3). For example, the signalling molecule used by *V. fischeri* to activate bioluminescence is 3-oxo-C6-HSL. As the population density increases, the signalling molecule binds to LuxR receptor. LuxR is the cytoplasmic receptor for 3-oxo-C6-HSL and also the transcriptional activator of the luciferase luxICDABE operon (Engebrecht et al. 1983; Ng and Bassler 2009). The structures of four LuxR type receptors are

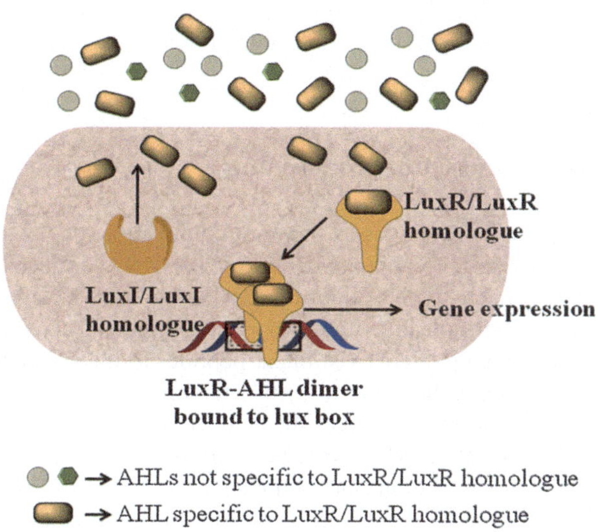

Fig. 2.3 Schematic representation of AHL-based QS system. AHLs are produced by LuxI/LuxI homologous enzymes present in bacterial cells. These AHLs are recognised by the receptor proteins LuxR/LuxR homologues specific to the bacterial cell. In a heterogeneous bacterial population, a variety of AHLs are present in the surroundings. At a threshold concentration of signalling molecule, the LuxR/LuxR homologue bound to AHL form a homodimer complex. This dimer complex then binds to the specific DNA sequence (lux box) that triggers gene expression to regulate group behaviour

known- TraR from *Agrobacterium tumefaciens* and *Rhizobium* sp. NGR234 (Vannini et al. 2002; Zhang et al. 2002), QscR from *P. aeruginosa* (Lintz et al. 2011) and CviR from *Chromobacterium violaceum* (Chen et al. 2011). The structure of the ligand-binding site of LasR from *P. aeruginosa* has also been studied (Bottomley et al. 2007). LuxR-proteins achieve AHL- specificity through variation in the amino acid side chain residues and also through flexibility of the binding pocket. Apart from the LuxR-receptors which work in partnership with LuxI-synthases, there exist solo LuxR-receptors as well. About 76% of the LuxR receptors do not have any accompanying LuxI or LuxI homologous synthases (Hudaiberdiev et al. 2015). QscR from *P. aeruginosa* is an example of LuxR-solo receptor. It has slightly relaxed ligand binding specificity as opposed to non-solo LuxR-type receptors. As a result of this, it activates gene expression at nanomolar concentrations of various AHLs like C8-HSL, C10-HSL, 3-oxo-C10-HSL, C12-HSL, 3-oxo-C12-HSL and C14-HSL (Lee et al. 2006). Because of this relaxation in specificity, QscR may be used by *P. aeruginosa* to detect autoinducers that are produced by other cohabiting species.

In addition to the cytoplasmic LuxR or LuxR-homologous receptors, one more class of receptors is found in Gram-negative bacteria. These are the two-component receptors which consist of membrane-bound histidine kinases that signal to cytoplasmic transcription factors through phosphorylation. An example of AHL based signalling molecule which is detected by a membrane bound receptor is 3-OH C4-HSL which is found in *Vibrio harveyi*. The receptor involved in the detection is LuxN (Swem et al. 2008). The receptor for this signalling molecule is a two-component system consisting of the periplasmic protein, LuxN and the cytoplasmic domains, LuxU and LuxO.

2.3 Peptide Based Quorum Sensing

Gram-positive bacteria use peptides as their communication language. The signalling peptides are called as autoinducing peptide (AIP) as they induce the communication process through activation of quorum sensing circuit. The structure and size of the peptide signalling molecules may vary depending on bacterial species and their functions. For example, there exist linear peptides as the signalling molecules that govern various group behaviours in certain Gram-positive bacteria. For example, *Streptococcus pneumoniae* regulates the competence by using a 17-residue linear peptide (Håvarstein et al. 1995). *Bacillus subtilis* also uses a series of linear peptides that regulate sporulation and competence (Lazazzera 2001). Similarly, *Bacillus cereus* and *Enterococcus faecalis* use various linear peptides to regulate the expression of virulence factors and plasmid-mediated conjugation respectively (Slamti and Lereclus 2002; Clewell et al. 2002). However important may the linear peptides be for certain bacteria, the more well-known class of peptide signalling molecules are cyclic in structure. A recent study suggested that the cyclic peptide based quorum sensing circuit is conserved among a large number of Gram-positive bacteria (Wuster and Babu 2008). Therefore, herein we focus our discussion on the cyclic class of peptide signalling molecules. We will discuss the structural variation in cyclic peptides and role of various parameters in guiding the signal induction. Biosynthesis of these peptides and their specificity towards receptors will be discussed in detail. To the end, we will narrate how the signal recognition and transduction take place that lead to gene expression and thereby regulate the group behaviour in bacterial world.

2.3.1 Peptide Signalling Molecules

Peptide signalling molecules regulate the autoinducing circuit that control bacterial communication in Gram-positive bacteria. The cyclic peptide signals were first discovered in *S. aureus* and later on similar molecules have been identified in many other bacterial species. These include *Streptococcus mutans* (Kaur et al. 2015), *Enterococcus faecalis* (Nakayama et al. 2001; Nishiguchi et al. 2009), *Lactobacillus plantarum* (Sturme et al. 2005; Fujii et al. 2008), *Listeria monocytogenes* (Autret et al. 2003; Riedel et al. 2009), *Clostridium perfringens* (Ohtani et al. 2009; Vidal et al. 2009), *Clostridium botulinum* (Cooksley et al. 2010) and many more. To date, four different AIPs are mostly characterized as chemical languages, namely AIP I-IV (Fig. 2.4). Despite the structural diversity of these peptides, they share some common structural features. For example, all the AIPs consist of Cys→*C*-terminus macrocyclic thiolactone core comprising a 5-amino acid sequence (shown in Red) along with an exocyclic tail of two to four amino acid residues (shown in Black). They may differ in amino acid sequence however; all of them possess an increased hydrophobicity from their *N*- to *C*-terminal end in the peptide structure. At the end of the sequence to the *C*-terminal positions, AIPs are found to have amino acids

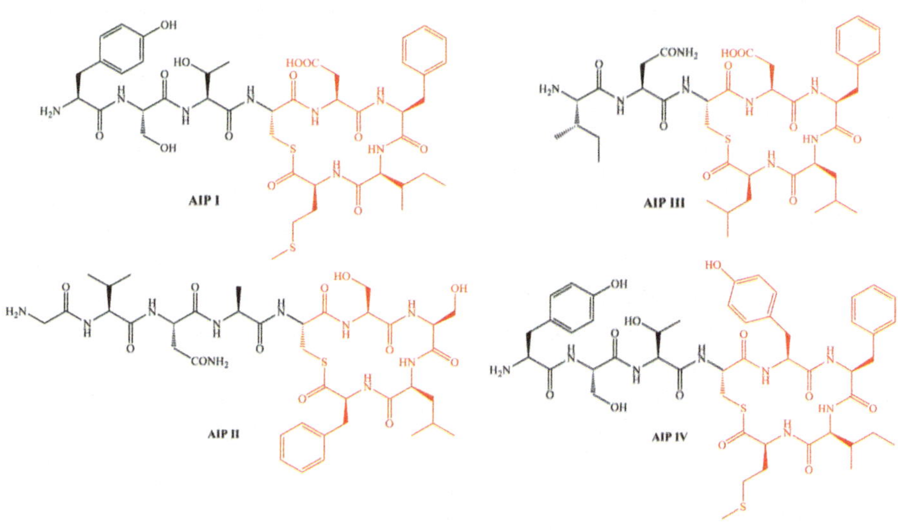

Fig. 2.4 Chemical structure of AIPs used by Gram-positive bacteria *S. aureus*

with hydrophobic side chains such as methionine (AIP I and IV), phenylalanine (AIP II) and leucine (AIP III). Detailed investigations have been performed to understand the essential structural motif through extensive structure-activity relationship (SAR) studies (Mayville et al. 1999; Lyon et al. 2000, 2002; MDowell et al. 2001; Tal-Gan et al. 2013, 2016; Yang et al. 2016). The cyclic ring structure is crucial for AIP to function as it has been found that corresponding linear peptide analogues or hydrolysis of the thioester moiety inactivates its functions as a signalling molecule (Ji et al. 1997). The thioester moiety present in the macrocyclic ring is also crucial as an ester moiety was found to inactivate the signalling capability of the peptide. Moreover, upon removal of the *N*-terminal exocyclic residues, the AIP loses its signalling capability. Rather it has been reported that macrocyclic ring of AIP-I, AIP-II or AIP-IV without the exocyclic tail functions as the signal inhibitor (Lyon et al. 2000, 2002). Therefore, both the macrocyclic ring as well as the exocyclic tail of the AIPs is essential for its signalling property.

2.3.2 Signal Production

The AIP based quorum sensing system is controlled by a regulatory cascade known as accessory gene regulator (*agr*). All the necessary genes are located in this *agr* locus that regulate the process of signal production, recognition and transduction. This locus consists of two divergent transcripts, referred to as RNAII and RNAIII (Janzon et al. 1989; Novick et al. 1995). The transcript RNAII which is an operon of four genes *agrBDCA* encodes the factor responsible for signal production and thereby switches on the response regulatory cascade (Fig. 2.5). Two components, AgrD and AgrB are involved in the biosynthesis of AIPs. While the AgrD functions as the precursor propeptide that gets converted into AIP after a series of transformations, the membrane bound protein AgrB plays an important role in this process by acting as an enzyme.

2.3.2.1 AgrD

The precursor propeptide AgrD can be structurally divided into three distinct regions; (1) the *N*-terminal residue, (2) the middle domain that encodes the residue for AIP and (3) the *C*-terminal residue. While the *N*-terminal domain is amphipathic in nature, the *C*-terminal domain is enriched with negatively charged residues. This structural feature of AgrD is well conserved among all the AIPs produced by various bacterial species (Novick and Geisinger 2008). Although the amino acid residues in *N*-terminal region can differ significantly, the amphipathic nature is conserved. This amphipathic region of AgrD propeptide is responsible for association with the cell membrane that in turn makes possible the interaction with AgrB (Zhang et al. 2002) leading to further processing for AIP biosynthesis. The middle region of AgrD comprising of AIP residues significantly differs among the bacterial species however, the cysteine residue is found to be conserved (Dufour et al. 2002).

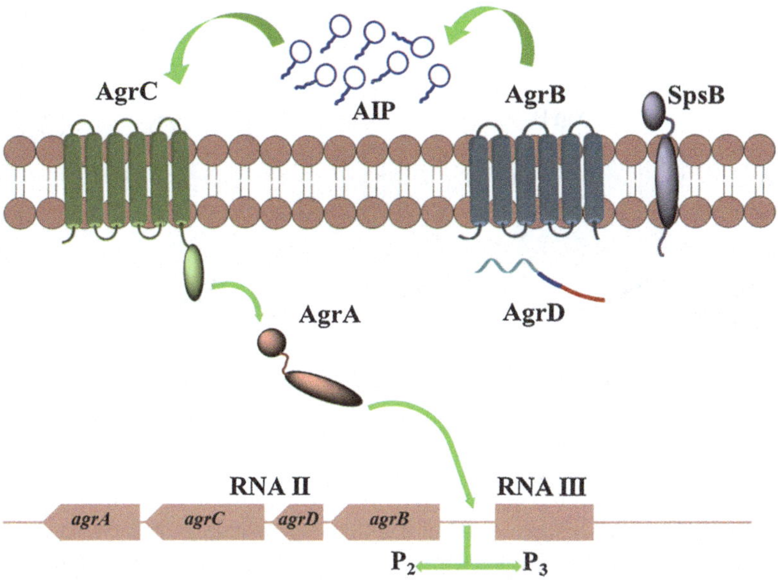

Fig. 2.5 Schematic representation of AIP-based QS system. The agr locus consists of two divergent transcripts referred to as RNAII and RNAIII, which are regulated by the promoters P2 and P3 respectively. The signalling molecule, AIP is produced from the precursor peptide AgrD with the help of membrane bound proteins AgrB and SpsB. Once the threshold concentration of signalling molecules is achieved in the bacterial population, binding of AIP with receptor protein AgrC leads to phosphorylation of the response regulator AgrA. This in turn activates the P2 and P3 promoters, which results in gene expression to regulate the collective behaviour of bacteria

This amino acid residue is responsible for the thiolactone ring structure of AIP upon cyclization. Exceptions also exist to this; serine is present in place of cysteine in several bacterial species such as *S. intermedius*, *S. pseudintermedius*, and *S. delphini*. These bacteria produce AIPs with lactone moiety instead of thiolactone. The *C*-terminal region is the most conserved region in AgrD.

Highly conserved, the first nine residues are necessary for the complete production of AIP (Dufour et al. 2002). The first two amino acid residues of this domain, namely aspartate and glutamate are essential that are found to be conserved among the Staphylococci species however; this conservation rule is not followed in other Gram-positive bacteria (Dufour et al. 2002). It has also been seen that mutations in the aspartate or glutamate residues prevent the AIP production in *S. aureus*, which highlights the importance of these residues.

2.3.2.2 AgrB

The membrane bound protein AgrB is the enzyme for processing the propeptide AgrD into AIP (Saenz et al. 2000). There exist many reports in literature that support the endopeptidase activity of AgrB (Zhang et al. 2002; Qiu et al. 2005; Thoendel

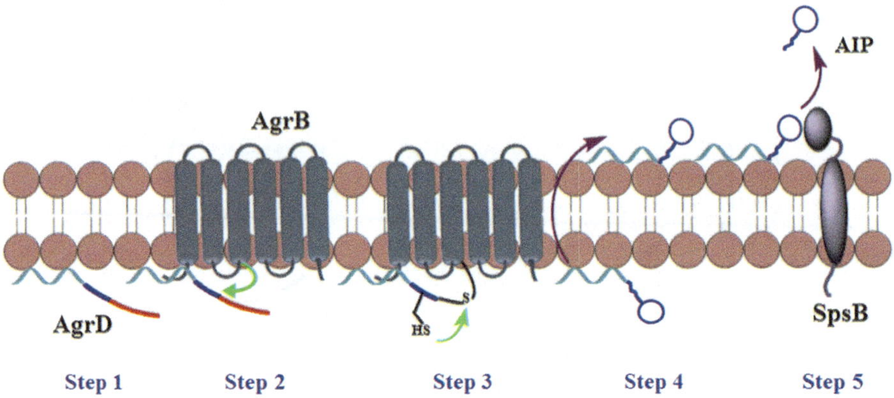

Fig. 2.6 The proposed mechanism of AIP biosynthesis pathway

and Horswill 2009). The enzymatic activity of AgrB results in removal of *C*-terminal region of AgrD, which leads to AIP biosynthesis proceeding in forward direction. Additionally, two amino acid residues His-77 and Cys-84 were reported to be essential for proteolysis activity (Qiu et al. 2005), which remain well conserved among all the Gram-positive bacterial species. Taken together, this suggests that AgrB plays an important role in AIP biosynthesis by functioning as a cysteine protease. Moreover, AgrB is a unique protein that does not possess any homologous similarity with the other proteins present in various quorum-sensing systems (Novick et al. 1995). Although the AgrB sequence significantly differs from bacterium to bacterium, the overall features remain conserved. For example, the hydrophobic region of AgrB that forms the transmembrane portion is maintained in all bacteria (Thoendel et al. 2011). Similarly, the *N*-terminal region (specially the first 34 residues) of AgrB is found to be highly conserved in all four *S. aureus* type AIP producing bacteria.

2.3.2.3 Mechanism of AIP Production

The mechanism of AIP biosynthesis pathway has been proposed through a series of experimental observations (Thoendel and Horswill 2009, 2010). The proposed mechanism consists of five consecutive steps as outlined in Fig. 2.6. In the first step, the precursor peptide AgrD associates with the membrane with the help of amphipathic *N*-terminal region. Then, in the second step AgrB accomplishes a nucleophilic reaction through a cysteine moiety, which results in removal of *C*-terminal residue of AgrD (Fig. 2.6). In this process AgrB gets covalently linked with AgrD via thioester bond. In the next step, the -SH group of cysteine residue in AgrD attacks on this covalently linked intermediate leading to a thiolactone ring structure followed by cleavage of the covalent linkage of AIP precursor and AgrB. The fourth step is associated with transportation of AIP precursor from inner leaflet to outer leaflet of the membrane. In the final step, SpsB cleaves the *N*-terminal region of AIP precursor to produce the AIP molecule.

2.3.3 Signal Recognition and Transduction

Signal recognition and response transduction of AIP based quorum sensing system consists of two components, AgrC and AgrA. Peptide signal is recognized through transmembrane receptor AgrC and signal transduction takes place through response regulator (RR) AgrA. Like the signal producing components (AgrD and AgrB), the signal recognizing and transducing components AgrC and AgrA are also regulated by the factor encoding in transcript RNAII (Fig. 2.5). It is noteworthy that there are some Staphylococcal species that possess AgrC homologue as the signal receiver instead of AgrC (Dufour et al. 2002). The other Gram positive bacteria *C. perfringens* (Lyristis et al. 1994; Ohtani et al. 2009), *L. monocytogenes* (Riedel et al. 2009), *E. faecalis* (Qin et al. 2000) and *L. plantarum* (Sturme et al. 2005) also possess AgrC homologues that are capable to sense cyclic peptides consisting of either lactone or thiolactone groups in the signalling molecule. At an optimum concentration of extracellular AIP, effective binding with the sensory domain in AgrC results in autophosphorylation (Lina et al. 1998). This process in turn activates the response regulator (RR) AgrA and the response transduces via phosphotransfer with aspartic acid residue of RR. At the end, this phosphorylated RR binds to DNA leading to regulation of necessary gene expression (Fig. 2.5).

2.3.3.1 AgrC

AgrC belongs to histidine protein kinases (HPKs) family. Structurally, it can be divided into several distinct regions. The *N*-terminal domain that spans across the membrane is referred to as the sensory domain (Tusnády and Simon 1998). This domain is actually the receiver of peptide signal. The specificity towards a particular AIP is brought about by the specific interaction of this transmembrane domain and AIPs. Followed by this sensory domain, an intracellular subdomain is present that comprises of a dimerization and histidine phosphotransfer site which is referred to as DHp. After that, a highly coiled region exists with an autophosphorylation site formed by a conserved histidine residue (Lupas 1996). Finally at the *C*-terminal end, a catalytic region is located consisting of ATP binding site denoted as CA. Both, the DHp and CA regulate the autokinase activity of AgrC, whereas the phosphotransfer and phosphatase activity is regulated only by DHp (Gao and Stock 2009).

2.3.3.2 AgrA

AgrA is the transcriptional regulator that transduces signal response through activation of *agr* P2 and P3 promoters (Novick et al. 1993, 1995). Structure of AgrA has been determined experimentally. DNA binding experiments with the *C*-terminal residue of AgrA suggested that a sequence of minimum 15 bp is required for optimum interaction between ArgA and DNA. This 15 bp sequence consists of consensus 9 bp sites along with flanking 3 bp on either side of DNA (Sidote et al. 2008). Moreover, the crystal structure of AgrA in complex with oligonucleotide duplex is solved that suggests a 10 β-stranded topology of AgrA along with a two-turn α helix. In the complex, AgrA bends about 38° that results in a concave

DNA-binding region. This slanted orientation of AgrA possibly leads to effective binding with the successive major grooves of DNA along the same face. Additional experimental studies confirm that AgrA interacts through dimer formation, where the *N*-terminal residues are involved in protein-protein interactions and the *C*-terminal residues remain apart from each other. Only two residues of *C*-terminal domain (H169 and R233), specifically bind with the major grooves of DNA and a third residue (N201) interacts with water in the minor groove, which were identified as essential for AgrA as mutations of these residues resulted in lesser DNA binding (Sidote et al. 2008).

2.4 Other Signalling Molecules

In addition to AHL-based and AIP-based quorum sensing, many other signalling systems are present in the bacterial world, which constitute an important role in regulating various biological processes in them (Fig. 2.7). In this section, we will briefly discuss the other signalling molecules that have been discovered in recent years.

2.4.1 Autoinducer 2 (AI-2)

AI-2 is a set of interconverting molecules derived from 4,5-dihydroxy-2,3-pentanedione (DPD) (Schauder and Bassler 2001). The structure and role of AI-2 has been explored well for the species *V. fischeri* (Fig. 2.7). LuxS is the synthase enzyme of DPD, which is present in many bacterial species (Pereira et al. 2013). The starting material for AI-2 synthesis is SAM. (Federle and Bassler 2003). DPD can cyclise to form different furanone molecules. Different forms of DPD act as AI-2 signals for different bacteria. The highly reactive nature of DPD leads to its quick inter-conversion into various moieties. The chemical nature of AI-2 is the same for many different bacteria. This suggests that AI-2 is used for inter-species communication to detect other bacterial populations. In *V. harveyi*, the active AI-2 signal is boronated whereas

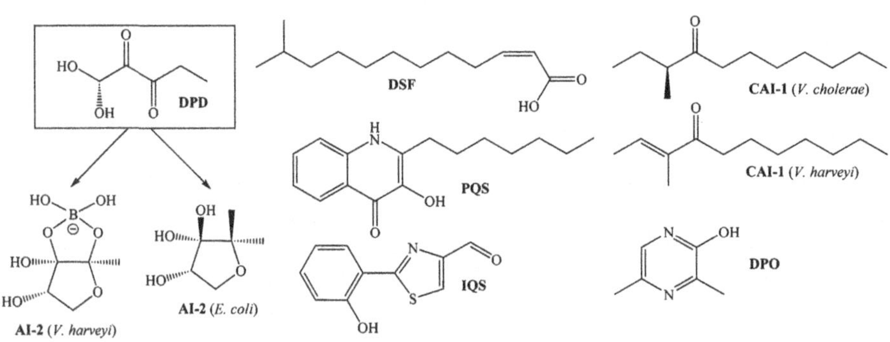

Fig. 2.7 Other signalling molecules used by bacteria

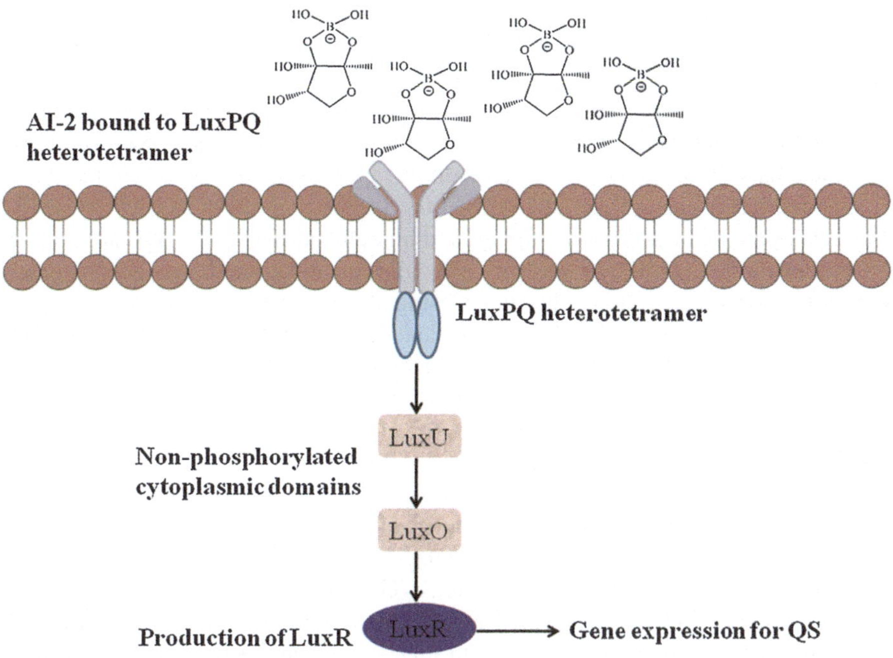

Fig. 2.8 QS system circuit involving AI-2 in Vibrio species. At a threshold population density, AI-2 binds to the LuxPQ heterotetramer. This prevents the phosphorylation of the cytoplasmic domains, LuxU and LuxO. This leads to the production of LuxR which results in gene expression guiding the collective behaviour of bacteria

in *Escherichia coli* and *Salmonella* species, the active AI-2 signal is non-boronated (Miller et al. 2004). In AI-2 based quorum sensing, the signal recognition and transduction take place through a two-component pathway. The first component consists of two proteins, LuxP and LuxQ. The second component constitutes the phosphotransferase protein LuxU and a cytoplasmic response regulator LuxO. There exist two LuxPQ complexes which form a symmetric heterotetramer in the absence of AI-2. Upon binding with AI-2, this tetramer undergoes a conformational change. This conformational change prevents the phosphorylation of the cytoplasmic proteins, LuxU and LuxO, which terminates the expression of quorum regulatory genes (*qrr* genes). Finally, LuxR is produced which leads to gene expression to guide quorum sensing behaviours (Fig. 2.8). This recognition system is used for the detection of AI-2 in Vibrio species which use boronated AI-2 as the signalling molecule.

2.4.2 DSF-Family

The diffusible signal factor (DSF) constitutes an interesting class of signalling molecules, mainly found in Gram-negative bacteria (Deng et al. 2011; Ryan et al. 2015; Zhou et al. 2017). Quorum sensing system based on this class of signalling molecules has emerged as another important communication tool with a distinct mechanism of signalling pathways. The signalling system was first identified in plant bacteria *Xanthomonas campestris* (Wang et al. 2004). This bacterium uses *cis*-11-methyl-2-dodecenoic acid as the signalling molecule which is the first member of the DSF-family (Fig. 2.7). In the recent past however, many other DSF-signalling molecules have been discovered as a result of rapid research progress in this field (Huang and Lee Wong 2007; Deng et al. 2008; Davies and Marques 2009; He et al. 2010). Till date, various classes of plant and human bacterial pathogens are known that produce DSF signalling molecules as communicating language. Like the other signalling systems, DSF based quorum sensing also regulates various essential processes in bacteria such as cell growth, virulence factor production and biofilm formation, which are critical for their survival and pathogenesis.

2.4.3 PQS and IQS

PQS is referred to pseudomonas quinolone signal. 2-heptyl-3-hydroxy-4-quinolone is used by *P. aeruginosa* as a chemical language (Pesci et al. 1999; Heeb et al. 2011) for QS (Fig. 2.7). The signalling molecule is generated by the proteins encoding in the genes *pqsABCDH*. However, the function of PQS quorum sensing is connected with AHL-based quorum sensing (Lee and Zhang 2015) in this bacterium. In combination with AHL-based quorum sensing, the PQS signal controls many essential biological processes in *P. aeruginosa*, which include virulence factor production, biofilm formation and so on. Additionally, *P. aeruginosa* uses another signalling molecule 2-(2-hydroxyphenyl)-thiazole-4-carbaldehyde (Fig. 2.7), also known as IQS which is synthesized by proteins encoding in the gene cluster *ambBCDE* (Lee et al. 2013). The signal receptor of IQS based quorum sensing is yet to be discovered.

2.4.4 CAI-1

The cholera autoinducer 1, CAI-1 was first discovered in cholera causing bacteria *Vibrio cholerae* (Higgins et al. 2007). This bacterium produces CAI-1 at its high cell density in order to terminate the virulence factor production, which is nothing but the indication to leave the host and enter in the environment to infect other hosts in a large number (Higgins et al. 2007). In the recent past, many other Vibrio species are reported to produce this class of signalling molecules. CAI-1 based quorum sensing is mainly used for the intra-species bacterial communication. Therefore, each Vibrio species produces structurally different CAI-1 signal as their secret language. While, *V. cholerae* uses (*S*)-3-hydroxytriecan-4-one as the signal, *V. harveyi* uses (*Z*)-3-aminoundec-2-en-4-one (Fig. 2.7). However, the presence of an acyl long chain is the main characteristic feature of this class of signalling molecules. The signalling molecule is produced by the synthase enzyme CqsA and the signal is received by the receptor protein CqsS, which are mostly conserved among Vibrio species (Ng et al. 2011).

2.4.5 DPO

DPO (3,5-dimethylpyrazin-2-ol) is the most recent class of signalling molecules (Papenfort et al. 2017) (Fig. 2.7). Like CAI-1, this signalling molecule was also discovered in *V. cholerae*. DPO-based signalling system functions through activation of the transcription factor VqmA. DPO binding regulates the VqmA activity leading to expression of the *vqmR* gene which controls biofilm formation and virulence factor production in this species of bacteria.

2.5 Conclusion

Quorum sensing, the bacterial communication system regulates the collective behaviour in bacterial population. Through this process, bacteria can function like multicellular organisms to take necessary action against harsh environmental conditions. They produce signalling molecules to communicate with other members of their own species as well as the members of other species. While, Gram-negative bacteria produce AHLs for intra-species communication, Gram-positive bacteria use AIP. For universal communication, only one class of signalling molecules, AI-2, is used by bacteria. In the recent past, many other interesting classes of signalling molecules have been discovered. DSF, which comprises of an aliphatic chain as a characteristic feature is mainly used by Gram-negative bacteria as communication signal. In addition to AHL-based signalling systems, *P. aeruginosa* uses other specific signalling systems (such as PQS and IQS) for their QS purpose. Similarly, human pathogenic bacteria *V. cholerae* is known to use CAI-1 and DPO-based signalling systems to control the pathogenicity.

2.6 Opinion

The distinct importance of QS is already established in regulating bacterial pathogenesis. In the production of virulence factors and biofilm formation by bacteria, QS constitutes an important role. In the recent past, research in this field has reached a new level of understanding. The application of QS has emerged in various biotechnological fields, which include bioenergy, bioremediation, bio-sensors, anti-cancer therapy, genetic engineering, marine industry, agricultural crops and so on. However, more investigation is required to identify the signalling molecules that still remain unrevealed. Additionally, the quorum sensing pathway of these unrevealed signalling molecules needs to be explored. Future research should focus more in understanding the pathogenic behaviour of disease-causing bacteria, especially the ones which cause a major threat to human health globally. A clear challenge in this field is to understand how quorum sensing regulates the disease-causing nature of pathogenic bacteria in real scenario under in-vivo settings. The field of inter-species quorum sensing should also be forayed into further. Other than AI-2, other signalling molecules which bacteria probably use for inter-species communication need to be investigated. Moreover, it is important to know whether pathogenic bacteria use any specific signalling molecules to modulate the behaviour of commensal bacteria present in the host that makes their disease-causing event successful. Probably, this understanding will result in a better strategy to prevent the diseased state caused by pathogenic bacteria. At present, there is no clear understanding of how quorum sensing regulates the Antimicrobial Resistance (AMR) development in bacteria. Future research should focus to find the link between quorum sensing and AMR. A detailed understanding will possibly lead to the development of effective strategies in resolving this unsolved problem, which has created a major threat to mankind causing an enormous rate of mortality and morbidity every year.

References

Atkinson S, Throup JP, Stewart GS, Williams P (1999) A hierarchical quorum-sensing system in Yersinia pseudotuberculosis is involved in the regulation of motility and clumping. Mol Microbiol 33:1267–1277. https://doi.org/10.1046/j.1365-2958.1999.01578.x

Autret N, Raynaud C, Dubail I, Berche P, Charbit A (2003) Identification of the agr locus of *Listeria monocytogenes*: role in bacterial virulence. Infect Immun 71:4463–4471. https://doi.org/10.1128/IAI.71.8.4463-4471.2003

Bainton NJ, Stead P, Chhabra SR, Bycroft BW, Salmond GP, Stewart GS, Williams P (1992) N-(3-oxohexanoyl)-L-homoserine lactone regulates carbapenem antibiotic production in Erwinia carotovora. Biochem J 288:997–1004 doi: Not available

Bassler BL (2002) Small talk: cell-to-cell communication in bacteria. Cell 109:421–424 doi: Not available

Bassler BL, Wright M, Showalter RE, Silverman MR (1993) Intercellular signalling in Vibrio harveyi: sequence and function of genes regulating expression of luminescence. Mol Microbiol 9:773–786 doi: Not available

von Bodman SB, Majerczak DR, Coplin DL (1998) A negative regulator mediates quorum-sensing control of exopolysaccharide production in Pantoea stewartii subsp. stewartii. Proc Natl Acad Sci U S A 95:7687–7692 doi: Not available

Bottomley MJ, Muraglia E, Bazzo R, Carfì A (2007) Molecular insights into quorum sensing in the human pathogen Pseudomonas aeruginosa from the structure of the virulence regulator LasR bound to its autoinducer. J Biol Chem 282:13592–13600. https://doi.org/10.1074/jbc.M700556200

Chen G, Swem LR, Swem DL, Stauff DL, O'Loughlin CT, Jeffrey PD, Bassler BL, Hughson FM (2011) A strategy for antagonizing quorum sensing. Mol Cell 42:199–209. https://doi.org/10.1016/j.molcel.2011.04.003

Chernin LS, Winson MK, Thompson JM, Haran S, Bycroft BW, Chet I, Williams P, Stewart GS (1998) Chitinolytic activity in Chromobacterium violaceum: substrate analysis and regulation by quorum sensing. J Bacteriol 180:4435–4441 doi: Not available

Clewell DB, Francia MV, Flannagan SE, An FY (2002) Enterococcal plasmid transfer: sex pheromones, transfer origins, relaxases, and the *Staphylococcus aureus* issue. Plasmid 48:193–201. https://doi.org/10.1016/S0147-619X(02)00113-0

Cooksley CM, Davis IJ, Winzer K, Chan WC, Peck MW, Minton NP (2010) Regulation of neurotoxin production and sporulation by a Putative agrBD signalling system in proteolytic *Clostridium botulinum*. Appl Environ Microbiol 76:4448–4460. https://doi.org/10.1128/AEM.03038-09

Davies DG, Marques CN (2009) A fatty acid messenger is responsible for inducing dispersion in microbial biofilms. J Bacteriol 191:1393–1403. https://doi.org/10.1128/JB.01214-08

Deng Y, Wu J, Eberl L, Zhang LH (2008) Structural and functional characterization of diffusible signal factor family quorum-sensing signals produced by members of the *Burkholderia cepacia* complex. Appl Environ Microbiol 76:4675–4683. https://doi.org/10.1128/AEM.00480-10

Deng Y, Wu J, Tao F, Zhang LH (2011) Listening to a new language: DSF-based quorum sensing in Gram-negative bacteria. Chem Rev 111:160–173. https://doi.org/10.1021/cr100354f

Dufour P, Jarraud S, Vandenesch F, Greenland T, Novick RP, Bes M, Etienne J, Lina G (2002) High genetic variability of the agr locus in *Staphylococcus species*. J Bacteriol 184:1180–1186. https://doi.org/10.1128/JB.184.4.1180-1186.2002

Eberl L, Winson MK, Sternberg C, Stewart GS, Christiansen G, Chhabra SR, Bycroft B, Williams P, Molin S, Givskov M (1996) Involvement of N-acyl-L-hormoserine lactone autoinducers in controlling the multicellular behaviour of Serratia liquefaciens. Mol Microbiol 20:127–136. https://doi.org/10.1111/j.1365-2958.1996.tb02495.x

Engebrecht J, Nealson K, Silverman M (1983) Bacterial bioluminescence: isolation and genetic analysis of functions from Vibrio fischeri. Cell 32:773–781 doi: Not available

Federle MJ, Bassler BL (2003) Interspecies communication in bacteria. J Clin Invest 112:1291–1299. https://doi.org/10.1172/JCI20195

Fujii T, Ingham C, Nakayama J, Beerthuyzen M, Kunuki R, Molenaar D, Sturme M, Vaughan E, Kleerebezem M, de Vos W (2008) Two homologous Agr-like quorum-sensing systems cooperatively control adherence, cell morphology, and cell viability properties in *Lactobacillus plantarum* WCFS1. J Bacteriol 190:7655–7665. https://doi.org/10.1128/JB.01489-07

Galloway WR, Hodgkinson JT, Bowden SD, Welch M, Spring DR (2011) Quorum sensing in Gram-negative bacteria: small-molecule modulation of AHL and AI-2 quorum sensing pathways. Chem Rev 111:28–67. https://doi.org/10.1021/cr100109t

Gao R, Stock AM (2009) Biological insights from structures of two-component proteins. Annu Rev Microbiol 63:133–154. https://doi.org/10.1146/annurev.micro.091208.073214

Givskov M, Eberl L, Molin S (1997) Control of exoenzyme production, motility and cell differentiation in *Serratia liquefaciens*. FEMS Microbiol Lett 148:115–122. https://doi.org/10.1111/j.1574-6968.1997.tb10276.x

Glessner A, Smith RS, Iglewski BH, Robinson JB (1999) Roles of Pseudomonas aeruginosa las and rhl quorum-sensing systems in control of twitching motility. J Bacteriol 181:1623–1629 doi: Not available

Håvarstein LS, Coomaraswamy G, Morrison DA (1995) An unmodified heptadecapeptide pheromone induces competence for genetic transformation in *Streptococcus pneumoniae*. Proc Natl Acad Sci U S A 92:11140–11144 doi: Not Available

He YW, Wu J, Cha JS, Zhang LH (2010) Rice bacterial blight pathogen *Xanthomonas oryzae pv. oryzae* produces multiple DSF-family signals in regulation of virulence factor production. BMC Microbiol 10:187. https://doi.org/10.1186/1471-2180-10-187

Heeb S, Fletcher MP, Chhabra SR, Diggle SP, Williams P, Cámara M (2011) Quinolones: from antibiotics to autoinducers. FEMS Microbiol Rev 35:247–274. https://doi.org/10.1111/j.1574-6976.2010.00247.x

Hentzer M, Givskov M (2003) Pharmacological inhibition of quorum sensing for the treatment of chronic bacterial infections. J Clin Invest 112:1300–1307. https://doi.org/10.1172/JCI20074

Higgins DA, Pomianek ME, Kraml CM, Taylor RK, Semmelhack MF, Bassler BL (2007) The major *Vibrio cholerae* autoinducer and its role in virulence factor production. Nature 450:883–886. https://doi.org/10.1038/nature06284

Huang TP, Lee Wong AC (2007) Extracellular fatty acids facilitate flagella-independent translocation by *Stenotrophomonas maltophilia*. Res Microbiol 158:702–711. https://doi.org/10.1016/j.resmic.2007.09.002

Hudaiberdiev S, Choudhary KS, Vera Alvarez R, Gelencsér Z, Ligeti B, Lamba D, Pongor S (2015) Census of solo LuxR genes in prokaryotic genomes. Front Cell Infect Microbiol 5:20. https://doi.org/10.3389/fcimb.2015.00020

Janzon L, Löfdahl S, Arvidson S (1989) Identification and nucleotide sequence of the delta-lysin gene, hld, adjacent to the accessory gene regulator (agr) of *Staphylococcus aureus*. Mol Gen Genet 219:480–485 doi: Not Available

Ji G, Beavis R, Novick RP (1997) Bacterial interference caused by autoinducing peptide variants. Science 276:2027–2030 doi: Not Available

Kaur G, Rajesh S, Adline Princy S (2015) Plausible Drug Targets in the Streptococcus mutans Quorum Sensing Pathways to Combat Dental Biofilms and Associated Risks. Indian J Microbiol 55(4):349–356

Koul S, Kalia VC (2017) Multiplicity of Quorum Quenching Enzymes: A Potential Mechanism to Limit Quorum Sensing Bacterial Population. Indian J Microbiol 57(1):100–108

Koul S, Prakash J, Mishra A, Kalia VC (2016) Potential Emergence of Multi-quorum Sensing Inhibitor Resistant (MQSIR) Bacteria. Indian J Microbiol 56(1):1–18

Lazazzera BA (2001) The intracellular function of extracellular signalling peptides. Peptides 22:1519–1527 doi: Not Available

Lee J, Zhang L (2015) The hierarchy quorum sensing network in *Pseudomonas aeruginosa*. Protein Cell 6:26–41. https://doi.org/10.1007/s13238-014-0100-x

Lee JH, Lequette Y, Greenberg EP (2006) Activity of purified QscR, a Pseudomonas aeruginosa orphan quorum-sensing transcription factor. Mol Microbiol 59:602–609. https://doi.org/10.1111/j.1365-2958.2005.04960.x

Lee J, Wu J, Deng Y, Wang J, Wang C, Wang J, Chang C, Dong Y, Williams P, Zhang LH (2013) A cell-cell communication signal integrates quorum sensing and stress response. Nat Chem Biol 9:339–343. https://doi.org/10.1038/nchembio.1225

Lewenza S, Conway B, Greenberg EP, Sokol PA (1999) Quorum sensing in Burkholderia cepacia: identification of the LuxRI homologs CepRI. J Bacteriol 181:748–756 doi: Not available

Lina G, Jarraud S, Ji G, Greenland T, Pedraza A, Etienne J, Novick RP, Vandenesch F (1998) Transmembrane topology and histidine protein kinase activity of AgrC, the agr signal receptor in *Staphylococcus aureus*. Mol Microbiol 28:655–662. https://doi.org/10.1046/j.1365-2958.1998.00830.x

Lintz MJ, Oinuma K, Wysoczynski CL, Greenberg EP, Churchill ME (2011) Crystal structure of QscR, a Pseudomonas aeruginosa quorum sensing signal receptor. Proc Natl Acad Sci U S A 108:15763–15768. https://doi.org/10.1073/pnas.1112398108

Lupas A (1996) Prediction and analysis of coiled-coil structures. Methods Enzymol 266:513–525 doi: Not Available

Lupp C, Ruby EG (2005) Vibrio fischeri uses two quorum-sensing systems for the regulation of early and late colonization factors. J Bacteriol 187:3620–3629. https://doi.org/10.1128/JB.187.11.3620-3629.2005

Lyon GJ, Mayville P, Muir TW, Novick RP (2000) Rational design of a global inhibitor of the virulence response in *Staphylococcus aureus*, based in part on localization of the site of inhibition to the receptor-histidine kinase. AgrC Proc Natl Acad Sci U S A 97:13330–13335. https://doi.org/10.1073/pnas.97.24.13330

Lyon GJ, Wright JS, Muir TW, Novick RP (2002) Key determinants of receptor activation in the agr autoinducing peptides of *Staphylococcus aureus*. Biochemistry 41:10095–10104. https://doi.org/10.1021/bi026049u

Lyristis M, Bryant AE, Sloan J, Awad MM, Nisbet IT, Stevens DL, Rood JI (1994) Identification and molecular analysis of a locus that regulates extracellular toxin production in *Clostridium perfringens*. Mol Microbiol 12:761–777. https://doi.org/10.1111/j.1365-2958.1994.tb01063.x

Mayville P, Ji G, Beavis R, Yang H, Goger M, Novick RP, Muir TW (1999) Structure-activity analysis of synthetic autoinducing thiolactone peptides from *Staphylococcus aureus* responsible for virulence. Proc Natl Acad Sci U S A 96:1218–1223. https://doi.org/10.1046/j.1365-2958.2001.02539.x

McClean KH, Winson MK, Fish L, Taylor A, Chhabra SR, Camara M, Daykin M, Lamb JH, Swift S, Bycroft BW, Stewart GS, Williams P (1997) Quorum sensing and Chromobacterium violaceum: exploitation of violacein production and inhibition for the detection of N-acylhomoserine lactones. Microbiology 143:3703–3711. https://doi.org/10.1099/00221287-143-12-3703

MDowell P, Affas Z, Reynolds C, Holden MT, Wood SJ, Saint S, Cockayne A, Hill PJ, Dodd CE, Bycroft BW, Chan WC, Williams P (2001) Structure, activity and evolution of the group I thiolactone peptide quorum-sensing system of *Staphylococcus aureus*. Mol Microbiol 41:503–512. https://doi.org/10.1046/j.1365-2958.2001.02539.x

Miller ST, Xavier KB, Campagna SR, Taga ME, Semmelhack MF, Bassler BL, Hughson FM (2004) Salmonella typhimurium recognizes a chemically distinct form of the bacterial quorum-sensing signal AI-2. Mol Cell 15:677–687. https://doi.org/10.1016/j.molcel.2004.07.020

Milton DL, Hardman A, Camara M, Chhabra SR, Bycroft BW, Stewart GS, Williams P (1997) Quorum sensing in Vibrio anguillarum: characterization of the vanI/vanR locus and identification of the autoinducer N-(3-oxodecanoyl)-L-homoserine lactone. J Bacteriol 179:3004–3012. https://doi.org/10.1128/jb.179.9.3004-3012.1997

Moore GF, Audrey S, Barker M, Bond L, Bonell C, Hardeman W, Moore L, O'Cathain A, Tinati T, Wight D, Baird J (2015) Process evaluation of complex interventions: Medical Research Council guidance. BMJ 350:h1258. https://doi.org/10.1136/bmj.h1258

Nakayama J, Cao Y, Horii T, Sakuda S, Akkermans AD, de Vos WM, Nagasawa H (2001) Gelatinase biosynthesis-activating pheromone: a peptide lactone that mediates a quorum sensing in *Enterococcus faecalis*. Mol Microbiol 41:145–154 doi: Not Available

Nasser W, Bouillant ML, Salmond G, Reverchon S (1998) Characterization of the Erwinia chrysanthemi expI-expR locus directing the synthesis of two N-acyl-homoserine lactone signal molecules. Mol Microbiol 29:1391–1405 doi: Not available

Ng WL, Bassler BL (2009) Bacterial quorum-sensing network architectures. Annu Rev Genet 43:197–222. https://doi.org/10.1146/annurev-genet-102108-134304

Ng WL, Perez LJ, Wei Y, Kraml C, Semmelhack MF, Bassler BL (2011) Signal production and detection specificity in Vibrio CqsA/CqsS quorum-sensing systems. Mol Microbiol 79:1407–1417. https://doi.org/10.1111/j.1365-2958.2011.07548.x

Nishiguchi K, Nagata K, Tanokura M, Sonomoto K, Nakayama J (2009) Structure-Activity Relationship of Gelatinase Biosynthesis-Activating Pheromone of Enterococcus faecalis. J Bacteriol 191(2):641–650

Novick RP, Geisinger E (2008) Quorum sensing in Staphylococci. Annu Rev Genet 42:541–564. https://doi.org/10.1146/annurev.genet.42.110807.091640

Novick RP, Ross HF, Projan SJ, Kornblum J, Kreiswirth B, Moghazeh S (1993) Synthesis of staphylococcal virulence factors is controlled by a regulatory RNA molecule. EMBO J 12:3967–3975 doi: Not Available

Novick RP, Projan SJ, Kornblum J, Ross HF, Ji G, Kreiswirth B, Vandenesch F, Moghazeh S (1995) The agr P2 operon: an autocatalytic sensory transduction system in *Staphylococcus aureus*. Mol Gen Genet 248:446–458 doi: Not Available

Ohtani K, Yuan Y, Hassan S, Wang R, Wang Y, Shimizu T (2009) Virulence gene regulation by the agr system in *Clostridium perfringens*. J Bacteriol 191:3919–3927. https://doi.org/10.1128/JB.01455-08

Papenfort K, Bassler BL (2016) Quorum sensing signal-response systems in Gram-negative bacteria. Nat Rev Microbiol 14:576–588. https://doi.org/10.1038/nrmicro.2016.89

Papenfort K, Silpe JE, Schramma KR, Cong JP, Seyedsayamdost MR, Bassler BL (2017) A *Vibrio cholerae* autoinducer-receptor pair that controls biofilm formation. https://doi.org/10.1038/nchembio.2336

Passador L, Cook JM, Gambello MJ, Rust L, Iglewski BH (1993) Expression of Pseudomonas aeruginosa virulence genes requires cell-to-cell communication. Science 260:1127–1130 doi: Not available

Pereira H, Azevedo F, Rego A, Sousa MJ, Chaves SR, Côrte-Real M (2013) The protective role of yeast cathepsin D in acetic acid-induced apoptosis depends on ANT (Aac2p) but not on the voltage-dependent channel (Por1p). FEBS Lett 587:200–205. https://doi.org/10.1016/j.febslet.2012.11.025

Pesci EC, Milbank JB, Pearson JP, McKnight S, Kende AS, Greenberg EP, Iglewski BH (1999) Quinolone signalling in the cell-to-cell communication system of *Pseudomonas aeruginosa*. Proc Natl Acad Sci U S A 96:11229–11234 doi: Not Available

Pierson LS 3rd, Keppenne VD, Wood DW (1994) Phenazine antibiotic biosynthesis in Pseudomonas aureofaciens 30–84 is regulated by PhzR in response to cell density. J Bacteriol 176(13):3966–3974 doi: Not available

Piper KR, Beck Von Bodman S, Hwang I, Farrand SK (1999) Hierarchical gene regulatory systems arising from fortuitous gene associations: controlling quorum sensing by the opine regulon in Agrobacterium. Mol Microbiol 32:1077–1089 doi: Not available

Pirhonen M, Flego D, Heikinheimo R, Palva ET (1993) A small diffusible signal molecule is responsible for the global control of virulence and exoenzyme production in the plant pathogen Erwinia carotovora. EMBO J 12:2467–2476 doi: Not available

Puskas A, Greenberg EP, Kaplan S, Schaefer AL (1997) A quorum-sensing system in the free-living photosynthetic bacterium Rhodobacter sphaeroides. J Bacteriol 179:7530–7537. https://doi.org/10.1128/jb.179.23.7530-7537.1997

Qin X, Singh KV, Weinstock GM, Murray BE (2000) Effects of *Enterococcus faecalis* fsr genes on production of gelatinase and a serine protease and virulence. Infect Immun 68:2579–2586 doi: Not Available

Qiu R, Pei W, Zhang L, Lin J, Ji G (2005) Identification of the putative staphylococcal AgrB catalytic residues involving the proteolytic cleavage of AgrD to generate autoinducing peptide. J Biol Chem 280:16695–16704. https://doi.org/10.1074/jbc.M411372200

Rasmussen TB, Givskov M (2006) Quorum-sensing inhibitors as anti-pathogenic drugs. Int J Med Microbiol 296:149–161. https://doi.org/10.1016/j.ijmm.2006.02.005

Riedel CU, Monk IR, Casey PG, Waidmann MS, Gahan CG, Hill C (2009) AgrD-dependent quorum sensing affects biofilm formation, invasion, virulence and global gene expression profiles in *Listeria monocytogenes*. Mol Microbiol 71:1177–1189. https://doi.org/10.1111/j.1365-2958.2008.06589.x

Ryan RP, An SQ, Allan JH, McCarthy Y, Dow JM (2015) The DSF family of cell-cell signals: an expanding class of bacterial virulence regulators. PLoS Pathog 11:e1004986. https://doi.org/10.1371/journal.ppat.1004986

Saenz HL, Augsburger V, Vuong C, Jack RW, Götz F, Otto M (2000) Inducible expression and cellular location of AgrB, a protein involved in the maturation of the staphylococcal quorum-sensing pheromone. Arch Microbiol 174:452–455. https://doi.org/10.1007/s002030000223

Sanchez-Contreras M, Bauer WD, Gao M, Robinson JB, Allan Downie J (2007) Quorum-sensing regulation in rhizobia and its role in symbiotic interactions with legumes. Philos Trans R Soc Lond Ser B Biol Sci 362(1483):1149–1163. https://doi.org/10.1098/rstb.2007.2041

Schaefer AL, Greenberg EP, Oliver CM, Oda Y, Huang JJ, Bittan-Banin G, Peres CM, Schmidt S, Juhaszova K, Sufrin JR, Harwood CS (2008) A new class of homoserine lactone quorum-sensing signals. Nature 454:595–599. https://doi.org/10.1038/nature07088

Schauder S, Bassler BL (2001) The languages of bacteria. Genes Dev 15:1468–1480. https://doi.org/10.1101/gad.899601

Sidote DJ, Barbieri CM, Wu T, Stock AM (2008) Structure of the *Staphylococcus aureus* AgrA LytTR domain bound to DNA reveals a beta fold with an unusual mode of binding. Structure 16:727–735. https://doi.org/10.1016/j.str.2008.02.011

Slamti L, Lereclus D (2002) A cell-cell signalling peptide activates the PlcR virulence regulon in bacteria of the *Bacillus cereus* group. EMBO J 21:4550–4559. https://doi.org/10.1093/emboj/cdf450

Smith D, Wang JH, Swatton JE, Davenport P, Price B, Mikkelsen H, Stickland H, Nishikawa K, Gardiol N, Spring DR, Welch M (2006) Variations on a theme: diverse N-acyl homoserine lactone-mediated quorum sensing mechanisms in gram-negative bacteria. Sci Prog 89:167–211 doi: Not available

Sturme MH, Nakayama J, Molenaar D, Murakami Y, Kunugi R, Fujii T, Vaughan EE, Kleerebezem M, de Vos WM (2005) An agr-like two-component regulatory system in *Lactobacillus plantarum* is involved in production of a novel cyclic peptide and regulation of adherence. J Bacteriol 187:5224–5235. https://doi.org/10.1128/JB.187.15.5224-5235.2005

Swem LR, Swem DL, Wingreen NS, Bassler BL (2008) Deducing receptor signalling parameters from in vivo analysis: LuxN/AI-1 quorum sensing in Vibrio harveyi. Cell 134:461–473. https://doi.org/10.1016/j.cell.2008.06.023

Swem LR, Swem DL, O'Loughlin CT, Gatmaitan R, Zhao B, Ulrich SM, Bassler BL (2009) A quorum-sensing antagonist targets both membrane-bound and cytoplasmic receptors and controls bacterial pathogenicity. Mol Cell 35:143–153. https://doi.org/10.1016/j.molcel.2009.05.029

Swift S, Winson MK, Chan PF, Bainton NJ, Birdsall M, Reeves PJ, Rees CE, Chhabra SR, Hill PJ, Throup JP, Bycroft BW, Salmond GPC, Williams P, Stewart GSAB (1993) A novel strategy for the isolation of luxI homologues: evidence for the widespread distribution of a LuxR:LuxI superfamily in enteric bacteria. Mol Microbiol 10:511–520 doi: Not available

Swift S, Karlyshev AV, Fish L, Durant EL, Winson MK, Chhabra SR, Williams P, Macintyre S, Stewart GS (1997) Quorum sensing in Aeromonas hydrophila and Aeromonas salmonicida: identification of the LuxRI homologs AhyRI and AsaRI and their cognate N-acylhomoserine lactone signal molecules. J Bacteriol 179:5271–5281 doi: Not available

Tal-Gan Y, Ivancic M, Cornilescu G, Cornilescu CC, Blackwell HE (2013) Structural characterization of native autoinducing peptides and abiotic analogues reveals key features essential for activation and inhibition of an AgrC quorum sensing receptor in *Staphylococcus aureus*. J Am Chem Soc 135:18436–18444. https://doi.org/10.1021/ja407533e

Tal-Gan Y, Ivancic M, Cornilescu G, Blackwell HE (2016) Characterization of structural elements in native autoinducing peptides and non-native analogues that permit the differential modulation of AgrC-type quorum sensing receptors in *Staphylococcus aureus*. Org Biomol Chem 14:113–121. https://doi.org/10.1039/c5ob01735a

Thoendel M, Horswill AR (2009) Identification of *Staphylococcus aureus* AgrD residues required for autoinducing peptide biosynthesis. J Biol Chem 284:21828–21838. https://doi.org/10.1074/jbc.M109.031757

Thoendel M, Horswill AR (2010) Biosynthesis of peptide signals in gram-positive bacteria. Adv Appl Microbiol 71:91–112. https://doi.org/10.1016/S0065-2164(10)71004-2

Thoendel M, Kavanaugh JS, Flack CE, Horswill AR (2011) Peptide signalling in the staphylococci. Chem Rev 111:117–151. https://doi.org/10.1021/cr100370n

Throup JP, Camara M, Briggs GS, Winson MK, Chhabra SR, Bycroft BW, Williams P, Stewart GS (1995) Characterisation of the yenI/yenR locus from Yersinia enterocolitica mediating the synthesis of two N-acylhomoserine lactone signal molecules. Mol Microbiol 17:345–356. https://doi.org/10.1111/j.1365-2958.1995.mmi_17020345.x

Tusnády GE, Simon I (1998) Principles governing amino acid composition of integral membrane proteins: application to topology prediction. J Mol Biol 283:489–506. https://doi.org/10.1006/jmbi.1998.2107

Vannini A, Volpari C, Gargioli C, Muraglia E, Cortese R, De Francesco R, Neddermann P, Marco SD (2002) The crystal structure of the quorum sensing protein TraR bound to its autoinducer and target DNA. EMBO J 21:4393–4401 doi: Not available

Vidal JE, Chen J, Li J, McClane BA (2009) Use of an EZ-Tn5-based random mutagenesis system to identify a novel toxin regulatory locus in *Clostridium perfringens* strain 13. PLoS One 4:e6232. https://doi.org/10.1371/journal.pone.0006232

Wang LH, He Y, Gao Y, Wu JE, Dong YH, He C, Wang SX, Weng LX, Xu JL, Tay L, Fang RX, Zhang LH (2004) A bacterial cell-cell communication signal with cross-kingdom structural analogues. Mol Microbiol 51:903–912. https://doi.org/10.1046/j.1365-2958.2003.03883.x

Whitehead NA, Barnard AML, Slater H, Simpson NJL, Salmond GPC (2001) Quorum-sensing in Gram-negative bacteria. FEMS Microbiol Rev 25:365–404. https://doi.org/10.1111/j.1574-6976.2001.tb00583.x

Williams P, Winzer K, Chan WC, Cámara M (2007) Look who's talking: communication and quorum sensing in the bacterial world. Phil Trans R Soc B 362:1119–1134. https://doi.org/10.1098/rstb.2007.2039

Wood DW, Gong F, Daykin MM, Williams P, Pierson LS 3rd (1997) N-acyl-homoserine lactone-mediated regulation of phenazine gene expression by Pseudomonas aureofaciens 30-84 in the wheat rhizosphere. J Bacteriol 179(24):7663–7670 doi: Not available

Wuster A, Babu MM (2008) Conservation and evolutionary dynamics of the agr cell-to-cell communication system across firmicutes. J Bacteriol 190:743–746. https://doi.org/10.1128/JB.01135-07

Yang T, Tal-Gan Y, Paharik AE, Horswill AR, Blackwell HE (2016) Structure-function analyses of a *Staphylococcus epidermidis* autoinducing peptide reveals motifs critical for AgrC-type receptor modulation. ACS Chem Biol 11:1982–1991. https://doi.org/10.1021/acschembio.6b00120

Zhang L, Gray L, Novick RP, Ji G (2002) Transmembrane topology of AgrB, the protein involved in the post-translational modification of AgrD in *Staphylococcus aureus*. J Biol Chem 277:34736–34742. https://doi.org/10.1074/jbc.M205367200

Zhou L, Zhang LH, Cámara M, He YW (2017) The DSF family of quorum sensing signals: diversity, biosynthesis, and turnover. Trends Microbiol 25:293–303. https://doi.org/10.1016/j.tim.2016.11.013

Fundamentals of Bacterial Biofilm: Present State of Art

3

Soumya Pandit, Shruti Sarode,
and Kuppam Chandrasekhar

Abstract

Bacteria are ubiquitous and it is reported that among them surface attached bacteria is over 99% population wise. However, the attached growth or sessile life of bacteria poses a different problem in human life. Chronic infection, contamination in food industries and biofouling in industrial materials are serious challenges which need to overcome. It is imperative to understand the bacterial adhesion and subsequent biofilm development to study biofilm-associated deceases and mitigating biofouling. In the present book chapter, the technical know-how of bacterial biofilm development has been depicted. A thorough understanding of the fundamental principles of bacterial biofilm would help to perceive new aspects of biofouling and pathogenicity associated with biofilm and how fouling could be controlled and contamination can be prevented. In this chapter, parameters affecting biofilm formation, the role of extracellular polymeric substances (EPS) and eDNA, the factors of materials which influence attachment and detachment of biofilm have been clearly described. The present chapter highlights the major factors involved in the signaling system to promote Biofilm formation.

Authors Soumya Pandit and Shruti Sarode have equally contributed for this chapter.

S. Sarode · S. Pandit (✉)
Department of biotechnology, IIT, Kharagpur, India

K. Chandrasekhar (✉)
Bio-Engineering and Environmental Science (BEES), CSIR-IICT, Hyderabad, India

School of Applied Bioscience, Kyungpook National University (KNU), Daegu, South Korea

V. C. Kalia (ed.), *Quorum Sensing and its Biotechnological Applications*,
https://doi.org/10.1007/978-981-13-0848-2_3

3.1 Introduction

Biofilm is a thin but strong layer of mucilage adhering to a surface which contains an organized community of microorganisms. The cells in biofilms differ from free-living cells for the reason that their low growth rate, extracellular matrix formation, and type of gene regulation. Free bacteria and bacteria growing in a biofilm express different sets of genes. Both extracellular, as well as intracellular signals are needed to express the genes needed to construct the biofilm. Studies have shown that attachment of microbes to a solid surface enhances their rate of growth, a phenomenon known as "bottle effect". This shows that a biofilm confers several advantages on bacteria (Donlan 2002). The formation of biofilm is a multistage process. In the first stage, a conditioning layer is formed that provides nutrients to the microbes in the biofilm. This is followed by bacterial adhesion, growth, and expansion. It can exist on various kinds of surfaces like soil, metal, wood, glass, plastic, food products, etc. A single or multiple species can form biofilms in monolayer or multilayer with three-dimensional structure.

3.2 Biofilm Structure

Biofilms are comprised of microbes on a matrix of extracellular polymeric substances. Biofilm development in flow cells is monitored by Confocal Scanning Laser Microscope (CSLM), which aids in examining the biofilm without disrupting the microbial community. The composition of biofilm has been determined successfully using CSLM. 15% of the biofilm is composed of microbial communities and the rest 85% is extracellular polymers. The extracellular polysaccharides (EPS) are mainly comprised of polysaccharides, which are neutral or anionic. The anionic properties of EPS are confirmed by the presence of uronic acids (mannuronic; D-galacturonic; D-galacuronic) or ketal linked pyruvate. These anionic linkages in combination with divalent cations like calcium or magnesium help in crosslinking and confer greater binding force in a developed biofilm. These extracellular polymeric substances (EPS) backbone comprises of 1, 3 or 1, 4 β linked hexose residues (Bishop 1997). The quantity of EPS production is directly related to the age of the biofilm, that is, more the age of a biofilm, greater is the amount of EPS produced. EPS synthesis is upsurge by the limitation of nitrogen, phosphate, and potassium.

3.3 Role of Biofilms in Microbial Communities

Microbial communities adopt biofilm strategy because of the following reasons:

3.3.1 Protection from Environment

The bacteria which inhabit the biofilm are sheltered by the EPS. The EPS matrix acts as an anion exchanger hence preventing the contact of antimicrobial substances.

It also limits the diffusion of harmful substances from the environment into the biofilm. Furthermore, PS sequesters toxins, metal ions, and cations. It also protects the microbes from different environmental stresses like UV radiation, pH shift and osmotic shock (Flemming 1993).

3.3.2 Nutrient Availability

The water channels are an effective means for nutrient exchange and the transfer of metabolic intermediates with the main part of the aqueous phase, which enhances the nutrient accessibility and elimination of potential toxins. The small colonies in biofilm usually comprise a wide variety of microbial populations, which can be metabolically cooperative, therefore, their closeness enables interspecies substrate transport, exchange, and removal of metabolic intermediates (Bishop 1997). Biofilm offers an ideal situation for the formation of syntrophic relations.

3.3.3 Acquisition of New Genetic Trait

Horizontal gene transfer is a vital feature in a biofilm where the transmission of DNA between different genomes occurs in a manner other than traditional reproduction. It is vital for growth and genetic variety of a microbial population. Acquiring new genetic traits increases the probability of a microbial community to copy the essential genes to become a lively member of biofilm populations. It is well established that the biofilm forming communities transcribe different genes and the phenotypic character is the expression of specific genotypic characters. For example, alginate production is increased four times in biofilm-associated cells in which algC gene was transcribed than that in planktonic cells. Studies have shown that pulmonary isolates are mucoid because of the synthesis of large amounts of alginate. Furthermore, sigma factor undesirably controls the synthesis of flagellum but it positively regulates alginate synthesis. Hence, when the synthesis of EPS and alginate are increased in the microbes associated with the biofilm, the synthesis of flagella decreases. Accordingly, the bacteria must differentiate into biofilm-associated cells to become an effective member of biofilm community (Christensen et al. 1998). Hence, the synthesis of flagella must be repressed because it destabilizes the biofilm.

3.4 Factors Favoring Biofilm Formation

Following are the various factors which affect the formation of biofilms:

3.4.1 Substratum Effect

The microbial colonies upsurge faster in rough surfaces due to high surface area. The wettability of surfaces indicated high surface free energy. It was reported that low surface energy of the surface prevents bacterial attachment. However, surfaces with high surface free energies tend to be more hydrophilic. Usually, these surfaces demonstrate more bacterial attachment than hydrophobic surfaces (Watnick and Kolter 2000).

3.4.2 Conditioning Film

Conditioning film is the coating of polymers on a solid surface when it is exposed to the aqueous medium. The rate and range of bacterial attachment are affected by this chemical modification. "Acquired pellicle" formed on tooth enamel surface consists of albumin, lysozyme, glycoprotein, phosphoproteins, and lipids. The pellicle conditioned surfaces are colonized by the bacteria residing in the oral cavity within hours of contact with these surfaces. The attachment of bacteria to biomaterials is affected by host produced conditioning films like blood, saliva, urine, tears and respiratory secretions (Costerton et al. 1999).

3.4.3 Hydrodynamics

Studies have shown the biofilm response is altered by hydrodynamic conditions such as turbulent and laminar flow. Bacterial biofilms grown under both these conditions were found 'patchy'; nevertheless, the ones grew under laminar flow consisted of patchy cell masses parted by interstitial voids, and the ones grown under turbulent flow are elongated 'streamers' and they oscillate in the bulk fluid. Cell size and motility also affect the association of cells with the surface (Kostakioti et al. 2013).

3.4.4 Characteristics of Aqueous Medium

The biofilm formation and microbial attachment in various aqueous systems are affected by season. Several physicochemical characteristics of aqueous medium like pH, ionic strength, nutrient levels also play a vital role in the rate of bacterial attachment to the surfaces. Studies have shown that the increase in concentration of various cations like calcium, sodium, ferric ions helps in neutralization of negative charge in cell surface and EPS. This eventually reduces the repulsive forces between the cell surface and glass substrate; helps in the attachment of *Pseudomonas fluorescens* (Dufour et al. 2010).

3.4.5 Interaction of Bacterial Cells with Water Molecules

Various techniques are adopted to investigate the relative hydrophilicity of the microbial cells. They are as follows:

1. Adherence to polystyrene.
2. The separation of two-phase system containing hydrocarbon (hydrophobic part) and aqueous phase containing dextran polymer and polyethylene glycol clubbed with hydrophobic groups.
3. Hydrophobic interaction chromatography using octyl agarose gel preparations (Flemming 1993).

There is a high chance that none of the methods effectively measures the degree of hydrophobicity (or hydrophilicity). Charged or to neutral surface polymers can also cause hydrophobicity. For instance, the presence of neutral O-specific side-chain polysaccharides of the lipopolysaccharide dominating cell surface causes the R-forms of *Salmonella typhimurium* to indicate both hydrophobic and charged hydrophilic properties whereas the S-forms to indicate only extensive non-charged hydrophilicity. A significant range of relative surface hydrophobicity or hydrophilicity has been observed by the implementation of these methods to a several types of bacteria. The surface of oral *streptococci* is highly hydrophobic. Environmental conditions may play a major determinative role. Proteins and amphipathic polymers are the surface components which may contribute to hydrophobicity. Surface M-protein and lipoteichoic acid both are considered to be the primary components to both the hydrophobic properties and adherence of *streptococci* to eukaryotic cells. More exposure of lipid A at the surface of the outer membrane in R-forms of *S. typhimuriuma* rendered high degree of hydrophobicity in R-forms as compared to S-forms (Garrett et al. 2008).

3.4.6 Signaling System to Promote Biofilm Formation

A biofilm is made of a matrix of polysaccharides, glycopeptides and lipids onto which bacteria are embedded. The bacterial population in a biofilm can be either homogeneous or homogeneous. The formation of a biofilm occurs in four distinct stages and each stage requires the expression of a different set of genes depending on internal and environmental stimuli which includes physical parameters temperature and pressure of the surface and biological parameters like cell density. Hence, the cells in a biofilm need to communicate with each other to regulate gene expression and form a complex structure that can be either monolayered or multilayered. The external stimuli can be registered and transmitted into the cell via three systems: the two-component system (TCS), quorum sensing (QS) and extra cytoplasmic function (ECF). Triggering biofilm formation also involves secondary messengers like cyclic guanosine monophosphate (c-di-GMP).

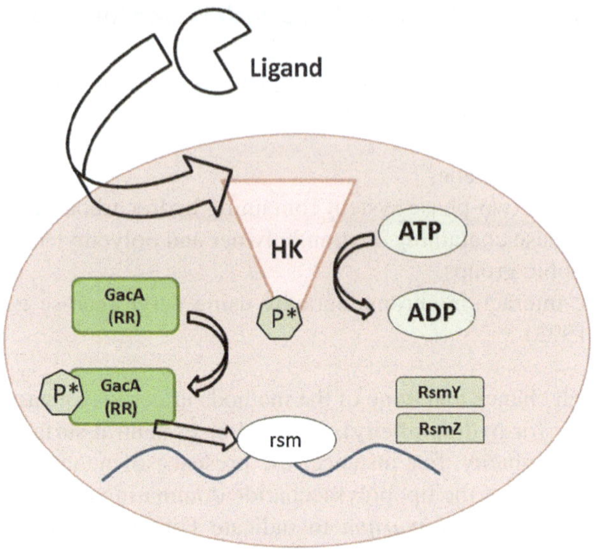

Fig. 3.1 Adhesion of bacteria and development of biofilm

Two-component signaling system comprises of GacS which is a histidine kinase (HK) that has a C-terminal kinase domain and an N-terminal ligand-binding domain and GacA which is a response regulator (RR) protein which is a transcriptional regulator (Fig. 3.1). The phosphorylation of a specific conserved histidine residue in HK is succeeded by the phosphorylation and activation of an aspartate residue in RR (Hall-Stoodley et al. 2004). The activated RR then encourages the expression of Rsm genes which code for RsmY and RsmZ which regulate the change between planktonic and sedentary forms. Two additional HKs associated with the Gac system are RetS, which suppresses formation of biofilm and LadS, which activates the formation of biofilms. Various organisms including *Pseudomonas aeruginosa*and *Staphylococcus auereus* have adopted TCS (Fig. 3.2). In gram-positive bacteria, biofilm formation is regulated by oligopeptides that do not travel into the cells but are recognized by sensor kinases present on the surface of the cell. An important protein involved in biofilm formation is PIA or polysaccharide intercellular adhesion which is encoded by the Ica operon and regulated by the regulator protein, IcaR, which suppresses PIA. The IcaABCD genes are vital for the development of biofilms and convey virulence to the bacteria.

The ECF signaling pathway comprises of a sigma factor and an antisigma factor which binds to and inhibits the sigma factor (Gupta et al. 2016). The periplasmic proteins, on receiving extracellular signals, degrade the antisigma factor that releases the free sigma factor that leads to the transcription of genes involved in biofilm formation (Sutherland 2001). In *P. aeruginosa*, the sigma factor is AlgU and the antisigma factor is Muc A whose C-terminal periplasmic domains are sliced by the act of a protease AlgW, thus releasing AlgU which activates the algUmucABCD operon. The algUmucABCD operon is involved in the production of EPS alginate and type IVA pili assembly that in turn facilitate in the proliferation of biofilm.

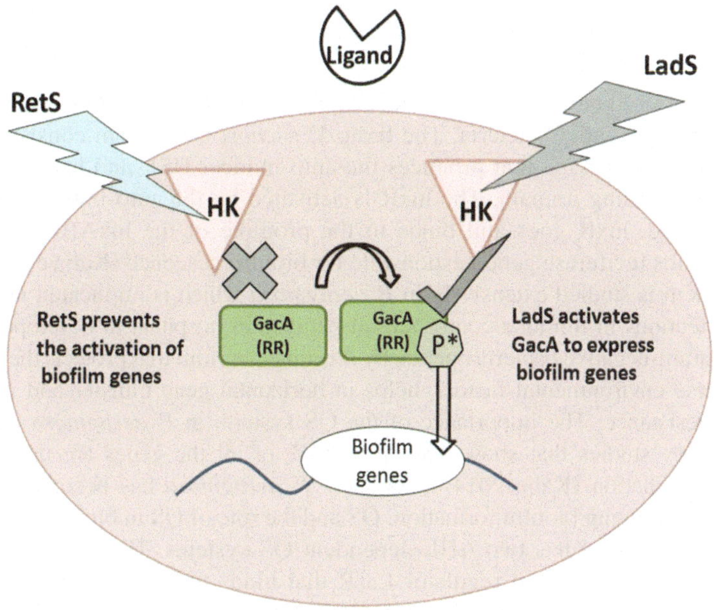

Fig. 3.2 GacS (HK)/GacA (RR) based 2-component signaling system for biofilm development in *P. aeruginosa*

3.4.6.1 Quorum Sensing (QS)

Biofilm formation takes place when bacteria present in a local community come together to form an adherent layer of cells. As such, a cell to cell communication protocol is necessary to regulate the genes involved in the formation of biofilms (Kim et al. 2011). QS is a mechanism, used by bacteria, for interspecies as well as intraspecies cell to cell communication. It plays an important role in many bacterial population behaviors like biofilm formation, virulence and antibiotic resistance in which bacteria need to alter their gene expression based on the local bacterial cell density. QS is mediated by chemical signals that are synthesized and secreted to the local environment by bacteria. These signaling molecules can pass down a concentration gradient, through the plasma membrane into the cytoplasm. The density of these chemical signals in the extracellular environment is proportional to the number of cells and acts as a measure of the cell density, allowing all the bacteria to express a specific gene at the same time, that is, in unison. This capability is crucial to certain process like virulence where the bacteria need to go undetected till they have reached a threshold concentration (Kumar and Anand 1998).

QS Systems were first observed in the marine bacteria *Vibrio fischeri*where they are involved in the regulation of bioluminescence. Since, then they have been observed in many other bacterial species, including, *Escherichia coli*, *Salmonella enterica*, *Streptococcus pneumoniae* and Yersinia. QS system can be broadly categorized into three types on the basis of the nature of the signaling molecules being used. Gram-positive bacteria use oligopeptides while gram-negative bacteria use AHLs (Huma et al. 2011). The third class of molecules called autoinducer 2 or AI-2, a furanosyl borate diester, is used by both the groups of bacteria (Vipin Chandra Kalia 2014a).

The *V. fischeri* QS system, which uses N-(3-oxohexanoyl)-homoserinelactone (HSL) as the QS signaling molecule, can be taken as model for the QS systems present in all gram-negative bacteria. The basic *V. fischeri* QS system consists of two proteins luxI, an enzyme that produces the auto-inducer HSL and luxR, a protein with a DNA-binding domain. The luxR is activated by the auto-inducer and after being activated, luxR goes and binds to the promote of the luxABCDE operon which contains luciferase gene, responsible for bioluminescence (Kalia et al. 2014).

QS System is studied extensively in *P. aeruginosa* which is implicated in causing chronic infections in immune-compromised patients in hospitals in developed countries. The gram-negative bacteria operate by forming a biofilm that protects the bacteria from adverse environmental factors, helps in horizontal gene transfer and improves antibiotic resistance. The importance of the QS system, in *P. aeruginosa,* has been established by studies that show that up to 11% of all the genes are under AHL-dependent regulation (Kalia 2014b). As such, *P. aeruginosa* has becomes a model organism for studying biofilm formation, QS and the role of QS in biofilm formation.

The *P. aeruginosa* has two AHL-dependent QS systems. The las system comprises of the transcriptional regulator LasR that binds to the corresponding AHL signaling molecule, N-(3-oxododecanoyl)-L-homoserine lactone (3-oxo-C12-HSL), while the RhI systems comprises of the transcriptional regulator RhI that binds to the corresponding AHL signaling molecule, N-butyryl-L-homoserine lactone (C4-HSL). Studies have shown that the biofilm formed by lasI^{-}strains was flat, undifferentiated and highly unstable, dispersing from the substratum on exposure to Sodium Dodecyl Sulphate (SDS) while the biofilm formed by the wild type strain remained intact. While, no such effect was observed in RhI^{-} strains, subsequent studies have shown that RhI has a role in biofilm formation but the exact mechanism is still unknown (Kalia 2013).

The cells are held on the biofilm by a matrix extracellular polymeric substance (EPS) that consists of polysaccharides, proteins and DNA. Out of the five gene clusters in *P. aeruginosa* that are assumed to function in EPS synthesis, only one, the pel biosynthetic operon has been shown definitively to be under QS regulation. The pel operon has seven genes pelABCDEFG that code for proteins accountable for the formation of glucose-rich biofilm exo-polysaccharide which imparts a wrinkled colony phenotype to the biofilm (Kalia et al. 2014). The *P. aeruginosa* cells grown in the presence of DNAse I completely dissolved and could not form biofilm. Biofilms formed by lasI and RhI QS mutants' showed lower extracellular DNA levels and improved susceptibility to SDS treatment. This shows that, while polysaccharides are an important structural section of the biofilm matrix, extracellular DNA is a crucial component that holds the bacterial cells together. In *P. aeruginosa* biofilms, the extracellular DNA comes from the lysis of cells which takes place by a QS regulated pathway. There is a second QS independent pathway but, it generates substantially less DNA (Miller and Bassler 2001).

Rhamnolipids are amphipathic glycolipids that have multiple roles in formation and maintenance of *P. aeruginosa* biofilm. Their production is under the control of their genes, RhlA, RhlB and RhlC all which are under the control of the QS system. They help in the formation and maintenance of the open channels surrounding micro-colonies that provide access to nutrients and oxygen and help in removal of waste products (Vipin Chandra Kalia 2014a). Therefore, inhibition of QS molecules and quorum quenching has become a potent approach to mitigate bacterial biofilm dispersion or preventing film formation (Koul and Kalia 2017).

3.4.7 Extracellular Polymeric Substances (EPS)

EPS is a high density complex compound released by bacteria to their surroundings. EPS endowed both structural and functional unity to bacterial biofilm. The quality of EPS determines the physiochemical properties. The EPS consists of major exo-polysaccharides along with protein, lipid, DNA and humic materials. EPS are considered as cement materials for biofilm which facilitate biofilm to hold firmly on the surface for sessile growth.

Bacteria produce a wide array of polysaccharides which includes exo-polysaccharides. It is composed of monosugar blended with a few non carbohydrate compounds like acetate, pyruvate, phosphate etc. EPS is therefore considered as an important raw material in different food processing and pharmaceutical manufacturing units. Efforts have been made to utilize gum like EPS which is potential enough to replace traditional plant and algal gums. Dextran, a major component in EPS has been used in panettone and other breads in the bakery industry (Flemming et al. 2007). EPS plays an important role in biofilm protection. Bacterial pathogenicity is bestowed with capsular EPS which protects biofilm from host immune system. Few strains of lactic acid bacteria, e.g., *Lactococcuslactis*sub sp. cremoris get a gelatinous texture to fermented milk products due to EPS. Rhizodeposits consists of EPS helps N_2 fixing bacteria to attach to plant roots and cause infection mediated by EPS (Flemming et al. 2007). Various roles of EPS in bacterial biofilm development have been tabulated elsewhere (Table 3.1). The di-valent cation is important for EPS to form biofilm structure which facilitates in binding opposite charged link of EPS and form flock in activated sludge. It was found that sulfate groups ($-SO_3^-$) and carboxylic groups ($-COO^-$) in polysaccharide and protein of EPS helps in bridging with divalent cations (Gacesa 1998). EPS has the significant function in bioaggregation, which is a common natural process and plays an important part in biological wastewater treatment technology. The flocculability, settle-ability, and dewater-ability for flocs, granules and shear resistance for biofilms largely depend on EPS mediating bioaggregation (Ding et al. 2015).

3.4.8 Role of eDNA

Extracellular DNA (eDNA) plays a significant part in biofilm formation. The eDNA produced due to autolysis of the cell as a consequence of controlled quorum sensing. The eDNA is the major component of biofilm matrix that helps in initial

Table 3.1 Different role of EPS in bacterial biofilm development

Nature of EPS component	Effect of EPS component	Role in biofilm	References
Neutral polysaccharides	Constructive	Structural component	Flemming et al. (2007)
Amyloids		Structural component	
Charged or hydrophobic polysaccharides	Sorptive	Ion exchange, sorption	Bishop (1997)
Amphiphilic	Surface-active	Interface interactions	Flemming et al. (2007)
Membrane vesicles		Export from cell, sorption	
Lectins	Informative	Specificity, recognition	Flemming et al. (2007)
Nucleic acids		Genetic information	
Extracellular enzymes	Active	Polymer degradation	Bishop (1997)
Bacterial refractory polymers	Redox active	Electron donor or acceptor	Ding et al. (2015)
Various polymers	Nutritive	Source of C, N, P	Ding et al. (2015)

adhesion of bacteria to surface. The eDNA was found in the biofilm of *P. aeruginosa*; *Enterococcus faecalis*; *Streptococcus mutans*. Researchers have found a new avenue to target eDNA in the matrix to dismantle biofilm. Most of the studies showed that DNAase treatment is useful in dispersing biofilm at the initial stage of its formation. DNAase I along with antibiotic treatment was found very effective in mitigating biofilm. However, production of DNAase is made from bovine pancreas. The cost of this enzyme limits its usage when the bulk amount of is required to get rid of bacterial biofilm. Currently, a cheaper and promising way has been adopted where DNAase production started using extracellular nucleases in *E. coli*. The long-term stability and efficacy of bacterial nuclease must be explored in order to its bulk use in industry (Montanaro et al. 2011).

3.5 Bacterial Adhesion- Steps Involved in Biofilm Development

3.5.1 The Sessile Mode of Bacterial Growth

In nature, bacteria usually grow as part of a rigid local community that forms an adherent layer called biofilm. The transition of from a free floating planktonic mode of growth to sessile mode of growth confers many benefits upon the bacteria (Watnick and Kolter 2000). The bacterial biofilm surface comprises of a very fluid and intricate interface with the environment which must be upheld by the bacterium for surviving amongst the omnipresent antibacterial agents and factors in nature.

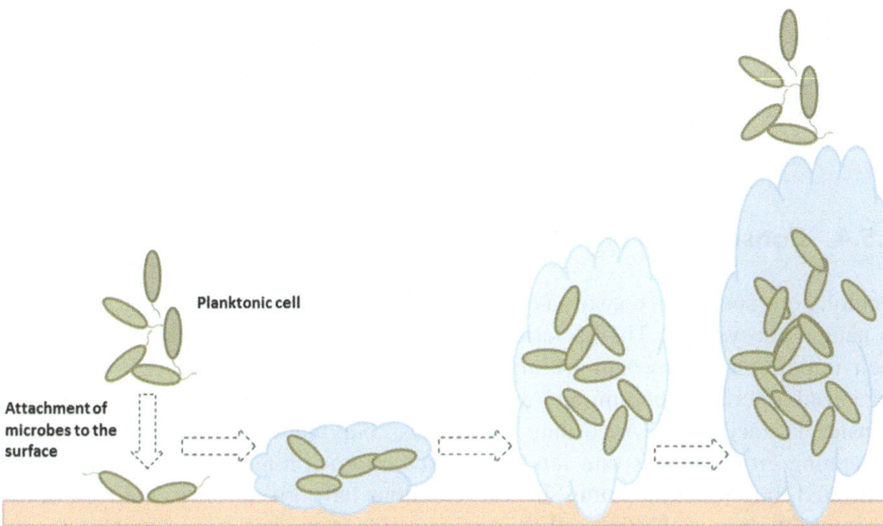

Fig. 3.3 RetS and LadS signaling system for *P. aeruginosa* biofilm formation

Such surfaces strengthen the differences between bacteria growing in natural and pathogenic ecosystems, therefore, novel and innovative techniques have been developed for the better understanding of these bacterial surfaces, i.e., for visualizing the bacterial surface structures as well as for stabilizing them (Fig. 3.3).

3.5.2 Bacterial Glycocalyx In Vivo In Vitro

The bacterial glycocalyces are complex substances containing polysaccharides moiety. They are present in the peptidoglycan layer in gram positive cells, whereas in gram negative cells, they are present outside the integral elements of the outer membrane. They have two subcategories, S layers and Capsules. S layers made up of regular arrays of glycoprotein subunits and while capsules are composed of fibrous matrix on the cell surface (McLean et al. 2001).

3.5.3 Microcolony Formation by Adherent Bacteria

Cells of the organisms in natural and pathogenic ecosystems most often replicate within a hydrated exo-polysaccharide matrix because of universal bacterial glycocalyx production, so that the daughter cells are trapped and forms micro-colonies of identical cells. The glycocalyxes facilitate both, micro-colony formation and adhesion, which results in the formation of micro-colonies on surfaces where they multiply and in due course come together to form an adherent biofilm (Fig. 3.2). If nutrients are present, they are trapped and used by a very efficient bacterial strategy and the bacterial cells flourish in the glycocalyx-enclosed microcolonies which are adherent as well as mature to bigger dimensions, but are seldom macroscopic. This approach also provides them safety from unwanted chemicals, antibodies, surfactants, antibacterial substances and phagocytic amoebae and leukocytes (Singh et al. 2006).

3.5.4 Consortium Formation by Adherent Bacteria

Homologous species in consortia contribute to a wide variety of bacterial processes in natural ecosystems. The bacterial members of the same consortium hold on to their initial substratum, and are also attached to form cluster, hence forming an organized microbial community within which substrate transfer and hydrogen transfer is facilitated. According to studies, butyrate-oxidizing and methane-generating group in consortia can form mixed micro-colonies, even when they are cultivated in a liquid medium. Such observations have given rise to the idea of an inherent interbacterial affinity. Bacteria that reside in the gastrointestinal tract usually attack the digestive substrata in a muchfocused manner. The bacteria within an adherent micro-colony that are in an environment which has more nutrition produce a "critical mass" of cells and exoenzymes and products in order to generate a microniche and erode the substratum immediately beneath the micro-colony. Because of this topographically focused nature of natural bacterial processes, pits arise in the digestible substrata. The cellulose and metal plays important role in developing a micro-colony that sinks into the pit as it develops (Garrett et al. 2008).

3.6 Occurrence of Bacterial Biofilm: Predominance in Aquatic System

According to studies, a stream can be considered as a "microbiological reactor" which flows past bacterial biofilm populations on submerged surfaces and carries a few detached bacteria and dissolved nutrients. Sessile bacteria are predominant, they are also not limited to only natural aquatic ecosystems, they have also been discovered in hundreds of industrial systems from heat exchangers to the injection tubes (McLean et al. 2001).

3.7 Physiology of Biofilm Population

Bacterial cell tends to grow on suitable moister surfaces, if the nutrient is available on the solid-aqueous interface; they are suitable for bacterial colonization. Hence, surfaces in aquatic environments are rapidly deposited with an organic coating which is rich in polysaccharides of microbial origin and this organic layer is retained by the surface which is colonized by adherent bacteria. Enormous quantities of fibrous glycocalyx material are produced by the bacteria soon after initial adsorption to a surface, and to convert their reversible adhesion to irreversible adhesion, they fix themselves onto the surface. Most of the chemical structures of glycocalyces are observed to comprise polyanionic matrices which might behave like ion-exchange resins, which attract Ca^{+2} type cations and molecules. Glycocalyces of biofilm bacteria makes up the ion-exchange matrix and it "loads" from its exposed surface, hence to have favored access to the rare ligands, the cells are expected to get closer to the exposed surface (de la Fuente-Núñez et al. 2013). Nonetheless, in this mode of growth, large amounts of the fibrous matrix make up individual biofilms and deeper cells are not hindered by overlying cells, hence it encourages the access of the cells preset farther from the surface to common nutrients. It can be concluded from several experiments that an adherent biofilm called as "quasitissue" which is comprised of micro-colonies of different types of aquatic bacteria that may have measurable rates of respiration and nutrient uptake.

3.8 Effects of Material Properties on Bacterial Adhesion

Properties of material surface have profound influence on initial bacterial adhesion and biofilm development. Surface charge, hydrophobicity, topology, roughness are important parameters affecting bacterial adhesion and biofilm formation. It was noticed that surfaces having negatively charged or super hydrophobic surfaces, super hydrophilic surfaces, and nm-scale surface roughness has the tendency to repel bacteria and prevent fouling (Song et al. 2015). However, several other factors like effect of surface stiffness and topography (except for roughness) on bacterial adhesion is still not unveiled clearly.

3.8.1 Surface Charge

Surface charge plays an important role in determining the binding force between bacteria and the surface, and it has long been known to affect biofilm formation. Most bacterial cells are negatively charged; thus, in general, a positively charged surface is more prone to bacterial adhesion, and a negatively charged surface is more resistant to bacterial adhesion. Meanwhile, surfaces presenting certain cationic groups, such as quaternary ammonium and polyethylenimines, have antimicrobial activities and thus can kill the attached cells.

In principle, controlling bacterial adhesion with surface charge may not work in static systems since the dead cells present a barrier that reduces the charge and facilitates the adhesion of other bacterial cells. It was observed that some positively charged surfaces like quaternary ammonium ion showed antimicrobial properties. Coating of these materials on the material surface is therefore a strategy to avoid biofouling. Nevertheless, bacteria have remarkable ability to break these resistances by producing EPS to negate repelling effect and subsequently form biofilms.

3.8.2 Surface Energy

The surface energy indicated hydrophobicity of the surface. In most cases hydrophobicity of the surface helps bacteria to develop biofilm on surface. Hydrophilic surface brings wettability which sometimes hindered film formation. Recently, it was found that both super-hydrophobic and super-hydrophilic surfaces can facilitate in resisting bacterial attachment and biofilm development. For an example, lotus leaf with a high contact angle of 170° due to wax coating can resist adhesion. The low surface energy of waxes renders super-hydrophobicity to protect leaves from all kind of unwanted attachment.

3.8.3 Roughness and Topography

The topography and surface roughness plays important role in bacterial binding to surface. It is assumed that more surface roughness can increase effect contact area of material surface to attach bacteria. Roughness also helps bacteria to protect it from shear force. Henceforth, smooth surface can prevent biofilm formation. A roughness Ra of 0.2 μm was found threshold to prevent attachment of bacteria. Nevertheless, the efficacy of designed roughness on bacterial attachment can vary with size and shape of bacterial cells and other physicochemical factor.

3.9 Measuring Biofilm

Several methods have been adopted to study the bacterial adhesion and biofilm development. The development of biofilm and its structure, roughness, thickness can be analyzed using confocal laser scanning (CLSM), atomic force (AF), scanning electron (SE), and transmission electron (TE) microscopy. The ruthenium red can be applied to detect the specific polysaccharide in order to observe its association with cells. To understand the metabolism of biofilm, research explored the fluorescent in-situ hybridization (FISH) and quantified using CLSM. FISH has been widely used to check the viability of the cells in biofilm. Fluorescent stains are required to stain biofilm and different part associated with this. These stains are designed in such a manner that these will excite consequently emit light at a specific wavelength. Several of these stains like DAPI (4′, 6-diamidino-2-phenylindole), Syto 9 are targeted to probe DNAs and RNAs. The different fluorescent probes used in biofilm studies have been provided in Table 3.2. Customized flow cells are generally used to grow biofilm on different surface under varied physicochemical conditions (Gupta et al. 2016). Recently, researchers used two noninvasive technologies for study related to bacterial adherence and biofilm formation experiments. The quartz crystal microbalance (QCM) is a nanogram sensitive technique that utilizes acoustic waves generated by oscillating a piezoelectric, single crystal quartz plate to measure bacterial mass. QCM with dissipation monitoring (QCM-D) helps us to understand the viscoelastic properties of biomass adhered on the quartz plate or sensor (Dixon 2008). Optical coherence tomography (OCT) is other industrial

Table 3.2 Different fluorescent probe used in biofilm studies

Mode of action	Fluorescent probe	Probes excitation/ emission	Reference
RNA, DNA intercalating agent (dead live cell kit)	Propidium iodide	530/615	Stiefel et al. (2015)
conjugated to lectins, antibodies, ficols, dextrans; binds to proteins	Fluorescein isothiocyanate (FITC)	490/520	Karygianni et al. (2012)
Detects esterase activity	Fluoroscein diacetate	495/520	Feng et al. (2014)
Detects neutral lipid and phospholipids	Nile red	450/530	Kokare et al. (2009)
Stains DNA and RNA	Acridine orange	490/530	Neu et al. (2001)
Calcium indicator	Fluo-3	506/526	Kokare et al. (2009)
Binds to proteins, coupled with antibodies, lectins,	Tetramethylrhodamine isothiocyanate (TRITC)	540/572	Neu et al. (2001)
pH indicator	NCECF	500/530 or 620	Kokare et al. (2009)

nondestructive testing (NDT) imaging technique that uses coherent light to capture micrometer-resolution, two- and three-dimensional images from within optical scattering biofilm. The technology is based on low-coherence interferometry, typically employing near-infrared light (Wagner and Horn 2017).

3.10 Removal of Biofilms

Cellular proliferation and exopolysaccharide production increases the biomass of biofilm, whereas cell destruction, dispersion of biofilm, and "grazing" by benthic microorganisms decreases the biomass, which gives rise to cyclic life of biofilms. It is very difficult to get rid of a biofilm; biofilm cannot be dispersed spontaneously. The applications of antimicrobial agent or biocide to eradicate bacterial biofilm from a colonized surface are often end with failure. Biofilms can develop continuously for very prolonged periods of time with a favourable environment, even with unfavorable condition, it can thrive (Sharma and Lal 2017). The main strategy to prevent biofilm formation is to clean and disinfect regularly before bacteria attach firmly to surfaces, therefore prevention of biofilm formation is a more rational option over its treatment. Chlorine, ozone, hydrogen peroxide are some chemical agents acts effectively to remove biofilms, these agents acts as an oxidizing antimicrobial agents which often generates free radicals which can have the ability to destroy both bacterial cell membrane biofilm matrix, hence physically removing the biofilm from the surface. UV radiation along with ozone was found efficient in getting rid of bacteria and dissolving biofilm matrices in studies of desalination plants (Kalia et al. 2017).

3.10.1 Removal of Biofilms

Cellular proliferation and exopolysaccharide production increases the biomass of biofilm, whereas cell death, objectivity at the smooth exterior, and "feeding" by benthic microorganisms decreases the biomass, which gives rise to cyclic life of biofilms. It is very difficult to get rid of a biofilm; neither natural nor biocide-driven slaying of microorganisms inside a biofilm spontaneously separates the biofilm from a populated surface. Biofilms occupies underwater planes in marine environs for extended eras of time. It has also been reported that some widespread biofilms comprise of lifeless microorganisms inside their wide matrices (Sharma and Lal 2017). Chlorine is effective in the removal of biofilms, it works as an oxidizing biocide which can together suppress bacterial growth and eliminate the polymeric components of the biofilm background that anchor this construction to the populated exterior, hence physically removing the biofilm from the surface. Bleach treatment method (using 5% NaClO) has been very effective in getting rid of microorganisms and softening biofilm backgrounds.

3.10.2 Regulation of Biofilm Formation

The chemical nature of the surface is one of the major determinants for the rate of bacterial adhesion for initial biofilm formation in experimental aquatic environments. Nonetheless, this noticeable influence of the chemical composition of the surface on the biofilm development is noticeably altered by the fact that all such planes are covered by a sheet of organic elements and also by other circumstances, that consequent biofilm growth is heavily dependent on the bacterial cell proliferation and cell to cell bond inside the biofilm background (Kalia et al. 2017). On comparing different surfaces within an aquatic system that have developed thick mature bacterial biofilms, hardly noticeable variances are observed among supporter cell quantities and thickness of the biofilm on the different surfaces like wood, rock, and plastic.

3.11 Conclusions

Understanding of biofilms plays a vital role in antimicrobial drug resistance, which is a very useful contribution towards public health perspective. Research on biofilms has risen steeply in recent years thanks to the increased awareness of the extensiveness and impact of biofilms on natural and industrial systems, also on human health. Pharmaceutical industries are threatened by the resistance of microbial communities that dwell in biofilms towards various types of antimicrobial agents because biofilms cost billions of dollars every year in product contamination, energy losses, equipment damage and medical infections (Kalia 2014). Therefore, it is advised to prevent its formation than treatment. But biofilms are not just harmful, they can also offer huge potential for biofiltering municipal and industrial water and wastewater, bioremediation of various waste sites, also they can be used to form bio-barriers to protect soil and groundwater from contamination. The study of biofilm activity and behavior is complex; hence it entails interdisciplinary research from biochemistry, engineering, mathematics and microbiology. Further studies are being done in order to come up with most effective control strategies to prevent biofilm formation, methods for total eradication of biofilms, using biofilms for bioremediation and a complete understanding of the differences between biofilm and planktonic phenotypes.

References

Bishop PL (1997) Biofilm structire and kinetics. Water Sci Technol, Biofilm Systems IIISelected Proceedings of the IAWQ 3rd International Specialised Conference on Biofilm Systems 36, 287–294. https://doi.org/10.1016/S0273-1223(97)00355-7

Christensen BB, Sternberg C, Andersen JB, Eberl L, Møller S, Givskov M, Molin S (1998) Establishment of new genetic traits in a microbial biofilm community. Appl Environ Microbiol 64:2247–2255

Costerton JW, Stewart PS, Greenberg EP (1999) Bacterial biofilms: a common cause of persistent infections. Science 284:1318–1322. https://doi.org/10.1126/science.284.5418.1318

Ding Z, Bourven I, Guibaud G, van Hullebusch ED, Panico A, Pirozzi F, Esposito G (2015) Role of extracellular polymeric substances (EPS) production in bioaggregation: application to wastewater treatment. Appl Microbiol Biotechnol 99:9883–9905. https://doi.org/10.1007/s00253-015-6964-8

Dixon MC (2008) Quartz crystal microbalance with dissipation onitoring: enabling real-time characterization of biological materials and their interactions. J Biomol Technol JBT 19:151–158

Donlan RM (2002) Biofilms: microbial life on surfaces. Emerg Infect Dis 8:881–890. https://doi.org/10.3201/eid0809.020063

Dufour D, Leung V, Lévesque CM (2010) Bacterial biofilm: structure, function, and antimicrobial resistance. Endod Top 22:2–16

Feng J, Wang T, Zhang S, Shi W, Zhang Y (2014) An optimized SYBR green I/PI assay for rapid viability assessment and antibiotic susceptibility testing for Borrelia burgdorferi. Wooten RM, ed. PLoS One 9(11):e111809. https://doi.org/10.1371/journal.pone.0111809

Flemming H-C (1993) Biofilms and environmental protection. Water Sci Technol 27:1–10

Flemming H-C, Neu TR, Wozniak DJ (2007) The EPS matrix: the "house of biofilm cells". J Bacteriol 189:7945–7947. https://doi.org/10.1128/JB.00858-07

Gacesa P (1998) Bacterial alginate biosynthesis – recent progress and future prospects. Microbiology 144:1133–1143. https://doi.org/10.1099/00221287-144-5-1133

Garrett TR, Bhakoo M, Zhang Z (2008) Bacterial adhesion and biofilms on surfaces. Prog Nat Sci 18:1049–1056. https://doi.org/10.1016/j.pnsc.2008.04.001

Gupta P, Sarkar S, Das B, Bhattacharjee S, Tribedi P (2016) Biofilm, pathogenesis and prevention – a journey to break the wall: a review. Arch Microbiol 198:1–15. https://doi.org/10.1007/s00203-015-1148-6

Hall-Stoodley L, Costerton JW, Stoodley P (2004) Bacterial biofilms: from the natural environment to infectious diseases. Nat Rev Microbiol 2:95–108. https://doi.org/10.1038/nrmicro821

Huma N, Shankar P, Kushwah J, Bhushan A, Joshi J, Mukherjee T, Raju S, Purohit HJ, Kalia VC (2011) Diversity and polymorphism in AHL-lactonase gene (aiiA) of Bacillus. J Microbiol Biotechnol 21:1001–1011

Kalia VC (2013) Quorum sensing inhibitors: an overview. Biotechnol Adv 31:224–245. https://doi.org/10.1016/j.biotechadv.2012.10.004

Kalia VC (2014) Microbes, antimicrobials and resistance: the battle goes on. Indian J Microbiol 54:1–2. https://doi.org/10.1007/s12088-013-0443-7

Kalia VC (2014a) Quorum sensing vs quorum quenching: a battle with no end in sight. Springer, New Delhi

Kalia VC (2014b) In search of versatile organisms for quorum-sensing inhibitors: acyl homoserine lactones (AHL)-acylase and AHL-lactonase. FEMS Microbiol Lett 359:143–143. https://doi.org/10.1111/1574-6968.12585

Kalia VC, Wood TK, Kumar P (2014) Evolution of resistance to quorum-sensing inhibitors. Microb Ecol 68:13–23. https://doi.org/10.1007/s00248-013-0316-y

Kalia VC, Prakash J, Koul S, Ray S (2017) Simple and rapid method for detecting biofilm forming bacteria. Indian J Microbiol 57:109–111. https://doi.org/10.1007/s12088-016-0616-2

Karygianni L, Follo M, Hellwig E et al (2012) Microscope-based imaging platform for large-scale analysis of oral biofilms. Appl Environ Microbiol 78(24):8703–8711. https://doi.org/10.1128/AEM.02416-12

Kim J-H, Choi D-C, Yeon K-M, Kim S-R, Lee C-H (2011) Enzyme-immobilized nanofiltration membrane to mitigate biofouling based on ouorum quenching. Environ Sci Technol 45:1601–1607. https://doi.org/10.1021/es103483j

Kokare CR, Chakraborty S, Khopade AN, Mahadik KR (2009) Biofilm: importance and applications. Indian J Biotechnol 8(2):159–168 http://nopr.niscair.res.in/handle/123456789/3883

Kostakioti M, Hadjifrangiskou M, Hultgren SJ (2013) Bacterial biofilms: development, dispersal, and therapeutic strategies in the dawn of the postantibiotic era. Cold Spring Harb Perspect Med 3:a010306–a010306. https://doi.org/10.1101/cshperspect.a010306

Koul S, Kalia VC (2017) Multiplicity of quorum quenching enzymes: a potential mechanism to limit quorum sensing bacterial population. Indian J Microbiol 57:100–108. https://doi.org/10.1007/s12088-016-0633-1

Kumar CG, Anand SK (1998) Significance of microbial biofilms in food industry: a review. Int J Food Microbiol 42:9–27. https://doi.org/10.1016/S0168-1605(98)00060-9

de la Fuente-Núñez C, Reffuveille F, Fernández L, Hancock RE (2013) Bacterial biofilm development as a multicellular adaptation: antibiotic resistance and new therapeutic strategies. Curr Opin Microbiol Antimicrob Genomics 16:580–589. https://doi.org/10.1016/j.mib.2013.06.013

McLean RJC, Cassanto JM, Barnes MB, Koo JH (2001) Bacterial biofilm formation under microgravity conditions. FEMS Microbiol Lett 195:115–119. https://doi.org/10.1016/S0378-1097(00)00549-8

Miller MB, Bassler BL (2001) Quorum sensing in bacteria. Annu Rev Microbiol 55:165–199. https://doi.org/10.1146/annurev.micro.55.1.165

Montanaro L, Poggi A, Visai L, Ravaioli S, Campoccia D, Speziale P, Arciola CR (2011) Extracellular DNA in biofilms. Int J Artif Organs 34:824–831. https://doi.org/10.5301/ijao.5000051

Neu TR, Swerhone GDW, Lawrence JR (2001) Assessment of lectin-binding analysis for in situ detection of glycoconjugates in biofilm systems. Microbiology 147:299–313. https://doi.org/10.1099/00221287-147-2-299

Sharma A, Lal R (2017) Survey of (Meta)genomic approaches for understanding microbial community dynamics. Indian J Microbiol 57:23–38. https://doi.org/10.1007/s12088-016-0629-x

Singh R, Paul D, Jain RK (2006) Biofilms: implications in bioremediation. Trends Microbiol 14:389–397. https://doi.org/10.1016/j.tim.2006.07.001

Song F, Koo H, Ren D (2015) Effects of material properties on bacterial adhesion and biofilm formation. J Dent Res 94:1027–1034 0022034515587690

Stiefel P, Schmidt-Emrich S, Maniura-Weber K, Ren Q (2015) Critical aspects of using bacterial cell viability assays with the fluorophores SYTO9 and propidium iodide. BMC Microbiol 15:36. 1186/s12866-015-0376-x

Sutherland IW (2001) Biofilm exopolysaccharides: a strong and sticky framework. Microbiology 147:3–9. https://doi.org/10.1099/00221287-147-1-3

Wagner M, Horn H (2017) Optical coherence tomography in biofilm research: a comprehensive review. Biotechnol Bioeng 114:1386–1402. https://doi.org/10.1002/bit.26283

Watnick P, Kolter R (2000) Biofilm, city of microbes. J Bacteriol 182:2675–2679. https://doi.org/10.1128/JB.182.10.2675-2679.2000

Phylogenomics and Evolutionary Perspective of Quorum Sensing Regulators (LuxI/LuxR) in Prokaryotes

4

Manoj Kumar and Akanksha Rajput

Abstract

The quorum sensing (QS) is a well-characterized phenomenon in the microbial world for cellular communication and being exploited for intraspecies, interspecies and even interkingdom interactions. This robust and widely distributed social networking cascade assists the microbes to emerge as strong parasites for the hosts. Moreover, the colonization with more efficient cross talk among the microbes further intensifies their infections. Here, we will be focusing to decipher the evolutionary status of the QS regulators (LuxI and LuxR) in the prokaryotic world. LuxI is a signal synthase, while LuxR is the recipient for sensing internal (cognate LuxR/solos LuxR) and external (solos LuxR) signals. These regulators are reported to evolve vertically as well as borrowed through horizontal gene transfer *w.r.t.* ecological niche. Their universal distribution in the microbial world further corroborates the need for targeting multiple signaling system regulators.

Keywords

Quorum sensing · Phylogenomics · LuxI/LuxR regulators · Bacteria · Archaea · Horizontal gene transfer (HGT) · Interspecies communication

M. Kumar (✉) · A. Rajput
Bioinformatics Centre, Institute of Microbial Technology, Council of Scientific and Industrial Research, Chandigarh, India
e-mail: manojk@imtech.res.in

V. C. Kalia (ed.), *Quorum Sensing and its Biotechnological Applications*,
https://doi.org/10.1007/978-981-13-0848-2_4

4.1 Introduction

Quorum sensing (QS) allows the phenotypic transformation of bacteria from unicellular to multicellular lifestyle (Atkinson and Williams 2009). It emerged as an imperative cell density dependent social network among prokaryotes, which accomplished through the help of various signaling molecules termed as quorum sensing signaling molecules (QSSMs) (Rajput et al. 2015 2016). The QSSMs are considered to be species-specific i.e. Acyl-homoserine lactones (AHLs) among Gram-negative bacteria (Parsek and Greenberg 2000; Rajput et al. 2016), quorum sensing peptides (QSPs) in Gram-positive bacteria (Wynendaele et al. 2013), etc. Due to the difference in their composition and conformation, they exhibit different mechanisms for secretion and perception among bacteria *via* one-component (AHLs) and two-component systems (QSPs) (Santos et al. 2012). The majorly used signaling molecule is AHLs with the alkyl group (varied length and saturation) attached to homoserine lactone ring (Churchill et al. 2011).

In 1965, Tomasz et al. firstly discovered the *cell-to-cell* communication in *Streptococcus pneumoniae*, a Gram-positive bacteria (Rajput et al. 2015). Later on, in 1970 the phenomenon was established in *Vibrio fischeri*, a marine luminescent Gram-negative bacterium. The *lux* operon is responsible for synthesis and perception of AHLs in Gram-negative bacteria. It contains regulatory (*luxIR*) and structural genes (*luxCDABFE*) (Rajput and Kumar 2017a). The operon encodes the key proteins LuxI and LuxR, which are responsible for synthesis (autoinducer synthase) and perception of AHLs (transcriptional regulator) respectively. The mechanism of QS in bacteria is provided in Fig. 4.1.

The LuxI is an autoinducer synthase *(InterPro ID: IPR001690)* that transfers the acyl moiety between acyl-acyl carrier protein (acyl-ACP) and S-adenosylmethionine (SAM) and led to the formation of AHLs with different degree of length and

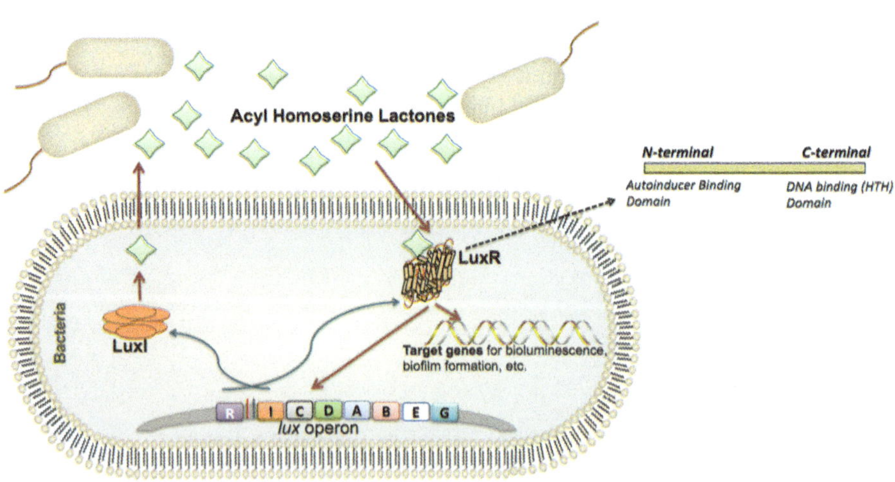

Fig. 4.1 Mechanism of quorum sensing among bacteria

saturation. While the LuxR regulators are responsible for perceiving and responding to AHLs. It is ~250 amino acid long cytosolic protein with two modules i.e. N-terminal (the 2/3rd portion of LuxR) or autoinducer-binding domain (ABD) *(InterPro ID: IPR005143)* and C-terminal (1/3rd portion of LuxR) or DNA binding domain *(InterPro ID: IPR000792)*. The C-terminal domain is accountable for DNA binding, transcriptional activation, and expression of *luxICDABEG*. Moreover, it also constitutes about 60 amino acids, helix-turn-helix (HTH) motif that is responsible for binding to promoters. In general, N-terminal domain is a repressor of the C-terminal domain in absence of AHLs' quorum in the vicinity. Although, the interaction of LuxR-AHL is important for bacteria to discriminate AHLs produced by its own species from others. LuxR solos (orphan or bachelor LuxR) are the separate class of response regulator, which does not have LuxI in their vicinity (~3000 bp or 3400 bp) (Hudaiberdiev et al. 2015). They are reported to possess same topology as cognate LuxR with ligand binding and DNA binding domains (Patankar and Gonzalez 2009). Unlike cognate LuxR, they are responsible for responding to external and internal signals (Subramoni and Venturi 2009). Intriguingly, solos LuxR are reported to be involved in interspecies and interkingdom communication through various AHL and non-AHL ligands (Subramoni and Venturi 2009; Rajput and Kumar 2017a, b).

Phylogenomics is the intersection field of evolution and genomics, which is helpful in deciphering the evolution pattern of gene family, prediction of horizontal gene transfer (HGT), gene functions, establishing the evolutionary relationships etc. (Pennisi 2008; Subramoni and Venturi 2009; Gupta et al. 2016; Rajput and Kumar 2017a). Therefore, reconstructing the evolutionary relationship of the particular gene among the species is important to complete our knowledge about the function of that gene. Moreover, the actual information about the evolutionary history of the gene is accomplished by comparing with the phylogenetic tree of conserved housekeeping genes like 16S rRNA. The phylogenetic tree of the conserved genes like 16S rRNA, *rpoB*, etc. clearly showed the distant positioning of all the groups (Rajput and Kumar 2017a). However, the intermixing of the sequences from different groups with good statistical bootstrapping support showed the possibility of HGT among the groups.

The distribution of QS regulators i.e. LuxI and LuxR among Gram-negative bacteria is well characterized. However, their presence in other groups of microbes like Gram-positive bacteria, and Archaea strengthen the concept of multilevel communication among the microbial world. Moreover, it poses several questions regarding the extent of occurrence, mode of transfer, and evolution.

4.2 LuxI and LuxR Regulators Among Gram-Negative Bacteria

The Gram-negative bacteria are characterized to possess LuxI/LuxR regulators for synthesizing and utilizing AHLs for completing QS cascade. A major phylum of Gram-negative bacteria is Proteobacteria, which is further sub-divided in alpha, beta, gamma, delta, and epsilon. In 2001, Gray and Garey checked the evolutionary pattern of LuxI and LuxR in Proteobacteria. The results showed that these regulators were evolved as regulatory cassettes in Proteobacteria along with the instances of HGT in a few species where multiple copies of LuxIR regulators were reported. The coevolved LuxI/LuxR homologs include TraI and TraR proteins of *Agrobacterium tumefaciens*; RhiI and RhiR of *Rhizobium leguminosarum* despite being non-adjacent in location. While SdiA regulator (a LuxR homolog) of *Escherichia coli* and *Salmonella typhimurium* seemed to be evolved through HGT from RhlR of *P. aeruginosa* (Rajput and Kumar 2017b).

In 2004, Lerat and Moran revealed the evolutionary history of QS systems (LuxI/LuxR and LuxS) among bacteria by checking the frequency of transfer and extent of the exchange of genes in this QS system *via* HGT or coevolution. They found that *luxR* gene transferred horizontally to several lineages of bacteria from Firmicutes e.g. *carR* genes of *Erwinia carotovora* and *Serratia marcescens*. The LuxI/LuxR system evolved together along with few instance of HGT especially in the case of gamma- Proteobacteria (Rajput and Kumar 2017b).

Albeit checking the evolutionary status of QS regulators in Proteobacteria, many groups scanned specific clades of Proteobacteria like *Aeromonas* (Jangid et al. 2007), Vibrionaceae (Purohit et al. 2013; Rasmussen et al. 2014), Roseobacteriacea (Cude and Buchan 2013), etc. In the individual studies, they found that the QS regulators are distributed in entire clades, with the instances of their horizontal and vertical evolution. Jangid et al. confirm the universality of LuxI and LuxR among the *Aeromonas* genus, which is known as "emerging pathogens". They confirm the presence of AHLs mediated response in all the 73 strains when tested *via* dot blot hybridization method. Further, the phylogenetic tree reconstruction was done to check their evolutionary status (Jangid et al. 2007). Purohit et al. checked the presence of AHLs in 57 members of Vibrionaceae family, through high-performance liquid chromatography-tandem mass spectrometry (HPLC-MS) and mapped on the 16 s rDNA phylogeny (Purohit et al. 2013). Rasmussen and coworkers decipher the global and phylogenetic distribution of AHLs in Vibrionaceae, and concluded that AHLs are responsible for intra- and inter-species communication (Rasmussen et al. 2014). Clude and Buchan reviewed the AHL-based QS among Roseobacter clade through 43 genomes and decipher their evolutionary pattern and found their overall conserved gene topologies with complex functional roles (Cude and Buchan 2013).

In 2015, Subramoni et al. performed bioinformatics survey of LuxR solos on the basis of their distribution, conservation, and probable function among bacteria (majorly Proteobacteria). They found that multiple LuxR solos from the same genome possess different level of the conservation of the invariant amino acids for AHL binding, thus exhibits a wide range of sensing ligands (Subramoni et al. 2015). Therefore, the LuxR solos emerged as an important component for multi level communication.

Thus, on checking the distribution and evolutionary context of QS regulators in Proteobacteria confer their universal distribution among clades and sub-clades of the phylum along with the instance of their evolution and lateral transfer from other member groups. Moreover, the ligands specificity of the LuxR solos, varies due to substitution at invariant amino acids for AHL binding.

4.3 LuxI and LuxR Regulators Among Gram-Positive Bacteria

The cell density based signaling in Gram-positive bacteria is accomplished through QSPs. Interestingly, the presence of LuxI/LuxR regulators in few studies were reported (Biswa and Doble 2013; Bose et al. 2017). Further, in 2012 Santos et al. described the phylogenomics distribution and functional diversity of LuxR regulators in Actinobacteria (a phylum of Gram positive bacteria). Among 53 genomes, only 991 proteins were reported to possess LuxR domains, moreover, 59% (maximally REC, receiver domain) possess extra domains. However, it was suggested that the occurrence of LuxR domains depends on genetic, ecological, and metabolic variables. Moreover, the evolution of LuxR regulators was considered though gene fusion/fission and duplication events especially through HGT and gene loss (Santos et al. 2012). In 2015, Polkade et al. reviewed the Actinobacteria phylum and decipher the functionality of gamma-butyrolactone signaling system (structural homolog of AHLs) along with methylenomycin furans, AI-2, and AHLs as the source of interspecies communication signals (Polkade et al. 2016).

Recently, our group studied overall Gram-positive bacteria (Actinobacteria and Firmicutes) for the presence of LuxI/LuxR system using conservational, functional, and phylogenetic aspects (Rajput and Kumar 2017b). Our study revealed the presence of putative LuxI/LuxR, having outnumbered LuxR as compared to LuxI, which possess similarity with Gram-negative bacteria, and are the result of HGT between Gram-negative and Gram positive bacteria. The similarity-based analyses were performed using amino acid composition, domain analyses, and multiple sequence alignment. All the similarity profiles confirm the likeness of LuxI and LuxR's among Gram positive with Gram-negative bacteria. Functional annotation of QS regulators of the Gram-positive bacteria showed that they have potential to synthesize and respond to ubiquitous signaling molecules including AHLs, gamma-butyrolactones, c-di-GMP, peptides, and many more. Moreover, authors also checked the transfer pattern of individual LuxI or LuxR as well as the complete cassette of LuxI and LuxR among Gram-positive bacteria and found that they possess both cases of individual transfer and complete cassette transfer between both groups of bacteria. Moreover, the instances of HGT are ecological niche specific among the groups for LuxI and LuxR.

The evolutionary history of LuxI/LuxR regulators in Gram positive bacteria suggests that they share the lateral gene transfer events with Gram-negative bacteria and thus possess the ability of undergo interspecies communication.

4.4 LuxI and LuxR Regulators Among Archaea

The Archaea kingdom was also scanned for the putative LuxI/LuxR regulators. Our group accomplished the study to decipher their presence in Archaea (Rajput and Kumar 2017a). The Archaea was explored computationally through taxonomic, probable function, distribution, and evolutionary characteristics. Interestingly, the Archaea reported to possess only LuxR regulators, which are similar to Gram-negative bacteria. Moreover, the functional characterization studies revealed that they hold the ability to sense and respond to AHLs, peptides, gamma-butyrolactones, nucleic acids, and many more. Additionally, the functional annotation using Gene Ontology, showed their involvement in the signal transduction cascade. Furthermore, the transfer of LuxR regulators in Archaea is the result of HGT events with Gram-negative bacteria.

Therefore, the Archaea able to receive and respond the signals from another group of microbes like Gram-negative bacteria, rather than secrete them. Thus, involved in inter kingdom cross-talk with bacteria.

4.5 LuxI and LuxR Regulators Among Prokaryotic Kingdom

In natural conditions, the prokaryotes like bacteria and Archaea are the inhabitant of same ecological niche, like in biofilms (Rajput et al. 2018). The individual clade specific studies of the evolutionary distribution of QS regulators were performed in the literature. While the exact status of QS regulators among the microbial world can be deciphered through phylogenomics study of individual LuxI and LuxR regulators.

Interestingly, the LuxI regulators are exclusively present in bacteria and absent in Archaea. The LuxI containing proteins among Gram-negative and Gram-positive bacteria showed the lateral exchange of LuxI between both the groups. For example *Asanoa ferruginea* and *Methylobacterium* spp.; *Streptomyces purpurogeneiscleroticus* and *Methylobacterium* spp.; *Mumia flava* with *Burkholderia* sp.; etc. The detailed overview of the phylogenetic tree of LuxI containing proteins is provided in Fig. 4.2.

The LuxR regulators are found distributed in the entire prokaryotic world. Both cognate and solos LuxR are reported in bacteria (Gram-positive and Gram-negative bacteria). While the Archaea possess solos LuxR regulators rather than cognate.

The phylogenomics analyses showed that instances of HGT among all the three groups of microbes with good bootstrap support. For example, *Euryarchaeota archaeon*, *Ruminococcaceae bacterium* D16 and *Clostridium* sp.; *Pseudoxanthobacter soli* and *Candidatus Nomurabacteria bacterium*; *Asanoa ferruginea* and *Methylobacterium* spp.; *Mumia flava* with *Burkholderia* sp.;and many more.

The diagrammatic representation of phylogenetic tree for LuxR containing proteins among Gram-negative bacteria, Gram-positive bacteria, and Archaea are provided in Fig. 4.3.

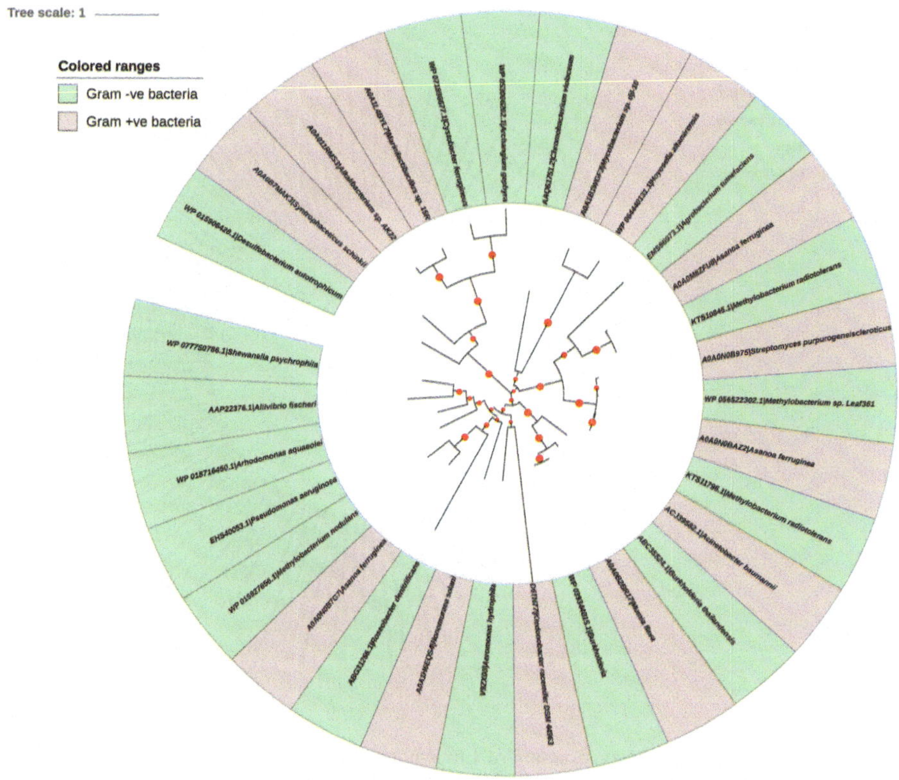

Fig. 4.2 Phylogenetic tree reconstruction of LuxI containing protein of Gram-negative and Gram-positive bacteria using Maximum likelihood method through 1000 bootstrap support

4.6 Application

Coexistence of different species in the same community is a natural process. The evolutionary pattern of QS regulators would be helpful in understanding the concept of their occurrence in prokaryotes. Although, the LuxI/LuxR system is responsible for synthesis and responds against AHL signaling molecule in Gram-negative bacteria, but their presence in other groups strengthen the concept of multilevel communication like intraspecies, interspecies, and interkingdom. Previously, it was considered that LuxR regulators are only responsible for causing virulence in the hosts by Gram-negative bacteria (Dubern and Diggle 2008). But, their presence and functionality in Gram-positive bacteria and Archaea decipher them as a potential target against various diseases caused by *Mycobcterium* spp., *Clostridium* spp., etc. (Santos et al. 2012). Moreover, the multilevel communication among various species, led to change the gene expression of the neighboring organism, and is one amongst the reason for antibiotic resistance (Kroger et al. 2016). However, the knowledge of LuxR regulator among the consortium of microbes would help in designing the inhibitors that in turn impede signaling cascade by blocking LuxR regulators.

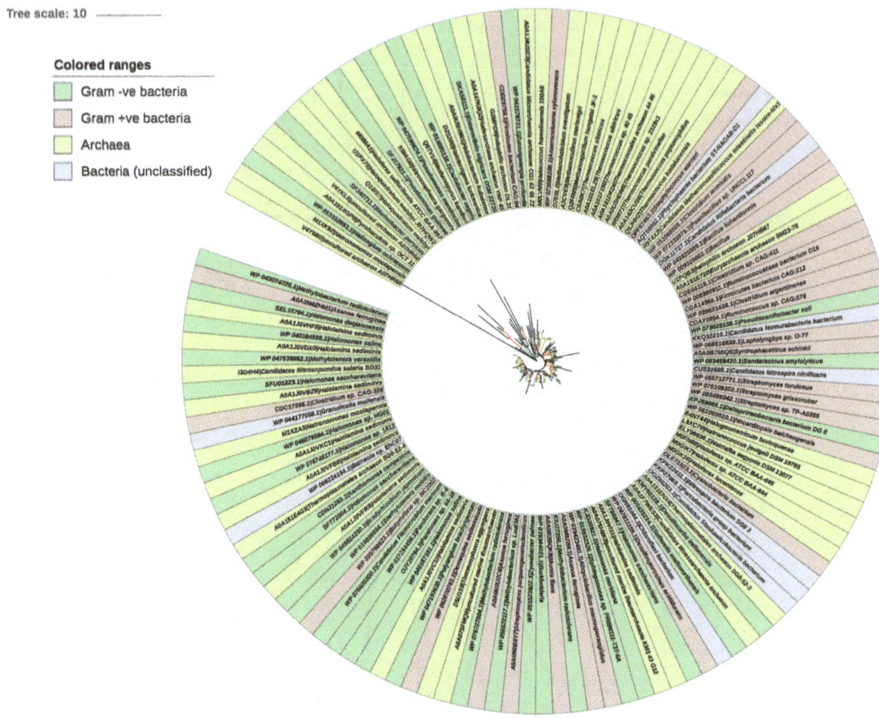

Fig. 4.3 Phylogenetic tree reconstruction of LuxR containing protein of Gram-negative, Gram-positive bacteria, and Archaea using Maximum likelihood method through 1000 bootstrap support

4.7 Conclusion

This book chapter is mainly focused on deciphering the evolutionary status of QS regulators (LuxI/LuxR) in whole prokaryotic kingdom. Putative LuxI homologs are distributed in both Gram-negative as well as in Gram-positive bacteria but absent in Archaea. While potential LuxR homologs are present in the complete microbial world and outnumbered LuxI. Phylogenetic analysis revealed the vertical evolution of these regulators with a number of instances of horizontal gene transfers among prokaryotes. These regulators assist in the establishment of QS-based biological processes like biofilm formation and virulence etc. (Kalia and Purohit 2011; Kalia et al. 2014; Sharma and Lal 2017). These, in turn, make the organisms more pathogenic towards human or plant hosts. Therefore, the therapeutic strategies need to be updated by not only targeting the specific signaling system for a particular group of organism but also universal regulators like LuxR, which is able to sense a broad range of ligands.

Acknowledgement Department of Biotechnology, Government of India (GAP0001) and Council of Scientific and Industrial Research (CSIR) supported this work.

References

Atkinson S, Williams P (2009) Quorum sensing and social networking in the microbial world. J R Soc Interface 6:959–978. https://doi.org/10.1098/rsif.2009.0203

Biswa P, Doble M (2013) Production of acylated homoserine lactone by gram-positive bacteria isolated from marine water. FEMS Microbiol Lett 343:34–41. https://doi.org/10.1111/1574-6968.12123

Bose U, Ortori CA, Sarmad S, Barrett DA, Hewavitharana AK, Hodson MP, Fuerst JA, Shaw PN, Boden R (2017) Production of N-acyl homoserine lactones by the sponge-associated marine actinobacteria *Salinispora arenicola* and *Salinispora pacifica*. FEMS Microbiol Lett 364. https://doi.org/10.1093/femsle/fnx002

Churchill ME, Sibhatu HM, Uhlson CL (2011) Defining the structure and function of acyl-homoserine lactone autoinducers. Methods Mol Biol 692:159–171. https://doi.org/10.1007/978-1-60761-971-0_12

Cude WN, Buchan A (2013) Acyl-homoserine lactone-based quorum sensing in the Roseobacter clade: complex cell-to-cell communication controls multiple physiologies. Front Microbiol 4:336. https://doi.org/10.3389/fmicb.2013.00336

Dubern JF, Diggle SP (2008) Quorum sensing by 2-alkyl-4-quinolones in *Pseudomonas aeruginosa* and other bacterial species. Mol BioSyst 4:882–888. https://doi.org/10.1039/b803796p

Gupta AK, Kaur K, Rajput A, Dhanda SK, Sehgal M, Khan MS, Monga I, Dar SA, Singh S, Nagpal G, Usmani SS, Thakur A, Kaur G, Sharma S, Bhardwaj A, Qureshi A, Raghava GP, Kumar M (2016) ZikaVR: an integrated Zika virus resource for genomics, proteomics, phylogenetic and therapeutic analysis. Sci Rep 6:32713. https://doi.org/10.1038/srep32713

Hudaiberdiev S, Choudhary KS, Vera Alvarez R, Gelencser Z, Ligeti B, Lamba D, Pongor S (2015) Census of solo LuxR genes in prokaryotic genomes. Front Cell Infect Microbiol 5:20. https://doi.org/10.3389/fcimb.2015.00020

Jangid K, Kong R, Patole MS, Shouche YS (2007) *luxRI* homologs are universally present in the genus *Aeromonas*. BMC Microbiol 7:93. https://doi.org/10.1186/1471-2180-7-93

Kalia VC, Purohit HJ (2011) Quenching the quorum sensing system: potential antibacterial drug targets. Crit Rev Microbiol 37:121–140. https://doi.org/10.3109/1040841x.2010.532479

Kalia VC, Wood TK, Kumar P (2014) Evolution of resistance to quorum-sensing inhibitors. Microb Ecol 68:13–23. https://doi.org/10.1007/s00248-013-0316-y

Kroger C, Kary SC, Schauer K, Cameron AD (2016) Genetic regulation of virulence and antibiotic resistance in *Acinetobacter baumannii*. Genes (Basel) 8:93. https://doi.org/10.3390/genes8010012

Parsek MR, Greenberg EP (2000) Acyl-homoserine lactone quorum sensing in gram-negative bacteria: a signaling mechanism involved in associations with higher organisms. Proc Natl Acad Sci U S A 97:8789–8793. https://doi.org/10.1073/pnas.97.16.8789

Patankar AV, Gonzalez JE (2009) Orphan LuxR regulators of quorum sensing. FEMS Microbiol Rev 33:739–756. https://doi.org/10.1111/j.1574-6976.2009.00163.x

Pennisi E (2008) Evolution. Building the tree of life, genome by genome. Science 320:1716–1717. https://doi.org/10.1126/science.320.5884.1716

Polkade AV, Mantri SS, Patwekar UJ, Jangid K (2016) Quorum sensing: an under-explored phenomenon in the Phylum Actinobacteria. Front Microbiol 7:131. https://doi.org/10.3389/fmicb.2016.00131

Purohit AA, Johansen JA, Hansen H, Leiros HK, Kashulin A, Karlsen C, Smalas A, Haugen P, Willassen NP (2013) Presence of acyl-homoserine lactones in 57 members of the Vibrionaceae family. J Appl Microbiol 115:835–847. https://doi.org/10.1111/jam.12264

Rajput A, Gupta AK, Kumar M (2015) Prediction and analysis of quorum sensing peptides based on sequence features. PLoS One 10:e0120066. https://doi.org/10.1371/journal.pone.0120066

Rajput A, Kaur K, Kumar M (2016) SigMol: repertoire of quorum sensing signaling molecules in prokaryotes. Nucleic Acids Res 44:D634–D639. https://doi.org/10.1093/nar/gkv1076

Rajput A, Kumar M (2017a) Computational exploration of putative LuxR solos in Archaea and their functional implications in quorum sensing. Front Microbiol 8:798. https://doi.org/10.3389/fmicb.2017.00798

Rajput A, Kumar M (2017b) In silico analyses of conservational, functional and phylogenetic distribution of the LuxI and LuxR homologs in Gram-positive bacteria. Sci Rep. https://doi.org/10.1038/s41598-017-07241-5

Rajput A, Thakur A, Sharma S, Kumar M (2018) aBiofilm: a resource of anti-biofilm agents and their potential implications in targeting antibiotic drug resistance. Nucleic Acids Res 46:D894–d900. https://doi.org/10.1093/nar/gkx1157

Rasmussen BB, Nielsen KF, Machado H, Melchiorsen J, Gram L, Sonnenschein EC (2014) Global and phylogenetic distribution of quorum sensing signals, acyl homoserine lactones, in the family of Vibrionaceae. Mar Drugs 12:5527–5546. https://doi.org/10.3390/md12115527

Santos CL, Correia-Neves M, Moradas-Ferreira P, Mendes MV (2012) A walk into the LuxR regulators of Actinobacteria: phylogenomic distribution and functional diversity. PLoS One 7:e46758. https://doi.org/10.1371/journal.pone.0046758

Sharma A, Lal R (2017) Survey of (Meta)genomic approaches for understanding microbial community dynamics. Indian J Microbiol 57:23–38. https://doi.org/10.1007/s12088-016-0629-x

Subramoni S, Florez Salcedo DV, Suarez-Moreno ZR (2015) A bioinformatic survey of distribution, conservation, and probable functions of LuxR solo regulators in bacteria. Front Cell Infect Microbiol 5:16. https://doi.org/10.3389/fcimb.2015.00016

Subramoni S, Venturi V (2009) LuxR-family 'solos': bachelor sensors/regulators of signalling molecules. Microbiology 155:1377–1385. https://doi.org/10.1099/mic.0.026849-0

Wynendaele E, Bronselaer A, Nielandt J, D'Hondt M, Stalmans S, Bracke N, Verbeke F, Van De Wiele C, De Tre G, De Spiegeleer B (2013) Quorumpeps database: chemical space, microbial origin and functionality of quorum sensing peptides. Nucleic Acids Res 41:D655–D659. https://doi.org/10.1093/nar/gks1137

Technology Platforms for Bioanalysis of Quorum Sensing System and Exploiting It as Biomonitors and Biosensors

5

Santoshkumar N. Patil and Swapnil C. Kamble

Abstract

Quorum sensing (QS) is a bacterial signaling phenomenon wherein bacteria regulate gene expression as per the concentration of signaling molecule. In the microbial milieu, bacteria use QS to sense their immediate environment and in-turn adjust QS genes. Our knowledge of this continuous process of biosensing and biomonitoring of QS signaling molecule and QS circuits has evolved over a period of time. Herein, we attempt to follow impact of newer bioanalysis techniques in understanding this QS phenomenon based on only recent technology platforms. Some of the technology platforms are at proof of concept stage wherein feasibility for QS studies is being demonstrated. We attempt to understand the enormous possibilities/potential these technologies withhold. Advancements in QS systems led researchers to attempt potential application of QS systems itself as technology platform. In this book chapter, few specific applications of QS system towards biosensing and biomonitoring are explored and covers above mentioned topics in four sections: (a) advanced structural based techniques involved in QS study (b) advanced biosensing and biomonitoring technologies (c) microarray technology (d) QS technologies for biosensing and biomonitoring activity. Specific examples are elaborated in details and for comprehensive reading on technology platform reader could refer to the references. In summary, advanced technology towards bioanalysis and applications of QS itself as biosensing and biomonitoring technology are discussed. The critical analysis, current trends, potential technology applications and the path forward are touched upon in key opinion and conclusion.

S. N. Patil (✉)
Sai Life Sciences Ltd, BTS, Chrysalis Enclave, International Bio Tech Park, Hinjawadi Phase II, Pune, India
e-mail: santosh.p@sailife.com

S. C. Kamble
Department of Technology, Savitribai Phule Pune University, Pune, Maharashtra, India

V. C. Kalia (ed.), *Quorum Sensing and its Biotechnological Applications*,
https://doi.org/10.1007/978-981-13-0848-2_5

Keywords
Bioanalysis · Quorum sensing · Biomonitors · Biosensors · Technology platform · Biofilms · Quorum sensing circuits

Abbreviations

AHL	Acylhomoserine lactone
AI	Autoinducer
AI-2	Autoinducer -2
CDA	Cis-2-decenoic acid
CRM	Confocal Raman microspectroscopy
DESI	Desorption electrospray ionization
EPS	Extracellular polymeric substances
GABA	Gamma-aminobutyric acid
GRAS	Generally regarded as safe
IMS	Imaging mass spectrometry
IR	Infrared
MALDI	Matrix-assisted laser desorption–ionization
MS	Mass spectroscopy
NMR	Nuclear magnetic resonance
OC8-HSL	N-(3-oxooctanoyl) homoserine lactone
OdDHL	*N*-(3-oxododecanoyl) homoserine lactone
PDMS	Polydimethylsiloxane
QQ	Quorum quenching
QS	Quorum sensing
RE	Restriction endonuclease
SDP	Sporulation delaying protein
SECM	Scanning electrochemical microscopy
SIMS	Secondary-ion mass spectrometry
SKF	Sporulation killing factor
SWV	Square wave voltammetry
TOF	Time of flight
UME	Ultramicroelectrode
WHO	World Health Organization

5.1 Introduction

In a complex natural habitat, bacteria continuously engage in biosensing and biomonitoring of their microenvironment. They accomplish this activity by employing different signaling molecules, called auto-inducers (AIs) that diffuse into the surroundings. Upon reaching a critical density, these AIs induce gene expression in the AI receiving cells that may be of same species or different species. This entire phenomenon of sending and receiving signals culminating in determination of bacterial population size is referred to as quorum sensing (QS).

Early travel logs of Greek and Roman sea travelers gestated the bioluminescence phenomenon, which was later attributed to marine microorganisms and was deciphered to as autoinduction by microbiologist Kenneth H. Nealson, Terry Platt and J. Woodland Hastings in 1970. During this deciphering phase, the role of tools and methodologies employed for understanding has played a critical role (Kalia 2015). Various microorganisms and different AIs have been discovered since then using classical microbial methodologies as well as recent methods of cell confinement technologies among others. The cell confinement technology provides insight into the role of QS amongst small groups of cells as well as in study of uncultivable bacteria. We employed microdroplet technology, a technology wherein picoliter droplets generated from water in oil emulsion act as a test tube and is used for interrogations of various biochemical phenomenon, to study QS diffusion of signaling molecule *N*-(3-oxododecanoyl) homoserine lactone (OdDHL) and intraspecies QS phenomenon (Shim et al. 2011; Bai et al. 2013).

5.2 Recent Microbiological Techniques for Deciphering QS

From early discovery till today, researchers are engaged in deciphering QS phenomenon with many more aspects of this bacterial communication still being explored. Some of these AIs maybe multifunctional in nature like imparting color to bacterial colonies. These pigment producing bacteria have been identified and suggested to have potential antimicrobial, anti-cancer, and anti-malarial activity as well. Shiva Krishna and colleagues combined the classical methods of isolation of marine bacteria, biochemical characterization with newer tools of molecular identification using 16S rRNA for strain identification to isolate novel pigments (Shiva Krishna et al. 2015). The potential AI was extracted by pigment extraction method and its antibacterial activity was determined. The simplicity involved in execution of this work underlies the wide variety of sources for AI that may interact with equally varied species. This can be exemplified with gamma-aminobutyric acid (GABA) – a plant hormone essential for cell-cell signaling. Its concentration increases as soon as a wound is inflicted in plant tissue. Interestingly, GABA inactivates N-(3-oxooctanoyl)homoserine lactone (OC8-HSL) – the AI of *Agrobacterium lactonase* AttM (Chevrot et al. 2006). As an experimental proof, transgenic tobacco plants with enhanced GABA expression had higher resistance to infection by *Agrobacterium tumefaciens* C58. Thus, application of GABA on plants can be utilized for designing of inhibitory methods for *A. tumefaciens*. Sanchart et al. identified Kung-Som as a substrate for biosynthesis of GABA using lactic acid bacteria like *Lactobacillus futsaii* (Sanchart et al. 2017). A detailed description on various plant QS inhibitors have been reviewed by Kalia (2013).

The presence of diversity of quorum quenching genes is remarkable. An experimental validation was executed by Huma et al. for acylhomoserine lactone (AHL) (Huma et al. 2011). The group screened nearly 800 bacterial isolates and identified quorum quenching (QQ) AHL-lactonase gene (aiiA) to be present in 42 strains. 16s rRNA sequencing method revealed the dominance of *Bacillus* species in this cohort. In silico restriction endonuclease (RE) digestion using 14 different Type II REs on AHL lactonase gene (aiiA) sequences led to identification of nucleotide fragments of varying sizes. Analyses of only four REs *viz.* AluI, DpnII, RsaI, and Tru9I led to generation of possible information while data from other REs was not analyzable. Further, polymorphism in AHL lactonase was observed among the different *Bacillus* species. In *Bacillus* sp. strain MBG11, the unique polymorphism (115 Alanine > Valine) could augment stability to AHL lactonase. Highly stable AHL lactonase is preferable for designing quorum quenchers and this AHL lactonase could be an ideal candidate for large scale application. Further, Kalia and group identified *Hyphomonas neptunium* ATCC15444, *Deinococcus radiodurans* R1, and *Photorhabdus luminescens* subsp. *laumondii* TTO1 to possess genes that encode for both AHL-lactonase and –acylase (Kalia 2014). Presence of multiple copies of the QQ enzymes in bacteria would suggest application in industries involving Generally Regarded As Safe (GRAS). This was validated by a comparative genomic analysis using sequences of AHL-lactonase from *Bacillus* sp. SB4 and AHL-acylase from *Ralstonia* sp. XJ12B (Koul and Kalia 2017).

In these communications, the key components are small signaling molecules (bacterial language) and QS circuits (language sensing and producing machinery). The bacterial language comprises predominately homoserine lactone and peptides. Different bacterial species employ their own QS circuits and the complexity and interdependence of QS circuits is still being deciphered. Likewise, new bacterial languages are being discovered and are added in bacterial lexicon. Towards understanding of QS and its eventual application as a biosensor and biomonitoring system, various analytical techniques and technology platforms ranging from traditional shake flask culture to current more sophisticated analytical technique that senses heterogeneity are used towards deciphering QS phenomenon.

5.2.1 Structural: Determination Based Analytical Techniques for QS Study

The detection and identification of small signaling molecule was initially carried out using standard structural elucidation techniques such as Proton Nuclear magnetic resonance (1DNMR), mass spectrometry (MS), infrared (IR) etc. However, the identification and three dimensional structural elucidations of big biological macromolecules involved in QS circuits required much more sophisticated techniques such as X-ray crystallography and solution phase 2DNMR (Krishnan and Rupp 2012). Most of the proteins or macromolecules involved in QS circuits were crystalisable and the X-ray crystallographic techniques helped in deciphering structure of QS circuits. X-ray gives near perfect static picture of a protein at proximate atomic level precision and it has no size limitation. Nevertheless, it requires the protein in crystal form, whereas 2D NMR techniques can be used even when the protein is not in crystal. It also gives information on dynamic state of proteins but it is limited to size below 50 kDa. The following table gives summary of X-ray crystallography and NMR techniques that were used to study QS systems (Table 5.1).

Sometimes structures of protein are predicted using homology model. For example, proteins involved in QS circuits of *A. baumannii* were predicted using homology modeling (Bhargava et al. 2015). AHL synthase (AbaI) of *A. baumannii* showed 46% similarity and 27.5% identity to LasI autoinducer synthase of *Pseudomonas aeruginosa*. Likewise, the tertiary structure of *Acinetobacter baumannii* AbaR was predicated based on *P. aeruginosa* AHL receptor (LasR) using protein homology modeling server CPH models 3.2. For a detailed discussion on homology model as a tool in predicting structure of *Acinetobacter baumannii* QS protein please, refer to book (Kalia 2015).

5.2.2 Advanced Biosensing and Biomonitoring Technologies

In previous section, we saw that application of analytical techniques helped in deciphering the QS circuits and understand the structure of the QS proteins and signaling molecule. The traditional analytical techniques played important role in understanding QS at molecular and microscopic level under homogenous laboratory conditions. However, in order to understand QS in their microenvironments, which are at its heterogeneous natural state, requires advanced bio-sensing and biomonitoring technologies. The ideal technology platform should be able to

(a) Quantitatively analyze chemical environment of complex QS community,
(b) Provide new insights into QS bacterial habitat, and
(c) Detect single cell or single colony level

Towards, the above goal we will discuss pros and cons of few emerging technologies. Advanced techniques include electrochemical techniques such as scanning electrochemical microscopy (SECM), mass spectrometry based techniques such as imaging mass spectrometry, matrix-assisted laser desorption–ionization (MALDI)–mass spectrometry, desorption electrospray ionization (DESI)–mass spectrometry and secondary-ion mass spectrometry (SIMS).

5.2.2.1 Scanning Electrochemical Microscopy (SECM)

SEM initially applied to probe the topography and surface reactivity of solid-state materials recently employed in bioanalysis of heterogeneous quorum sensing system. The SECM works on redox principle wherein ultramicroelectode (UME) is held at biofilm mounted on SECM stage. The potential of electrode is held at the standard potential of the analyte, leading to reduction or oxidation of analyte at electrode tip. Redox potential is recorded and data transformed into real time three-dimensional analyte concentration readout.

Table 5.1 Structures of macromolecules involved in QS

Title	Ligand name	PDB ID
3-D structure of the QS protein TraR bound to its AI and target DNA	Homoserine lactone, selenomethionine, 3-Oxooctanoic acid	1H0M
Crystal structure of the QS protein TraM from *Agrobacterium tumefaciens*	–	1US6
Crystal structure of the master transcriptional regulator, SmcR, in *Vibrio vulnificus:* DNA recognition mechanism	Selenomethionine, sulfate ion	3KZ9
Crystal structure of the AI-2-bound form of *Vibrio harveyi* LuxP – periplasmic domain of LuxQ complex	3a-methyl-5,6-dihydro-furo[2,3-D][1,3,2] dioxaborole-2,2,6,6a-tetraol	2HJ9
	Nickel (II) ion	2HJE
Crystal structure of *E. coli* LsrG	–	3QMQ
Crystal structure of RpfF	–	3M6N
Crystal structure of LasR LBD-QslA complex from *Pseudomonas aeruginosa*	N-3-oxo-dodecanoyl-L-homoserine lactone	4NG2
Crystal structure of QS transcriptional activator from *Yersinia enterocolitica*	Acetic acid, 1,2-ethanediol, selenomethionine, sulfate ion	5L07
Bacillus subtilis LuxS – 1.2 Å structure	Cysteinesulfonic acid, Zn ion	1J98
Crystal structure of the AHL synthase, EsaI	–	1KZF
Crystal structure of QS antiactivator TraM	–	1RFY
Regulatory mechanism of the QS repressor RsaL in *P. aeruginosa*	–	5J2Y
Crystal structure of luxp from *V. harveyi* complexed with AI-2	3a-methyl-5,6-dihydro-furo[2,3-D][1,3,2] dioxaborole-2,2,6,6a-tetraol, calcium ion	1JX6
QS signal integrator LuxO – catalytic domain	Acetate ion	5EP1
Structure, regulation, and inhibition of the QS signal integrator LuxO	Adenosine-5′-triphosphate, 1,2-ethanediol, 4-(2-hydroxyethyl)-1-piperazine ethanesulfonic acid	5EP4
High resolution crystal structure of LuxS – quorum sensor molecular complex from *Salmonella typhi* at 1.58 Å	3-sulfinoalanine,(2r,4s)-2-methyl-2,3,3,4-tetrahydroxytetrahydrofuran, zinc ion, methionine	5E68
SdiA in complex with 3-oxo-C6-homoserine lactone	3-oxo-N-[(3S)-2-oxotetrahydrofuran-3-Yl] hexanamide, sulfate ion	4Y15
Crystal structure of *Helicobacter pylori* LuxS	Methionine, selenomethionine, zinc ion	1J6X

(continued)

Table 5.1 (continued)

Title	Ligand name	PDB ID
Structure of the *P. aeruginosa* LasR ligand-binding domain bound to its AI	Selenomethionine, N-3-oxo-dodecanoyl-L-homoserine lactone	2UV0
Crystal structure of an anti-activation complex in bacterial QS	3-oxo-octanoic acid (2-oxo-tetrahydro-furan-3-Yl)-amide	2Q0O
QS control repressor, QscR, bound to N-3-oxo-dodecanoyl-L-homoserine lactone	N-3-oxo-dodecanoyl-L-homoserine lactone, sodium ion	3SZT
Crystal structure of LUXS	Glycerol, cysteinesulfonic acid, zinc ion	1IE0
Bacillus subtilis luxs/ribosilhomocysteine complex: the 2.2 Å resolution	2-amino-4-mercapto-butyric acid, cysteinesulfonic acid, zinc ion	1JVI
Solution structure and dynamics of LuxU from *V. harveyi*, a phosphotransferase protein involved in bacterial QS	–	1Y6D
Crystal structure and catalytic mechanism of the QQAHL hydrolase	Glycerol, homoserine lactone, zinc ion	2BR6
Structure of apo form of *Vibrio cholera* CqsA	Sulfate ion	2WK7
Crystal structure of the periplasmic domain of *V. cholera* LuxQ	–	3C38
LasR-OC12 HSL complex	N-3-oxo-dodecanoyl-L-homoserine lactone	3IX3
Solution structure of *E. coli* SdiA1-171	N-(2-oxotetrahydrofuran-3-yl) octanamide	2AVX
NMR structure of TPC3 in TFE	Amino group	2I2H
NMR structure of UA159sp in TFE	Amino group	2I2J

Pyocyanin Concentration Determination in Biofilm

Pyocyanin is a multifaceted secondary metabolite produced by *P. aeruginosa*. It is known for imparting the blue coloration to *P. aeruginosa* colonies in cultures on agar plates. It is involved in inhibiting and restricting its microbial competitors as well as it participates in quorum signaling. Further, pyocyanin is zwitter ion and hence can act in multiple ways. Koley et al. used SECM to measure the concentration and redox state of pyocyanin present in the biofilm (Koley et al. 2011). They engineered a utility of square wave voltammetry (SWV) for detection of pyocyanin dependent current which had a high sensitivity and minimal detection limit of ≈ 0.7 pA (that corresponds to ≈ 0.6 μM pyocyanin). Extending the detection in the z axis using SECM, it was discovered that a gradation of pyocyanin concentration is present in the biofilm which ranges over 400 μm from the biofilm surface.

Interestingly, Pyocyanin receives the electron instead of oxygen in the usual electron flow from cytochrome bc1 to oxygen. In a similar way, pyocyanin can also reduce Fe^{3+} to Fe^{2+} (soluble form) even in presence of oxygen at pH 7. Koley et al.,

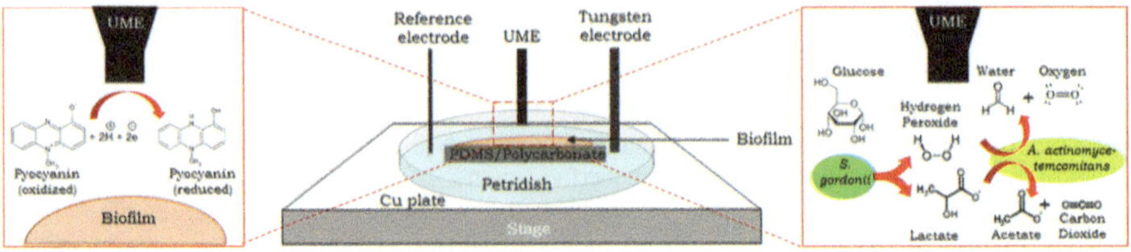

Fig. 5.1 The central figure shows generic schematic diagram of the SECM apparatus. The biofilm is grown on polydimethylsiloxane (PDMS)/polycarbonate material placed in petri dish. The change in redox is recorded with help of reference, UME and Tungsten electrodes. This can be applied for measurement of pyocyanin diffused from the biofilm (left) or for measurement of hydrogen peroxide released from the biofilm with *S. gordonii* and *A. actinomycestemcomitans* (right)

found that addition of Fe^{3+} in the environment allows pyocyanin gradient to change as soluble iron could be assimilated. Thus, the density gradient of pyocyanin is a function of external Fe. This also suggests preferential presence of *P. aeruginosa* in areas with Fe^{2+} over Fe^{3+} rich microenvironment. This report signifies the role of biofilm in nutrient assimilation and thus the regulation and dynamics of biofilm size (Fig. 5.1).

Hydrogen Peroxide Monitoring

Oral microflora has a plethora of microbes that is in a continuous dynamic flux. One of the common beneficial members of this flora is *Streptococcus gordonii*. It gives competition to pathogenic bacteria by its presence and production of inhibitory levels of hydrogen peroxide. *S. gordonii* utilizes the sugars to produce lactic acid as an end product. Further, opportunistic oral pathogen *Aggregatibacter actinomycetemcomitans* circumvents the hydrogen peroxide levels by utilizing lactic acid produced by *S. gordonii*. Hydrogen peroxide induces expression of katA and apiA in *A. actinomycetemcomitans* that make it resistant to host innate immunity response. Since hydrogen peroxide is promptly acted upon by catalase, it is necessary to study its effective concentration in a complex microenvironment.

Liu et al. quantified local hydrogen peroxide concentrations in a solution above a *S. gordonii* biofilm by SECM (Liu et al. 2011). The authors report the first such application of SECM in the real-time recording of hydrogen peroxide concentration present in a complex structure of a biofilm. Further, it was also used over a period of time for the identifying the consumption rate of hydrogen peroxide. The authors determined the concentration of hydrogen peroxide is in the range of 0.7 mM and 1.6 mM when bacteria were grown in 10 mM glucose for 2–8 h. The results were also validated using fluoremetric analyses. The significance of biofilm can be appreciated by the fact that planktonic *S. gordonii* had less hydrogen peroxide in vicinity in comparison to that in biofilm. When the two bacteria were present together in a biofilm contributed by both, then hydrogen peroxide levels varied with ratio and number of bacteria of each species. Usually, the concentration of hydrogen peroxide near the biofilm surface that is produced by *S. gordonii* was sufficient to prevent growth of various bacteria. Thus, the application of SECM gave insight in the mass transfer in the intricate dynamic system of biofilms that are otherwise difficult to experimentally determine.

5.2.2.2 Imaging Mass Spectrometry

Imaging mass spectrometry (IMS) technique was first demonstrated in 1960, however it has been underutilized and researchers have started applying it in microbiology very recently. IMS is a conglomeration of various ionization techniques and includes MALDI–mass spectrometry, DESI–mass spectrometry, SIMS etc. IMS provides two-dimensional information at atomic and molecular level and hence has the potential to analyze the bacterial microcolony at the molecular level in three dimensions. This technique is incredibly powerful, particularly in understanding chemical process involved in heterogeneous complex biofilms (Table 5.2). The working principle of IMS is simple. It first ionizes the material source that is operated by computer controlled X-Y motor stage. After raster, based upon single mass from mass spectrum, an ion image is displayed and its relative abundance is shown as a false color scaling wherein each sampling coordinate is indicated by color intensity within pixel (Fig. 5.2).

5.2.2.3 Microarray Technology

In the previous section we discussed how advanced analytical techniques helped in our understanding of both homogenous and heterogeneous QS circuits at molecular level. In this section, microarray technology enabled QS understating at genetic level is discussed. The microarray technology involves orderly arrangements of samples wherein base paring rule is used for matching known and unknown DNA samples. The focused discussion on recent valuable information generated using microarray technology is illustrated. A detailed discussion on strength and weakness of microarray technology in analysis of quorum sensing system and detail transcriptome studies of individual bacterial species is beyond the scope of this book chapter, please refer Vasil (2003). *P. aeruginosa* is one of the most widely studied QS microorganism. The *P. aeruginosa* biofilm is responsible for life threatening consequences for patients with chronic illness history. Rahmani-Badi and colleagues used microarray technology for examination of cis-2-decenoic acid (CDA) signaling network within *P. aeruginosa* (Rahmani-Badi et al. 2015). CDA has been implicated in inducing biofilm formation and is crucial for inter-kingdom signaling. Components of CDA signaling pathway within *P. aeruginosa* were determined through transcriptome analysis by a comparative pair of experiment – one with CDA and other lacking it. Protein-protein interaction linkage analyses as constructed using STRING and Cytoscape identified 666 genes are differentially expressed in presence of CDA. The functionality associated with these genes was performed using gene ontology that associated CDA with induced enhanced motility, metabolic activity and virulence. Further, CDA synthesis and its perception was mediated through a constellation of five genes (*viz.* PA4978, PA4979, PA4980, PA4982, PA4983).

Table 5.2 Recent imaging MS techniques used for QS

Imaging mass spectrometry technique	Molecule biosensed/finding	Invention summary/future trend	Reference
Imaging between liquid time-of flight (ToF)-SIMS and super resolution florescence microscopy	Extracellular polymeric substance (EPS) material: Fatty acids (e.g., palmitic acid), quinolone signal, and riboflavin fragments are found to respond after the biofilm is treated with Cr (VI)	First chemical mapping of EPS of *Shewanella oneidensis* biofilm	Ding et al. (2016)
		New avenue opened for mechanistic in-sight of QS communications using in situ IMS	
MALDI MS and confocal Raman microspectroscopy (CRM)	EPS materials such as glycolipids, rhamnolipids, polysaccharides and secreted proteins were identified	The comparative study of structural and chemical features of planktonic and biofilm cells of the bacterium *P. aeruginosa* and found out that three day old biofilm showed dramatic difference compared to planktonic culture	Masyuko et al. (2014)
MALDI-guided SIMS	Bioactive secondary metabolites, including rhamnolipids and quinolones, were detected and visualized on both macro- and microscopic size scales	The technological improvisation wherein challenges of locating microscopic chemical of interest was addressed	Lanni et al. (2014)
		First (MALDI) MSI used to obtain low-resolution molecular maps of a sample. The molecular map guided direct subsequent microscopic SIMS imaging and tandem mass spectrometry (MS/MS) experiments	
MALDI-TOF-IMS	Sporulation killing factor (SKF) and sporulation delaying protein (SDP) cannibalism peptides of *B. subtilis*	Specificity and heterogeneity associated with signaling molecule of the genetically identical *Bacillus subtilis* strains studied	Shank and Kolter (2011)
DESI-MS	Polyhydroxyanthraquinones secondary metabolite that inhibit QS	The spatial and temporal distribution of the polyhydroxyanthraquinone was examined	Figueroa et al. (2014)
		First example of employment of DESI-MS imaging technique to scan polyhydroxyanthraquinone in a guttate-forming fungus	

(continued)

Table 5.2 (continued)

Imaging mass spectrometry technique	Molecule biosensed/finding	Invention summary/future trend	Reference
MALDI-imaging HRMS	Spatial distribution of QS molecules AHL in the biosensor strain, *C. violaceum*	First to quantify and visualize the spatial distribution of the QS molecules in the biosensor strain, C. *violaceum*	Kusari et al. (2014)
MALDI TOF	Secondary metabolite	Studied interaction between *B. subtilis* and *Streptomyces coelicolor* using MALDI-TOF IMS	Yang et al. (2009)

The QS regulon of AHL functioning in acidophilic chemolithoautotrophic *Acidithiobacillus ferrooxidans* ATCC 23270 through transcriptomic analyses was performed by Mamani and group using AHL super agonist analog (Mamani et al. 2016). The tetrazole analog was used to stimulate adherence of bacteria on sulfur coupons and DNA microarray assays to determine genes involved in early biofilm formation. DNA microarray analyses revealed that nearly 60 genes are related to biofilm formation. These majorly include induction of phosphate and ammonium transporters and genes encoding F0-ATPase; and repression of genes for carbohydrate metabolism. Balasubramanian and Mathee reviewed comparative transcriptome analyses of *P. aeruginosa* (Balasubramanian and Mathee 2009). Recently, comparison of metabolic pathways for absorption of n-alkane (C_{10}–C_{16}) in *P. aeruginosa* strains *viz.* ATCC 33988 and PAO1 (with >99% average nucleotide identity) through an array of assays was performed to determine cause of how AT33988 consumed faster in comparison to PAO1 (Grady et al. 2017). These tests included proteomics, small-molecule LC-MS, Ribo-seq, RNA-seq and microarray as a single experiment. The integration of –omics data revealed lack of lasI/lasR arm of QS response in ATCC 33988 causing absence of rhamnolipid production, and as an alternative to expressing QS genes, it upregulates operons for alkaline proteases and sphingosine metabolism.

Overall, transcriptional analysis for confirmation of gene expression through usage of DNA microarrays has been generating valuable information over a period of time in various species and strains.

5.3 QS Technologies

The role of advanced analytical and molecular biological techniques used for biomonitoring and understanding of QS phenomenon were discussed in previous section. In turn, the QS phenomenon by itself serves as an interesting platform that is being exploited to be developed for industrial or commercial use. In this section, we will discuss recent focused applications of QS technology in biomonitoring and biosensing mechanisms.

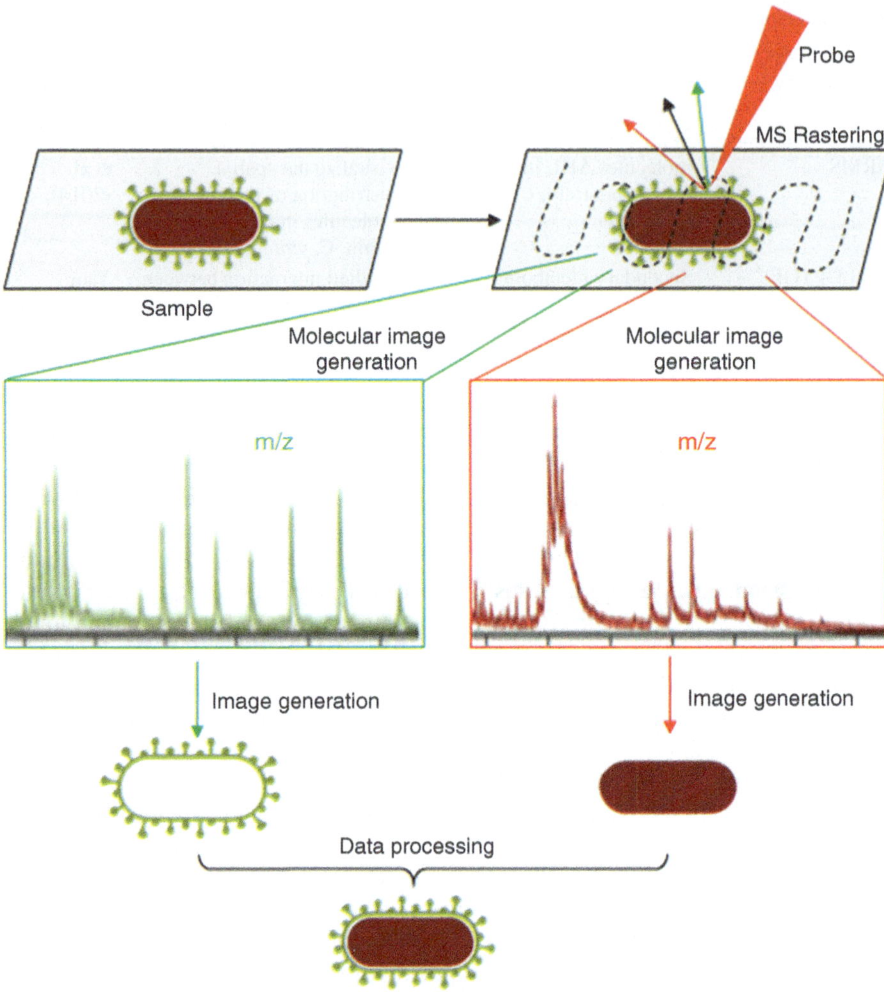

Fig. 5.2 Overview of the workflow in imaging mass spectrometry. Microbial samples are prepared by either as cryosection or isolated and fixed on slide. The area on the slide is selected and a raster grid is prepared for it. Mass spectra data is collected for the sample and all information is combined into a single image with pseudo colors (green or red). Algorithms are applied to merge the multiple images and consolidate it as a single image. *m*/*z*, mass-to-charge ratio

5.3.1 Population Level Oscillator for Arsenic Detection

Synthetic biology has been limited with robust circuits in microenvironment with various clatters of signals. Presence of noise lowers the specificity of signal over a long range. Microbial communities directed towards detection of low available metabolites and its simultaneously signalling can be better utilized by signal amplification. Synchronization of signals emitted by a small population when based on oscillation provides a scalable and future ready option.

An oscillatory fluorescence-generating circuit was developed by Prindle and colleagues that not only synchronized local but global sensing systems for variable arsenic concentration (Prindle et al. 2011). This was achieved by synchronization among oscillating colonies of modified *E. coli* arranged within 500 "biopixels", where each biopixel represented ~5000 individual *E. coli* cells as an array in an area of centimeter-length. This is a significant increase from previously established size of millimeter scale (Danino et al. 2010) which, if directly extrapolated over a longer dimension, is slower for generating macroscopic oscillation. Hence, these smaller oscillations by smaller colonies needed synchronization by designing another level of circuitry for faster communication. The oscillations among the biopixel were achieved by coupling AHL QS within each colony with gas-phase redox signaling (H_2O_2 based). Application of gas-phase signaling allowed detection of unsynchronized, low globally synchronized oscillations and high low globally synchronized oscillations for low (mild repression of the lux promoter), intermediate and very high H_2O_2 (permanent activation of the lux promoter) production respectively.

The gene circuit was coupled with arsenic sensor using variations of oscillation period. In essence, the presence of arsenic changed the oscillation period within the microbial colony present as biopixel. Consequentially, varying amount of arsenic caused proportional GFP oscillation period that propagated among the biopixels of the array. Thus, presence of two systems for communication within the device architecture permitted microbial colony as well as the arrays of colonies to be treated separately. This system had high sensitivity of 0.2 μM (as per World Health Organization, WHO that recommends detection of minimum 0.5 μM arsenite) and was integrated as a handheld sensor device. This device had liquid crystal display (LCD)– like macroscopic clock built on above mentioned stage to make it simple, handy and cost effective. Thus, a macroscopic biosensor involving a defined population of cells was made for arsenic detection within WHO sensitivity limits. Nonetheless, many more examples involving synthetic consortia or circuits for biosensing applications are required.

5.3.2 Genetic Clocks

Synchronized clocks form an underlying mechanism to coordinate rhythmic behavior amongst individual components in a community or a large complex system. Various biological functions are based on intercellular coupling mechanisms including respiration, cardiac activity, and circadian rhythms among others (Glass 2001; Young and Kay 2001). The rhythm generation by few thousand cells present in the mammalian suprachiasmatic nuclei successfully circadian clock of the body. Presence of autoinducer has been reported to induce synchronized oscillations in a cellular population (Garcia-Ojalvo et al. 2004).

In order to utilize quorum sensing systems for regulated clock work, a critical microbial density would be required. Danino and colleagues designed a synchronized oscillator to exemplify usage of microbes in making macroscopic biosensor (Danino et al. 2010). This was achieved by combining cellular features of *V. fisheri* and *B. thurigensis*. Specifically, luxI (origin: *V. fischeri*), aiiA (origin: *B. thurigensis*) and yemGFP genes were placed under the control of three same copies of the luxI promoter. LuxI synthase catalyzes production of AHL that is negatively regulated by aiiA. The network circuitry so formed with an activator causes activity of its own repressor that is similar to system used in synthetic oscillators and is necessity for circadian clock complex. In addition, the authors used variable sized microfluidic devices for environmental volume reduction. The devices incorporated designs for a nutrient channel to feed the confined cells with nutrients or inducers in their respective chamber. Further, the excess population of cells, if generated, would be pushed towards the channel leading to waste port. Thus, sustainable population with a uniform number of cells could be maintained within the device. The research concluded to a size 100 × (80–100) μm^2 was best suited for examining intercellular oscillation activity. The modified cells when allocated in their chambers showcased stable synchronized oscillations. Overall, coupling of quorum sensing with genetic clocks created synchronized oscillations in the controlled population. This approach could be used for increasing the sensitivity and identifying the signal from noise to be utilized as potential biosensor.

5.4 Conclusion

In this chapter we have attempted to review impact of newer bioanalysis techniques and microbiological techniques in understanding QS phenomenon. Recent microbiological assays and advanced analytical techniques such as X-ray, microarray, NMR etc. deciphered QS phenomenon in homogenous state. However, since QS phenomenon is heterogonous in natural environment involving biofilm that comprised of diverse bacterial population, it was an eminent requirement of sophisticated techniques for understanding biosensing and biomonitoring. Thus, newer sophisticated technologies such as SECM and imaging microscopy, traditionally applied in physical sciences, as a probe to topography and surface reactivity are currently being employed to understand heterogonous QS phenomenon. On one side technologies are unraveling QS mysteries, a new trend is also being observed wherein QS itself is used as a technology like in genetic clock and detection of arsenic level. In summation, the newer bioanalysis techniques are uncovering QS phenomenon like never before and QS technology matured to a point wherein it is being now pursued as a technology platform for biosensing and biomonitoring.

5.5 Opinion

The advanced biosensing and biomonitoring technologies capable of resolution in time and three-dimensional space domain are needed to unravel heterogeneous QS system. Towards this endeavor, we reviewed newer sophisticated technologies such as SECM and imaging microscopy. However, these technological applications are at proof of concept stage and we believe it is high time these technologies could be applied to answer fundamental question. The need of an hour is collaborative efforts between QS scientist and technologist towards breakthrough in understanding of heterogeneous QS system. We also feel that prospective are bright towards use of

QS system itself as a technology for biomonitoring and biosensing purpose. We are optimistic that newer orthogonal bioanalysis techniques could lead to major breakthrough in understanding of heterogeneous QS system. We also hope to see QS based biosensing and biomonitoring technology could become market reality.

Acknowledgment Authors would like to thank Sai Life Sciences management and leadership team and Dean, Faculty of Technology, Savitribai Phule Pune University for approval and support. A special thanks to Dr. Sarma, VP, Medicinal Chemistry, Sai Life Sciences, Pune, India for support and encouragement provided during writing this chapter and Dr. Joyita Sarkar for proof-reading. We also thank Prof. Kalia, Emeritus Scientist, CSIR-Institute of Genomics and Integrative Biology, for guidance and an invitation to write book chapter.

References

Bai Y, Patil SN, Bowden SD, Poulter S, Pan J, Salmond GPC, Welch M, Huck WTS, Abell C (2013) Intra-species bacterial quorum sensing studied at single cell level in a double droplet trapping system. Int J Mol Sci 14:10570–10581. https://doi.org/10.3390/ijms140510570

Balasubramanian D, Mathee K (2009) Comparative transcriptome analyses of *Pseudomonas aeruginosa*. Hum Genomics 3:349–361. https://doi.org/10.1186/1479-7364-3-4-361

Bhargava N, Singh SP, Sharma A, Sharma P, Capalash N (2015) Attenuation of quorum sensing-mediated virulence of *Acinetobacter baumannii* by *Glycyrrhiza glabra* flavonoids. Future Microbiol 10:1953–1968. https://doi.org/10.2217/fmb.15.107

Chevrot R, Rosen R, Haudecoeur E, Cirou A, Shelp BJ, Ron E, Faure D (2006) GABA controls the level of quorum-sensing signal in *Agrobacterium tumefaciens*. Proc Natl Acad Sci 103:7460–7464. https://doi.org/10.1073/pnas.0600313103

Danino T, Mondragón-Palomino O, Tsimring L, Hasty J (2010) A synchronized quorum of genetic clocks. Nature 463:326–330. https://doi.org/10.1038/nature08753

Ding Y, Zhou Y, Yao J, Szymanski C, Fredrickson J, Shi L, Cao B, Zhu Z, Yu X-Y (2016) In situ molecular imaging of the biofilm and its matrix. Anal Chem 88:11244–11252. https://doi.org/10.1021/acs.analchem.6b03909

Figueroa M, Jarmusch AK, Raja HA, El-Elimat T, Kavanaugh JS, Horswill AR, Cooks RG, Cech NB, Oberlies NH (2014) Polyhydroxyanthraquinones as quorum sensing inhibitors from the guttates of *Penicillium restrictum* and their analysis by desorption electrospray ionization mass spectrometry. J Nat Prod 77:1351–1358. https://doi.org/10.1021/np5000704

Garcia-Ojalvo J, Elowitz MB, Strogatz SH (2004) Modeling a synthetic multicellular clock: repressilators coupled by quorum sensing. Proc Natl Acad Sci U S A 101:10955–10960. https://doi.org/10.1073/pnas.0307095101

Glass L (2001) Synchronization and rhythmic processes in physiology. Nature 410:277–284. https://doi.org/10.1038/35065745

Grady SL, Malfatti SA, Gunasekera TS, Dalley BK, Lyman MG, Striebich RC, Mayhew MB, Zhou CL, Ruiz ON, Dugan LC (2017) A comprehensive multi-omics approach uncovers adaptations for growth and survival of *Pseudomonas aeruginosa* on n -alkanes. GMC Genomics 18:1–19. https://doi.org/10.1038/85686

Huma N, Shankar P, Kushwah J, Bhushan A, Joshi J, Mukherjee T, Raju SC, Purohit HJ, Kalia VC (2011) Diversity and polymorphism in AHL-lactonase gene (aiiA) of Bacillus. J Microbiol Biotechnol 21:1001–1011. https://doi.org/10.4014/jmb.1105.05056

Kalia VC (2013) Quorum sensing inhibitors: an overview. Biotechnol Adv 31:224–245. https://doi.org/10.1016/J.BIOTECHADV.2012.10.004

Kalia VC (2014) In search of versatile organisms for quorum-sensing inhibitors: acyl homoserine lactones (AHL)-acylase and AHL-lactonase. FEMS Microbiol Lett 359:143–143. https://doi.org/10.1111/1574-6968.12585

Kalia VC (2015) Quorum sensing vs quorum quenching: a battle with no end in sight. https://doi.org/10.1007/978-81-322-1982-8

Koley D, Ramsey MM, Bard AJ, Whiteley M (2011) Discovery of a biofilm electrocline using real-time 3D metabolite analysis. Proc Natl Acad Sci U S A 108:19996–20001. https://doi.org/10.1073/pnas.1117298108

Koul S, Kalia VC (2017) Multiplicity of quorum quenching enzymes: a potential mechanism to limit quorum sensing bacterial population. Indian J Microbiol 57:100–108. https://doi.org/10.1007/s12088-016-0633-1

Krishnan V, Rupp B (2012) Macromolecular structure determination: comparison of X-ray crystallography and NMR spectroscopy. eLS. https://doi.org/10.1002/9780470015902.a0002716.pub2

Kusari P, Kusari S, Lamshöft M, Sezgin S, Spiteller M, Kayser O (2014) Quorum quenching is an antivirulence strategy employed by endophytic bacteria. Appl Microbiol Biotechnol 98:7173–7183. https://doi.org/10.1007/s00253-014-5807-3

Lanni EJ, Masyuko RN, Driscoll CM, Aerts JT, Shrout JD, Bohn PW, Sweedler JV (2014) MALDI-guided SIMS: multiscale imaging of metabolites in bacterial biofilms. Anal Chem 86:9139–9145. https://doi.org/10.1021/ac5020222

Liu X, Ramsey MM, Chen X, Koley D, Whiteley M, Bard AJ (2011) Real-time mapping of a hydrogen peroxide concentration profile across a polymicrobial bacterial biofilm using scanning electrochemical microscopy. Proc Natl Acad Sci U S A 108:2668–2673. https://doi.org/10.1073/pnas.1018391108

Mamani S, Moinier D, Denis Y, Soulère L, Queneau Y, Talla E, Bonnefoy V, Guiliani N (2016) Insights into the quorum sensing regulon of the acidophilic *Acidithiobacillus ferrooxidans* revealed by transcriptomic in the presence of an acyl homoserine lactone superagonist analog. Front Microbiol 7:1–19. https://doi.org/10.3389/fmicb.2016.01365

Masyuko RN, Lanni EJ, Driscoll CM, Shrout JD, Sweedler JV, Bohn PW (2014) Spatial organization of *Pseudomonas aeruginosa* biofilms probed by combined matrix-assisted laser desorption ionization mass spectrometry and confocal Raman microscopy. Analyst 139:5700–5708. https://doi.org/10.1039/c4an00435c

Prindle A, Samayoa P, Razinkov I, Danino T, Tsimring LS, Hasty J (2011) A sensing array of radically coupled genetic "biopixels". Nature 481:39–44. https://doi.org/10.1038/nature10722

Rahmani-Badi A, Sepehr S, Fallahi H, Heidari-Keshel S (2015) Dissection of the cis-2-decenoic acid signaling network in *Pseudomonas aeruginosa* using microarray technique. Front Microbiol 6:1–13. https://doi.org/10.3389/fmicb.2015.00383

Sanchart C, Rattanaporn O, Haltrich D, Phukpattaranont P, Maneerat S (2017) Lactobacillus futsaii CS3, a new GABA-producing strain isolated from Thai fermented shrimp (Kung-Som). Indian J Microbiol 57:211–217. https://doi.org/10.1007/s12088-016-0632-2

Shank EA, Kolter R (2011) Extracellular signaling and multicellularity in *Bacillus subtilis*. Curr Opin Microbiol 14:741–747. https://doi.org/10.1016/j.mib.2011.09.016

Shim J, Patil SN, Hodgkinson JT, Bowden SD, Spring DR, Welch M, Huck WTS, Hollfelder F, Abell C (2011) Controlling the contents of microdroplets by exploiting the permeability of PDMS. Lab Chip 11:1132–1137. https://doi.org/10.1039/c0lc00615g

Shiva Krishna P, Sudheer Kumar B, Raju P, Murty MSR, Prabhakar Rao T, Singara Charya MA, Prakasham RS (2015) Fermentative production of pyranone derivate I from marine Vibrio sp. SKMARSP9: isolation, characterization and bioactivity evaluation. Indian J Microbiol 55:292–301. https://doi.org/10.1007/s12088-015-0521-0

Vasil ML (2003) DNA microarrays in analysis of quorum sensing: strengths and limitations. J Bacteriol 185:2061–2065. https://doi.org/10.1128/JB.185.7.2061

Yang Y-L, Xu Y, Straight P, Dorrestein PC (2009) Translating metabolic exchange with imaging mass spectrometry. Nat Chem Biol 5:885–887. https://doi.org/10.1038/nchembio.252

Young MW, Kay SA (2001) Time zones: a comparative genetics of circadian clocks. Nat Rev Genet 2:702–715. https://doi.org/10.1038/35088576

Application of Microbial Quorum Sensing Systems for Bioremediation of Wastewaters

6

Vipin Chandra Kalia, Jyotsana Prakash, Subhasree Ray, and Shikha Koul

Abstract

Environmental pollution due to the use of naturally occurring and manmade recalcitrant compounds is a major cause of worry among Environmental and Health Department managers. Incidentally, microorganisms have the unique ability to catabolize a wide range of such pollutants. Bioremediation with the help of microbes is limited by their slow growth rate within the bioreactor. On the other hand, in continuous culture digestions, the rate of effluent discharge is quite high and cause an imbalance of food to microbe ratio. In order to retain bacteria, immobilization on different support materials have been recommended. However, Biofilm formation by quorum sensing system (QSS) has been envisaged as a novel approach to entrap microbes within the bioreactor. A few bacteria possessing QSS and ability to catabolize pollutants have been exploited for bioremediation of waste waters.

Keywords

Bioremediation · Quorum sensing · Wastewater · Aromatic compounds · *Pseudomonas* · *Acinetobacter* · Biofilm

V. C. Kalia (✉)
Molecular Biotechnology Lab, Department of Chemical Engineering, Konkuk University, Seoul, Republic of Korea
e-mail: vckaliaku@gmail.com

J. Prakash · S. Ray · S. Koul
Microbial Biotechnology and Genomics, CSIR – Institute of Genomics and Integrative Biology (IGIB), Delhi University Campus, New Delhi, Delhi, India

Academy of Scientific & Innovative Research (AcSIR), New Delhi, Delhi, India
e-mail: jyotsana.prakash@igib.in

V. C. Kalia (ed.), *Quorum Sensing and its Biotechnological Applications*,
https://doi.org/10.1007/978-981-13-0848-2_6

6.1 Introduction

The role of bacterial metabolism in bioremediation process has been well established. The continuous culture process is limited by the slow rate of bacterial multiplication within the reactor and a much faster washout with the effluent. The efficiency of the bioreactor is greatly influenced and even enhanced by higher retention of bacterial population. At high Food to Microbe ratio, the degradation process is not at its best. A lowering the Food to Microbe ratio by higher retention of bacteria helps to enhance the degradation process. Biofilms are known to help dramatically in achieving this target. Within the biofilm diverse bacterial populations stay together as a community and are able to resist environmental stress. In fact, it has been reported that biofilm formation is initiated as a result of bacterial exposure to various environmental stresses such as those caused by antibiotics and pollutants (Ahmed et al. 2009; Vaysse et al. 2009; Kang and Park 2010a; Sharma and Lal 2017). Biofilm formation due to the expression of Quorum sensing (QS) phenomenon has been well documented (Niu et al. 2008). QS signal molecules, acylhomoserine lactones (AHLs) have been reported to contribute significantly in treating wastewaters. The various processes involved in this degradation are: (i) granule formation in aerobic sludge (Ren et al. 2010, 2013; Tan et al. 2014), (ii) stabilization of microbes within the community (Valle et al. 2004) and (iii) production of enzymes (Chong et al. 2012). QS system (QSS) has also been shown to control phenol degradation (Valle et al. 2004; Yong and Zhong 2010), hexadecane degradation (Kang and Park 2010a), ammonium oxidation (De Clippeleir et al. 2011) and denitrification (Toyofuku et al. 2007, 2008). Among the various organisms in which QS mediated biofilm formation has been reported include: (i) *Acinetobacter* (Baldi et al. 1999; Niu et al. 2008; Sarkar and Chakraborty 2008; Tomaras et al. 2008), (ii) *Pseudomonas* (Kang et al. 2007; Toyofuku et al. 2008; Schertzer et al. 2009; Yong and Zhong 2010; Yong et al. 2015), (iii) *Marinobacter hydrocarbonoclasticus* (Vaysse et al. 2009), (iv) *Ochrobactrum* sp. (Imran et al. 2014), (v) *Oleiphilus* (Golyshin et al. 2002), (vi) *Sphingomonas* (Willison 2004), (vii) *Mycobacterium* (Bastiaens et al. 2000), (viii) *Burkholderia* (Matsumiya et al. 2007; Wattanaphon et al. 2008). Although the phenomenon of biofilm formation is an expression of QS pathogenicity of bacteria associated (Kalia 2013), however, it can be exploited as a mechanism to treat waste water and pollutants present in industrial effluents (Stach and Burns 2002; Zhang et al. 2011).

6.2 Biodegradation and QSS

The beginning of the confirmed role of QSS was well addressed by looking into the bacterial community in an industrial wastewater treatment plant. Among the several proteobacterial strains, *Thauera, Comomonas* and *Pseudomonas* spp. were found to be involved in phenol degradation (Valle et al. 2004). A good correlation between QSS and prevalent of aromatic degraders was reported through in silico study (Yeon et al. 2008; Huang et al. 2013). Samples collected from soil, wetland and marine waters, revealed the presence of Sphingomonadales and Rhizobiales, which had abilities to degrade phenanthrene- or pyrene and possessed AHL-production system as well. The production of AHL by Sphingomonads was confirmed through gas-chromatography-mass-spectrum and thin-layer-chromatography. This report of combination of aromatics-degradation and QSS was envisaged to open new avenues for developing bioremediation technologies (Huang et al. 2013).

6.2.1 Pseudomonas

Pseudomonas is a genus comprised of genetically diverse organisms (Bhushan et al. 2013). This genetic diversity is reflected in their equally diverse and efficient metabolisms. The most important are their abilities to degrade pollutants. Incidentally, this bacterium also causes very lethal infectious diseases, which provoked researchers to look for the mechanisms to control their pathogenicity (Kalia and Purohit 2011; Kalia 2013). Among the various mechanisms, biofilm formation by *Pseudomonas* through QSS has been identified as the most relevant target for drugs. Here, is the opportunity to exploit the two characteristics of *Pseudomonas* and reduce pollution, especially that caused by waste waters.

Pseudomonas aeruginosa strain CGMCC 1.860 can produce AHLs and degrade aromatic compounds. AHL production continued while metabolism of aromatic compounds such as including salicylate, benzoate, naphthalene, *p*-hydroxybenzoate, and phenol was observed. The role of QS during phenol biodegradation was evident under diverse conditions: (i) exogenously addition of AHL extracts, (ii) endogenous over-production of QS signals, (iii) inhibition by abolishing the QS signal molecule production, and (iv) was not affected if extracts without AHLs were added. The results indicated that AHL was involved in the process of biodegradation of pollutants (Table 6.1) (Yong and Zhong 2010, 2013a). In addition to direct involvement of QSS, certain aromatic compounds are degraded through multiple steps. Here, QS plays an indirect role. Biodegradation of anthranilate was influenced by AHL molecules in a QSS, which was LuxR-independent (Chugani and Greenberg 2010). Addition of AHLs such as C4HSL, C8HSL and C10HSL helped in enhancing the transcription of *catB*, a gene responsible for encoding catechol-1, 2-dioxygenase (Chugani and Greenberg 2010). In a similar kind of study, meta-cleavage of catechol by catechol 2, 3-dioxygenase by *P. aeruginosa* strain CGMCC 1.860 (Yong and Zhong 2013b).

Naphthalene and its metabolic products such as reactive oxygen species are highly toxic to a wide range of organisms (Yong et al. 2015). It is thus necessary to search bacteria, which must be resistant to these compounds and degrade them as well (Park et al. 2004; Kang et al. 2006). Metabolism of naphthalene and related compounds was dramatically enhanced by *Pseudomonas* sp. strain As1, which contained plasmid constructs conferring expression of antioxidant enzymes and superoxidase dismutase (Kang et al. 2007). A positive correlation between QS mediated biofilm formation and degradation of polycyclic aromatic hydrocarbons (PAHs) such as pyrene and phenanthrene was recorded with *Pseudomonas mendocina* strain

Table 6.1 Applications of microbial quorum sensing systems for bioremediation of waste waters

Organisms	QSS/molecule involved	Approaches	Applications	References
Bioremediation				
Burkholderia sp. DW2–1, *B. cenocepacia* BSP3	CepI/CepR	QS mediated biosurfactant production in Glucose containing nitrate supplemented medium	Waste water treatment and pesticides removal from soil	Matsumiya et al. (2007) and Wattanaphon et al. (2008)
Aerobic and anoxic ammonium-oxidizing bacteria(AerAOB, AnAOB)	N-dodecanoyl-L-Homoserinelactone	Tuning of QSS by addition of synthetic AHLs in AerAOB and AnAOB	Enhanced waste water treatment	De Clippeleir et al. (2011)
Microbes from aerobic granular sludge(AGS)	AutoInducer-2	Boron supplementation for increased exopolysaccharides (EPS) production and faster AGS formation	Increasing the efficiency of wastewater treatment	Zhang et al. (2011)
Pseudomonas aeruginosa PAO	N-decanoyl-L-HSL (C10HSL)	Upregulation of C10HSL responsive anthranilate metabolism, independent of las/rhl QSS	Anthranilate biodegradation in waste water	Chugani and Greenberg (2010)
P. aeruginosa	Pseudomonas quinolone signal (PQS)	Transcription profiling revealing iron chelation role of PQS	Putative role in bioremediation	Schertzer et al. (2009)
P. aeruginosa	LasI/LasR; RhlI/RhlR	Heterologous gene expression strategies for rhamnolipid production	Biodegradable surfactants for pharmaceutical, cosmetic, food industries and for enhanced bioremediation of oil contaminant sites	Müller et al. (2012)
P. aeruginosa CGMCC1.860	RhlI/RhlR	Tuning catechol meta-cleavage pathway	Improved degradation of aromatic contaminants (phenol, benzoate, and phenanthrene)	Yong and Zhong (2013b)
Acinetobacter sp. DR1	AqsI/AqsR	Supplementation of *aqsI* mutants with C12HSL enhanced utilization of hydrocarbons and biofilm formation	Hexadecane degradation from waste	Kang and Park (2010a, c)
Ochrobactrum sp. Pv2Z2	Not identified	Tuning of QS to increase the bacterial rhizospheric colonization	Biocontrol agent and biofertilizer	Imran et al. (2014)

QS quorum sensing, *QSS* quorum sensing system

NR802 and *P. aeruginosa* strain N6P6 (Mangwani et al. 2013, 2015). Phenanthrene degradation of up to 86% was observed after a period of 7 days. This degradation was negatively influenced by QS inhibitor – tannic acid, which also affected biofilm formation and other QS mediated expressions such as pyocyanin production (Mangwani et al. 2015), Bioremediation of PAHs is enhanced by their bioavailability, which was achieved by biosurfactants. Microbes can be employed to produce these biosurfactants during the dispersion phase of biofilm (Makkar and Rockne 2003; Wattanaphon et al. 2008). Biosurfactants such as rhamnolipids can be produced by biofilm forming *P. aeruginosa* (Boles et al. 2004; Müller et al. 2012; Shukla et al. 2014).

Biological metabolism of nitrate and nitrite rich wastewater can be achieved through the phenomenon of denitrification (Yong et al. 2015). The basic enzymes involved in this process are nitrate reductase, nitrirte reductase, nitrous oxide reductase and nitrogen dioxide reductase. In *P. aeruginosa* PAO1, the denitrification enzymes are under the control of QSS system- *rhlR* (Yoon et al. 2002; Toyofuku et al. 2007). It was subsequently shown that *Pseudomonas* quinolone signal (PQS) – mediated QSS is also critical for denitrification process. PQS system chelates iron and inhibits NO_3^- dependent respiration and other reductases (Toyofuku et al. 2008).

6.2.2 Acinetobacter

Acinetobacter can metabolize a wide range of hydrocarbons (HCs) (Throne-Holst et al. 2007; Yoon et al. 2007; Fischer et al. 2008; Jung and Park 2015). The degradation of alkane HCs takes place by the adsorption of bacterial cells on HCs or by development of biofilm on the interface of HCs and water (Baldi et al. 1999). Biofilm formation helps in effective degradation of HCs (Kang and Park 2010b; Bhargava et al. 2012). A close correlation between QS mediated biofilm (Table 6.1) (Kang and Park 2010c), and the HC degradation ability of *Acinetobacter* was shown through the use of wild type strain DR1 and its variants, which included rifampin resistant strain DR1R and *aqsI* mutant, which has lost the ability to synthesize AHLs. *aqsI* mutant strain showed severe defects in growth and mineralization of hexadecane. Addition of QS signals to the medium where *aqsI* was mineralizing HC resulted in improved process efficiency. C12-AHL and its derivatives such as 3-oxo, 3-hydroxy, or other 3-substituted QS signal molecules (Kang and Park 2010a). Phenol degrading *Acinetobacter calcoaceticus* strain PHEA-2 and dye decolorizing strain of *A. calcoaceticus* YC210 were obtained from waste water samples (Chen et al. 2011; Zhan et al. 2012). *A. calcoaceticus* YC210 could efficiently (94.5%) decolorize VBR in the pH range of 5–7 and this ability was well maintained up to 450 mg/L (Chen et al. 2011). *Acinetobacter* sp. isolated from soil contaminated with petroleum was observed to produce biosurfactant, which proves helpful in efficient degradation of organic pollutants by increasing the bioavailability of the substrate (Chen et al. 2012). *Acinetobacter gerneri* strain P7 was characterized for the production of polyurethanase enzymes using p-nitrophenyl-propanate where it recorded an activity of 37.58 U mg/L. The organism formed strong complex with polyurethane (Howard et al. 2012). Comparative genomic analysis of three strains of *Acinetobacter* species with distinct characteristics: (i) *A. baumannii* AYE, a human pathogen, (ii) *A. baylyi* ADP1, a strain adapted to soils, and (iii) a phenol degrader *A. calcoaceticus* PHEA-2 revealed horizontal gene transfer among them and a larger number of transport-related proteins were found in PHEA-2 rather than in ADP1 and AYE. It implied higher adaptation of PHEA-2 towards phenol-contaminated waste waters (Zhan et al. 2012). Genome sequencing of hydrocarbon degrading *Acinetobacter venetianus* strains RAG-1 and VE-C3 revealed the genetic basis of their adaptation to these pollutants (Fondi et al. 2012, 2013).

6.2.3 Other Potential Organisms

6.2.3.1 Marinobacter

The role of biofilm formation and hydrocarbon degradation ability of *M. hydrocarbonoclasticus* strain SP17 was established through proteomic analysis of the compounds present at their interphase (Klein et al. 2008; Vaysse et al. 2009). Cells within the biofilm showed that out of 1144 proteins, 576 showed modulation in comparison to those observed in planktonic cells. Within the biofilm cells, the most remarkable was the over expression of protein encoded by MARHY0478 – responsible for transportation of hydrophobic compounds (Vaysse et al. 2009).

6.2.3.2 Ochrobactrum

Bacteria present in the rhizosphere have generally been found to prove beneficial to plants for their growth and development. *Ochrobactrum* strain Pv2Z2 is a rhizospheric isolate which showed plant growth promoting features, antipathogenic properties, ability to degrade phenol and possessed QSS as well (Imran et al. 2014). Another strain of *Ochrobactrum* sp. NW-3 showed growth enhancement of cucumber plants (Xu et al. 2015).

6.2.3.3 Sphingomonas

Sphingomonas members are known to catabolize a large number of naturally occurring recalcitrant and anthropogenic compounds such as furans, biphenyl, oestradiol, napthalenes, chlorinated phenols etc. (Gan et al. 2014; Verma et al. 2017). The presence of QSS in *Sphingomonas* was reported by comparative genomic analysis of 62 sequenced genomes. It revealed luxI/R type of QSS in *Sphingomonas japonicum* and *S. lactosutens* (Gan et al. 2014). However, a direct involvement of QS in bioremediation is still being elucidated.

6.3 Opinion

The concept of combining QSS and ability to catabolize pollutants has a very great potential as this will ensure availability of requisite microbial population within the bioreactor. Another possibility can be using two different microbes each with either of the property. It will also prove beneficial as they will be behaving in a complementary manner. The need is to evaluate the ability of a large number of biofilm formers to degrade pollutants and thus help in bioremediation of waste waters. Another important factor, which emerges from this QS mediated biofilm is the fact that the system has to be protected against bacterial and QS inhibitors (Kalia 2013, 2014; Gui et al. 2014; Begum et al. 2016; Koul et al. 2016; Jeyanthi and Velusamy 2016; Varsha et al. 2016; Wadhwani et al. 2016; Ahiwale et al. 2017; Azman et al. 2017; Kalia et al. 2017; Koul and Kalia 2017).

Acknowledgments This work was supported by Brain Pool grant (NRF-2018H1D3A2001746) by National Research Foundation of Korea (NRF) to work at Konkuk University.

References

Ahiwale SS, Bankar AV, Tagunde S, Kapadnis BP (2017) A bacteriophage mediated gold nanoparticle synthesis and their anti-biofilm activity. Indian J Microbiol 57:188–194. https://doi.org/10.1007/s12088-017-0640-x

Ahmed NA, Petersen FC, Schele AA (2009) AI-2/LuxS is involved in increased biofilm formation by *Streptococcus intermedius* in the presence of antibiotics. Antimicrob Agents Chemother 53:4258–4263. https://doi.org/10.1128/AAC.00546-09

Azman CA-S, Othman I, Fang C-M, Chan K-G, Goh B-H, Lee L-H (2017) Antibacterial, anticancer and neuroprotective activities of rare actinobacteria from mangrove forest soils. Indian J Microbiol 57:177–187. https://doi.org/10.1007/s12088-016-0627-z

Baldi F, Ivosevic N, Minacci A, Pepi M, Fani R, Svetlicic V, Zutic V (1999) Adhesion of *Acinetobacter venetianus* to diesel fuel droplets studied with in situ electrochemical and molecular probes. Appl Environ Microbiol 65:2041–2048 doi: NA

Bastiaens L, Springael D, Wattiau P, Harms H, DeWachter R, Verachtert H, Diels L (2000) Isolation of adherent polycyclic aromatic hydrocarbon (PAH)-degrading bacteria using PAH-sorbing carriers. Appl Environ Microbiol 66:1834–1843. https://doi.org/10.1128/AEM.66.5.1834-1843.2000

Begum IF, Mohankumar R, Jeevan M, Ramani K (2016) GC–MS analysis of bioactive molecules derived from *Paracoccus pantotrophus* FMR19 and the antimicrobial activity against bacterial pathogens and MDROs. Indian J Microbiol 56:426–432. https://doi.org/10.1007/s12088-016-0609-1

Bhargava N, Sharma P, Capalash N (2012) N-acyl homoserine lactone mediated interspecies interactions between *A. baumannii* and *P. aeruginosa*. Biofouling 28:813–822. https://doi.org/10.1128/IAI.01600-14

Bhushan A, Joshi J, Shankar P, Kushwah J, Raju SC, Purohit HJ, Kalia VC (2013) Development of genomic tools for the identification of certain *Pseudomonas* up to species level. Indian J Microbiol 53:253–263. https://doi.org/10.1007/s12088-013-0412-1

Boles BR, Thoendel M, Singh PK (2004) Self-generated diversity produces "insurance effects" in biofilm communities. Proc Natl Acad Sci U S A 101:16630–16635. https://doi.org/10.1073/pnas.0407460101

Chen CC, Chen CY, Cheng CY, Teng PY, Chung YC (2011) Decolorization characteristics and mechanism of Victoria Blue R removal by *Acinetobacter calcoaceticus* YC210. J Hazard Mater 196:166–172. https://doi.org/10.1016/j.jhazmat.2011.09.015

Chen J, Huang PT, Zhang KY, Ding FR (2012) Isolation of biosurfactant producers, optimization and properties of biosurfactant produced by *Acinetobacter* sp. from petroleum-contaminated soil. J Appl Microbiol 112:660–671. https://doi.org/10.1111/j.1365-2672.2012.05242

Chong G, Kimyon O, Rice SA, Kjellebrg S, Manefiled M (2012) The presence and role of bacterial quorum sensing in activated sludge. Microb Biotechnol 5:621–633. https://doi.org/10.1111/j.1751-7915.2012.00348.x

Chugani S, Greenberg EP (2010) LuxR homolog-independent gene regulation by acyl-homoserine lactones in *Pseudomonas aeruginosa*. Proc Natl Acad Sci U S A 107:10673–10678. https://doi.org/10.1073/pnas.1005909107

De Clippeleir H, Defoirdt T, Vanhaecke L, Vlaeminck SE, Carballa M, Verstraete W, Boon N (2011) Long-chain acylhomoserine lactones increase the anoxic ammonium oxidation rate in an OLAND biofilm. Appl Microbiol Biotechnol 90:1511–1519. https://doi.org/10.1007/s00253-011-3177-7

Fischer R, Bleichrodt FS, Gerischer UC (2008) Aromatic degradative pathways in *Acinetobacter baylyi* underlie carbon catabolite repression. Microbiology 154:3095–3103. https://doi.org/10.1099/mic.0.2008/016907-0

Fondi M, Orlandini V, Emiliani G, Papaleo MC, Maida I, Perrin E, Vaneechoutte M, Dijkshoorn L, Fani R (2012) Draft genome sequence of the hydrocarbon-degrading and emulsan-producing strain *Acinetobacter venetianus* RAG-1. J Bacteriol 194:4771–4772. https://doi.org/10.1128/JB.01019-12

Fondi M, Rizzi E, Emiliani G, Orlandini V, Berna L, Papaleo MC, Perrin E, Maida I, Corti G, De Bellis G, Baldi F, Dijkshoorn L, Vaneechoutte M, Fani R (2013) The genome sequence of the hydrocarbon-degrading *Acinetobacter venetianus* VE-C3. Res Microbiol 164:439–449. https://doi.org/10.1016/j.resmic.2013.03.003

Gan HM, Gan HY, Ahmad NH, Aziz NA, Hudson AO, Savka MA (2014) Whole genome sequencing and analysis reveal insights into the genetic structure, diversity and evolutionary relatedness of *luxI* and *luxR* homologs in bacteria belonging to the *Sphingomonadaceae* family. Front Cell Infect Microbiol 4:188. https://doi.org/10.3389/fcimb.2014.00188

Golyshin PN, Chemikova TN, Abraham WR, Lunsdorf H, Timmis KN, Yakimov MM (2002) Oleiphilaceae fam. Nov., to include *Oleiphilus messinensis* gen. Nov., sp. nov., a novel marine bacterium that obligates utilizes hydrocarbons. Int J Syst Evol Microbiol 52:901–911. https://doi.org/10.1099/00207713-52-3-901

Gui Z, Wang H, Ding T, Zhu W, Zhuang X, Chu W (2014) Azithromycin reduces the production of α-hemolysin and biofilm formation in *Staphylococcus aureus*. Indian J Microbiol 54:114–117. https://doi.org/10.1007/s12088-013-0438-4

Howard GT, Norton WN, Burks T (2012) Growth of *Acinetobacter gerneri* P7 on polyurethane and the purification and characterization of a polyurethanase enzyme. Biodegradation 23:561–573. https://doi.org/10.1007/s10532-011-9533-6

Huang Y, Zeng Y, Yu Z, Zhang J, Fenh H, Lin X (2013) In silico and experimental methods revealed highly diverse bacteria with quorum sensing and aromatics biodegradation systems – a potential broad application on bioremediation. Bioresour Technol 124:23–38. https://doi.org/10.1016/j.biortech.2013.08.155

Imran A, Saadalla MJA, Khan SU, Mirza MS, Malik KA, Hafeez FY (2014) *Ochrobactrum* sp. Pv2Z2 exhibits multiple traits of plant growth promotion, biodegradation and N-acyl-homoserine-lactone quorum sensing. Ann Microbiol 64:1797–1806. https://doi.org/10.1007/s13213-014-0824-0

Jeyanthi V, Velusamy P (2016) Anti-methicillin resistant *Staphylococcus aureus* compound isolation from halophilic *Bacillus amyloliquefaciens* MHB1 and determination of its mode of action using electron microscope and flow cytometry analysis. Indian J Microbiol 56:148–157. https://doi.org/10.1007/s12088-016-0566-8

Jung J, Park W (2015) *Acinetobacter* species as model microorganisms in environmental microbiology: current state and perspectives. Appl Microbiol Biotechnol 99:2533–2548. https://doi.org/10.1007/s00253-015-6439-y

Kalia VC (2013) Quorum sensing inhibitors: an overview. Biotechnol Adv 31:224–245. https://doi.org/10.1016/j.biotechadv.2012.10.004

Kalia VC (2014) Microbes, antimicrobials and resistance: the battle goes on. Indian J Microbiol 54:1–2. https://doi.org/10.1007/s12088-013-0443-7

Kalia VC, Purohit HJ (2011) Quenching the quorum sensing system: potential antibacterial drug targets. Crit Rev Microbiol 37:121–140. https://doi.org/10.3109/1040841X.2010.532479

Kalia VC, Prakash J, Koul S, Ray S (2017) Simple and rapid method for detecting biofilm forming bacteria. Indian J Microbiol 57:109–111. https://doi.org/10.1007/s12088-016-0616-2

Kang YS, Park W (2010a) Contribution of quorum-sensing system to hexadecane degradation and biofilm formation in *Acinetobacter* sp. strain DR1. J Appl Microbiol 109:1650–1659. https://doi.org/10.1111/j.1365-2672.2010.04793.x

Kang YS, Park W (2010b) Protection against diesel oil toxicity by sodium chloride-induced exopolysaccharides in *Acinetobacter* sp. strain DR1. J Biosci Bioeng 109:118–123. https://doi.org/10.1016/j.jbiosc.2009.08.001

Kang YS, Park W (2010c) Trade-off between antibiotic resistance and biological fitness in *Acinetobacter* sp. strain DR1. Environ Microbiol 12:1304–1318. https://doi.org/10.1111/j.1462-2920.2010.02175

Kang YS, Kim YJ, Jeon CO, Park W (2006) Characterization of naphthalene-degrading Pseudomonas species isolated from pollutant-contaminated sites: oxidative stress during their growth on naphthalene. J Microbiol Biotechnol 16:1819–1825 doi: NA

Kang YS, Lee Y, Jung H, Jeon CO, Madsen EL, Park W (2007) Overexpressing antioxidant enzymes enhances naphthalene biodegradation in *Pseudomonas* sp. strain As1. Microbiology 153:3246–3254. https://doi.org/10.1099/mic.0.2007/008896-0

Klein B, Grossi V, Bouriat P, Goulas P, Grimaud R (2008) Cytoplasmic wax ester accumulation during biofilm-driven substrate assimilation at the alkane-water interface by *Marinobacter hydrocarbonoclasticus* SP17. Res Microbiol 159:137–144. https://doi.org/10.1016/j.resmic.2007.11.013

Koul S, Kalia VC (2017) Multiplicity of quorum quenching enzymes: a potential mechanism to limit quorum sensing bacterial population. Indian J Microbiol 57:100–108. https://doi.org/10.1007/s12088-016-0633-1

Koul S, Prakash J, Mishra A, Kalia VC (2016) Potential emergence of multi-quorum sensing inhibitor resistant (MQSIR) bacteria. Indian J Microbiol 56:1–18. https://doi.org/10.1007/s12088-015-0558-0

Makkar RS, Rockne KJ (2003) Comparison of synthetic surfactants and biosurfactants in enhancing biodegradation of polycyclic aromatic hydrocarbons. Environ Toxicol Chem 22:228–2292. https://doi.org/10.1897/02-472

Mangwani N, Shukla SK, Rao TS, Das S (2013) Calcium-mediated modulation of *Pseudomonas mendocina* NR802 biofilm influences the phenanthrene degradation. Colloids Surf B 114:301–309. https://doi.org/10.1016/j.colsurfb.2013.10.003

Mangwani N, Kumari S, Das S (2015) Involvement of quorum sensing genes in biofilm development and degradation of polycyclic aromatic hydrocarbons by a marine bacterium *Pseudomonas aeruginosa* N6P6. Appl Microbiol Biotechnol 99:10283–10297. https://doi.org/10.1007/s00253-015-6868-7

Matsumiya Y, Wakita D, Kimura A, Sanpa S, Kubo M (2007) Isolation and characterization of a lipid-degrading bacterium and its application to lipid-containing wastewater treatment. J Biosci Bioeng 103:325–330. https://doi.org/10.1263/jbb.103.325

Müller MM, Kügler JH, Henkel M, Gerlitzki M, Hörmann B, Pöhnlein M, Syldatk C, Hausmann R (2012) Rhamnolipids—next generation surfactants? J Biotechnol 162:366–380. https://doi.org/10.1016/j.jbiotec.2012.05.022

Niu C, Clemmer KM, Bonomo RA, Rather PN (2008) Isolation and characterization of an autoinducer synthase from *Acinetobacter baumannii*. J Bacteriol 190:3386–3392. https://doi.org/10.1128/JB.01929-07

Park W, Jeon CO, Cadillo H, DeRito C, Madsen EL (2004) Survival of naphthalene-degrading *Pseudomonas putida* NCIB 9816-4 in naphthalene-amended soils: toxicity of naphthalene and its metabolites. Appl Microbiol Biotechnol 64:429–435. https://doi.org/10.1007/s00253-003-1420-6

Ren TT, Yu HQ, Li XY (2010) The quorum sensing effect of aerobic granules on bacterial adhesion, biofilm formation and sludge granulation. Appl Microbiol Biotechnol 88:789–797. https://doi.org/10.1007/s00253-010-2796-8

Ren TT, Yu HQ, Li XY (2013) Effect of N-acyl-homoserine lactones-like molecules from aerobic granules on biofilm formation by *Escherichia coli* K12. Bioresour Technol 129:655–658. https://doi.org/10.1016/j.biortech.2012.12.043

Sarkar S, Chakraborty R (2008) Quorum sensing in metal tolerance of *Acinetobacter junii* BB1A is associated with biofilm production. FEMS Microbiol Lett 282:160–165. https://doi.org/10.1111/j.1574-6968.2008.01080.x

Schertzer JW, Boulette ML, Whiteley M (2009) More than a signal: non-signaling properties of quorum sensing molecules. Trends Microbiol 17:189–195. https://doi.org/10.1016/j.tim.2009.02.001

Sharma A, Lal R (2017) Survey of (Meta)genomic approaches for understanding microbial community dynamics. Indian J Microbiol 57:23–38. https://doi.org/10.1007/s12088-016-0629-x

Shukla SK, Mangwani N, Rao TS, Das S (2014) Biofilm-mediated bioremediation of polycyclic aromatic hydrocarbons. In: Das S (ed) Microbial biodegradation and bioremediation. Elsevier, USA, pp. 203–232. https://doi.org/10.1016/b978-0-12-800021-2.00008-x

Stach JE, Burns RG (2002) Enrichment versus biofilm culture: a functional and phylogenetic comparison of polycyclic aromatic hydrocarbon-degrading microbial communities. Environ Microbiol 4:169–182. https://doi.org/10.1046/j.1462-2920.2002.00283.x

Tan CH, Koh KS, Xie C, Tay M, Zhou Y, Williams R, Ng WJ, Rice SA, Kjelleberg S (2014) The role of quorum sensing signalling in EPS production and the assembly of a sludge community into aerobic granules. ISME J 8:1186–1197. https://doi.org/10.1038/ismej.2013.240

Throne-Holst M, Wentzel A, Ellingsen TE, Kotlar HK, Zotchev SB (2007) Identification of novel genes involved in long-chain n-alkane degradation by *Acinetobacter* sp. strain DSM 17874. Appl Environ Microbiol 73:3327–3332. https://doi.org/10.1128/AEM.00064-07

Tomaras AP, Flagler MJ, Dorsey CW, Gaddy JA, Actis LA (2008) Characterization of a two-component regulatory system from *Acinetobacter baumannii* that controls biofilm formation and cellular morphology. Microbiology 154:3398–3409. https://doi.org/10.1099/mic.0.2008/019471-0

Toyofuku M, Nomura N, Fujii T, Takaya N, Maseda H, Sawada I, Nakajima T, Uchiyama H (2007) Quorum sensing regulates denitrification in *Pseudomonas aeruginosa* PAO1. J Bacteriol 189:4969–4672. https://doi.org/10.1128/JB.00289-07

Toyofuku M, Nomura N, Kuno E, Tashiro Y, Nakajima T, Uchiyama H (2008) Influence of Pseudomonas quinolone signal on denitrification in *Pseudomonas aeruginosa*. J Bacteriol 190:7947–7956. https://doi.org/10.1128/JB.00968-08

Valle A, Bailey MJ, Whiteley AS, Manefield M (2004) N-acyl-l-homoserine lactones (AHLs) affect microbial community composition and function in activated sludge. Environ Microbiol 6:424–433. https://doi.org/10.1111/j.1462-2920.2004.00581.x

Varsha KK, Nishant G, Sneha SM, Shilpa G, Devendra L, Priya S, Nampoothiri KM (2016) Antifungal, anticancer and aminopeptidase inhibitory potential of a phenazine compound produced by *Lactococcus* BSN307. Indian J Microbiol 56:411–416. https://doi.org/10.1007/s12088-016-097-1

Vaysse PJ, Prat L, Magenot S, Cruveiller S, Goulas P, Grimaus R (2009) Proteomic analysis of *Marinobacter hydrocarbonoclasticus* SP17 biofilm formation at the alkane water interface reveals novel proteins and cellular processes involved in hexadecane assimilation. Res Microbiol 160:829–837. https://doi.org/10.1016/j.resmic.2009.09.010

Verma H, Bajaj A, Kumar R, Kaur J, Anand S, Nayyar N, Puri A, Singh Y, Khurana JP, Lal R (2017) Genome organization of *Sphingobium indicum* B90A: an archetypal hexachlorocyclohexane (HCH) degrading genotype. Genome Biol Evol 9:2191–2197. https://doi.org/10.1093/gbe/evx133

Wadhwani SA, Shedbalkar UU, Singh R, Vashisth P, Pruthi V, Chopade BA (2016) Kinetics of synthesis of gold nanoparticles by *Acinetobacter* sp. SW30 isolated from environment. Indian J Microbiol 56:439–444. https://doi.org/10.1007/s12088-016-0598-0

Wattanaphon HT, Kerdsin A, Thammacharoen C, Sangvanich P, Vangnai AS (2008) A biosurfactant from *Burkholderia cenocepacia* BSP3 and its enhancement of pesticide solubilization. J Appl Microbiol 105:416–423. https://doi.org/10.1111/j.1365-2672.2008.03755.x

Willison JC (2004) Isolation and characterization of a novel sphingomonad capable of growth with chrysene as sole carbon and energy source. FEMS Microbiol Lett 241:143–150. https://doi.org/10.1016/j.femsle.2004.10.012

Xu S, Liu Y, Wang J, Yin TT, Han YF, Wang XB, Huang ZY (2015) Isolation and potential of *Ochrobactrum* sp. NW-3 to increase the growth of cucmber. Int J Ag Policy Res 3:341–350. https://doi.org/10.15739/IJAPR.059

Yeon KM, Cheong WS, Oh HS, Lee WN, Hwang BK, Lee CH, Beyenal H, Lewandowski Z (2008) Quorum sensing: a new biofouling control paradigm in a membrane bioreactor for advanced wastewater treatment. Environ Sci Technol 43:380–385. https://doi.org/10.1021/es8019275

Yong YC, Zhong JJ (2010) N-acylated homoserine lactone production and involvement in the biodegradation of aromatics by an environmental isolate of *Pseudomonas aeruginosa*. Process Biochem 45:1944–1948. https://doi.org/10.1016/j.procbio.2010.05.006

Yong YC, Zhong JJ (2013a) Impact of quorum sensing on microbial metabolism and human health. Future Trends Bioetchnol 131:25–61. https://doi.org/10.1007/10_2012_138

Yong YC, Zhong JJ (2013b) Regulation of aromatics biodegradation by *rhl* quorum sensing system through induction of catechol meta-cleavage pathway. Bioresour Technol 136:761–765. https://doi.org/10.1016/j.biortech.2013.03.134

Yong YC, Wu XY, Sun JZ, Cao YX, Song H (2015) Engineering quorum sensing signaling of *Pseudomonas* for enhanced wastewater treatment and electricity harvest: a review. Chemosphere 140:18–25. https://doi.org/10.1016/j.chemosphere.2014.10.020

Yoon SS, Hennigan RF, Hilliard GM, Ochsner UA, Parvatiyar K, Kamani MC, Allen HL, DeKievit TR, Gardner PR, Schwab U, Rowe JJ, Iglewski BH, McDermott TR, Mason RP, Wozniak DJ, Hancock RE, Parsek MR, Noah TL, Boucher RC, Hassett DJ (2002) *Pseudomonas aeruginosa* anaerobic respiration in biofilms: relationships to cystic fibrosis pathogenesis. Dev Cell 3:593–603. https://doi.org/10.1016/S1534-5807(02)00295-2

Yoon YH, Yun SH, Park SH, Seol SY, Leem SH, Kim SI (2007) Characterization of a new catechol branch of the beta-ketoadipate pathway induced for benzoate degradation in *Acinetobacter lwoffii* K24. Biochem Biophys Res Commun 360:513–519. https://doi.org/10.1016/j.bbrc.2007.05.132

Zhan Y, Yan Y, Zhang W, Chen M, Lu W, Ping S, Lin M (2012) Comparative analysis of the complete genome of an *Acinetobacter calcoaceticus* strain adapted to a phenol-polluted environment. Res Microbiol 163:36–43. https://doi.org/10.1016/j.resmic.2011.10.006

Zhang SH, Yu X, Guo F, Wu ZY (2011) Effect of interspecies quorum sensing on the formation of aerobic granular sludge. Water Sci Technol 64:1284–1290. https://doi.org/10.2166/wst.2011.723

lux Gene: Quorum Sensing, Engineering and Applications

7

Bhagwan Rekadwad

Abstract

Phenotypes and genotypes of the microorganisms develop over a period of time as per the environmental conditions. Numbers of reason are responsible to make change outlook and characteristics of microorganisms. These include natural phenomenon such as quorum sensing, mutations, horizontal gene transfer that plays a critical part in bacterial quorum sensing development and has major clinical significance in bacterial evolution. This is the key to comprehend the components and energy of hereditary changes. Common change is the driving component for horizontal gene transfer in various genera of microscopic organisms. These changes may be due the necessity feel by microorganism allows to express their genes/environmental factors triggers the activation of genes. This book chapter present straightforward applications of *lux-system* in biotechnology and bioprocesses with industrial value.

Keywords

Biophotonic imaging · Dip-sticks · Lux-system · QS system · Luciferase · Reporter genes

7.1 Introduction

The blend of latest imaging innovations and advancement of *luciferase* (*lux*)-based bioluminescent system give a touchy and basic non-obtrusive identification technique (biophotonic imaging) for the investigation of diverse biological procedures and remedial intercessions (Bruckbauer et al. 2015). Bioluminescence is the era of

B. Rekadwad (✉)
National Centre for Microbial Resource, National Centre for Cell Science, Pune, India

V. C. Kalia (ed.), *Quorum Sensing and its Biotechnological Applications*,
https://doi.org/10.1007/978-981-13-0848-2_7

light by living life forms subsequently of vitality discharged when the light emanating particle luciferin experiences oxidation catalyzed by *luciferase* within the sight of oxygen. Bioluminescence capacities for pulling in mates, catching prey, and for countershading in night time amphibian life forms to avoid predators. It happens over a wide scope of life forms include plants animals and microorganisms. In microorganisms, quorum sensing is the phenomenon that regulates biofilm formation (Koutsoudis et al. 2006; Vidal et al. 2011; Fatima et al. 2010; Kalia 2015; Koul et al. 2016; Koul and Kalia 2017) governs quorum sensing system (QS system) on a certain extent (Kalia 2014; Rekadwad and Khobragade 2017a, b). Bacteria such *Photobacterium* spp., *Photorhabsus* sp. and *Vibrio* sp. exists in soil as well as in water (both fresh and marine) as commensals or pathogens take part in QS system. Bioluminescent microorganisms or their *lux* system can be applicable as biosensors for detection of water quality, lethality testing, for the discovery of anti-toxin deposits and pathogens in sustenance. Therefore, *lux system* is the part of QS system as a result of luciferase enzyme quality (Molina et al. 2016). Hence, up to date, bioluminescent microscopic organisms are not yet investigated in the direction as potential new bioresources of anti-infection agents to battle with the pathogens.

7.2 *lux*-Based Natural Transformation

A very touchy framework permits the location of a change occasion straightforwardly from a bacterial populace with no partition step or choice of cells (Choi et al. 2012). The framework depends on the bacterial *luciferase operon* (*lux system*) from *Photorhabdus luminescens*. The unique sub-atomic assembly comprises of the useful modules *luxCDE* and *luxAB*, which include a replicative plasmid and an integrative quality tape. An entrenched host for bacterial hereditary examinations, *Acinetobacter baylyi* ADP1, is utilized as the model bacterium for quorum sensing (McConnell et al. 2013; Jung and Park 2015). A common change taken after by homologous recombination or plasmid recircularization can be promptly recognized in both effectively developing and static biofilm-like communities, including exceptionally uncommon change occasions (). The framework permits the discovery of common change inside 1 h of bringing test DNA into the way of life. This strategy gives a helpful intends to concentrate the energy of characteristic change under factor various conditions (Santala et al. 2016).

7.3 Vector for Gene Targeting

Uses of vector for gene targeting have biomedical and industrial applications viz. as biomarker, in curation of dental biofilm through product induced gene expression (Kalia 2015; Kalia and Kumar 2015; Kaur et al. 2015; Pooja et al. 2015; Ray and Kalia 2017). Genes such as py*r*G can be used a quorum sensing marker in fungi viz. *A. Niger*. The DNA sequence arrangement can be done in such a ways that two pieces of the pyrG gene permits the homologous recombination of the newly amplified recombinant DNA (rDNA) at the pyrG locus. The 5′ end of the strand focusing on a tape contains a nonutilitarian and truncated pyrG g open perusing outline (initial 112 nucleotides erased) and the 3′ untranslated area (3′ UTR). While at 3′ end of DNA strand, the focusing on tape comprises of the 3′ flanking area of the *pyrG* quality. A one of a kind *NotI* gene site between the flanks permits the inclusion of a quality of intrigue. The linearized focusing on tape is changed to the *pyrG* gene in *A. niger* mutant strain AB4.1 or a subsidiary thereof. The utilization is constitutively communicated by *luciferase* (*mluc*)-reporter gene. Few bacterial *luciferase* capable to tolerate high temperature. This thermostable bacterial luciferase can acts as a reporter in plants (Cui et al. 2014). A bacterial *luciferase* gene (*lux*) acts to measure, performance of transformants in which *mluc* correspondent was coordinated at the *pyrG* gene locus, appeared practically identical and reproducible *lux* exercises. Results indicate that the new *pyrG* focusing on vector is a vital change to the current strategy for quality focusing in *A. niger*. Despite the fact that the vector (vehicle) is particular for *A. niger*, the introduced outline and loom is effectively pertinent for building mix vehicles/vectors for other parasites (Arentshorst et al. 2015).

7.4 Dip-Stick Type Biosensor for the Detection of Water Toxicity

Microbial entire cell bioreporters are hereditarily adjusted microorganisms that deliver a quantifiable yield because of the nearness of dangerous chemicals or different anxiety elements (Yagur-Kroll and Belkin 2014). To make use of microorganisms many phenomena are exists in nature such as quorum sensing. Artificially quorum sensing can be visualised using different experiments. A dip-stick is a new platform developed has applications in classification and identification toxicants in aqueous environments. It is comprises eight (08) optically color-coded (fluorescence) functional alginate beads. Each bead acts as fluorescent microbead beads each encapsulate bioluminescent bacteria. The plasmids with specific stress promoter were taken either from *Photorhabdus luminescent* or *Vibrio fischeri* (Mitchell et al. 2005; Renoz et al. 2017). Specific stress promoter such as *luxCDABE* genes was constructed based on plasmids based either on pDEW201 or pUCD615. The color coded microbeads were prepared and fabricated. The optically coded functional microbead biosensor is 85.250 × 8.0 × 1.50 mm in size with ten apertures holes having diameter 1.25 mm. The reverse trapezoid top and bottom sections avoid are the escape of the microbeads. A microbead has its own color code generated due to bioluminescent nature of bacteria included as per the needs. It can be easily detected using the bioluminescent reader (Fig. 7.1).

When we add portable dip-stick in water in the environment. It helps to detect any toxicity caused by water pollutants or contaminants which resulted in cell surface damage, DNA damage, protein deterioration (damage) in the response of bacteria to the toxicants or toxic and heavy chemicals (Jung et al. 2014). Practically, it is a helpful tool for detecting the status of biodiversity in vitro as well as invivo.

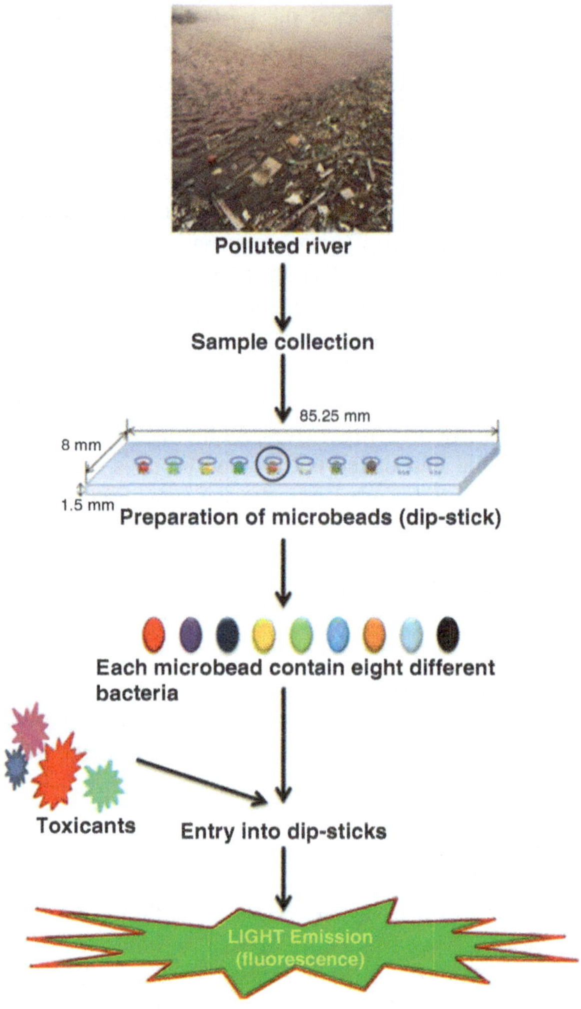

Fig. 7.1 Application of dip-stick type biosensor in quorum sensing and detection of environmental pollution. Fluorescence (bioluminescence) generated when microbeads on dip0stick come in contact with toxicants

7.5 Applications of Quorum Sensing (QS) in Synthetic Biology Systems

Quorum sensing (QS) manages numerous changes in microbiota and environment such as normal phenotypes, destructiveness, biofilm arrangement, anti-infection, pathogen resistance. All these applications nowadays are engineered and microorganisms made carrying their properties as hereditary character empowered by genes (Shukla et al. 2014; Siddiqui et al. 2015; Soma and Hanai 2015; Scott and Hasty 2016; Kalia et al. 2017). For instance, a library of *Escherichia coli lsr*-operon gave an organism with higher antibiotic resistance over the local promoter (*tet(C)* gene). This happens because of several changes in the QS repressor-*LsrR* gene (Hooshangi and Bentley 2011; Brito et al. 2013). The site-directed mutagenesis reestablishment in p-lsrR-box sites shows that two promoter regions EP01rec and EP14rec exhibits enhanced expression and their variants retained the LsrR-mediated QS switching activity. The fusion of these promoters will encourage future applications of QS-regulation in synthetic biology, protein expression and metabolic engineering (Tien et al. 2016; Hauk et al. 2016).

7.6 Regulation of Bioluminescence

Vibrionaceae is the family contains certain proteobacteria showing bioluminescence activity. In bioluminescence, pheromone signalling pathway plays a central role in regulation of production of light (Urbanczyk et al. 2010; Bjornsdottir-Butlera et al. 2016). But, certain microorganisms such as *Photobacterium* strains bypasses above regulation methods. *Photobacterium* strain possesses insertions in genes encoding important components which are necessary for the luciferase reaction -viz. operon- *lux*, *lum*, and *rib*- as well as other loci. This study can be performed using transposon mutagenesis and screening of intensity of luminescence which may be decreased (Aguirre-Ramirez et al. 2012; Bazire and Dufour 2014; Dunn et al. 2015; Shivak et al. 2016). In other review, it was accounted for that constitutive light expression is not metabolically expensive to *Citrobacter rodentium* and underpins the view that bioluminescent adaptations of organisms can be utilized as a substitute for their non-bioluminescent guardians to concentrate bacterial conduct in a wide assortment of situations (Read et al. 2016). Further in depth research is the necessity and needs to be carried out to discover roles and regulation of bioluminescence in *Photobacterium* strains.

References

Aguirre-Ramirez M, Medina G, Gonzalez-Valdez A, Grosso-Becerra V, Soberon-Chavez G (2012) The *Pseudomonas aeruginosa* rmlBDAC operon, encoding dTDP-L-rhamnose biosynthetic enzymes, is regulated by the quorum-sensing transcriptional regulator RhlR and the alternative sigma factor σS. Microbiology 158:908–916. https://doi.org/10.1099/mic.0.054726-0

Arentshorst M, Lagendijk EL, Ram AFJ (2015) A new vector for efficient gene targeting to the pyrG locus in *Aspergillus niger*. Fungal Biol Biotechnol 2:2. https://doi.org/10.1186/s40694-015-0012-4

Bazire A, Dufour A (2014) The *Pseudomonas aeruginosa* rhlG and rhlAB genes are inversely regulated and RhlG is not required for rhamnolipid synthesis. BMC Microbiol 14:160. https://doi.org/10.1186/1471-2180-14-160

Bjornsdottir-Butlera K, McCarthya SA, Dunlapb PV, Benner RA (2016) Photobacterium angustum and Photobacterium kishitanii, Psychrotrophic High-Level Histamine-Producing Bacteria Indigenous to Tuna. Appl Environ Microbiol 82:2167–2176. https://doi.org/10.1128/AEM.02833-15

Brito PH, Rocha EPC, Xavier KB, Gordo I (2013) Natural genome diversity of AI-2 quorum sensing in *Escherichia coli*: conserved signal production but labile signal reception. Genome Biol Evol 2013 5:16–30. https://doi.org/10.1093/gbe/evs122

Bruckbauer ST, Kvitko BH, Karkhoff-Schweizer RR, Schweizer HP (2015) Tn5/7-lux: a versatile tool for the dentification and capture of promoters in gram-negative bacteria. BMC Microbiol 15:17. https://doi.org/10.1186/s12866-015-0354-3

Choi H, Mascuch SJ, Villa FA, Byrum T, Teasdale ME, Smith JE, Preskitt LB, Rowley DC, Gerwick L, Gerwick WH, Honaucins AC (2012) Potent inhibitors of eukaryotic inflammation and bacterial quorum sensing: synthetic derivatives and structure-activity relationships. Chem Biol 19:589–598. https://doi.org/10.1016/j.chembiol.2012.03.014

Cui B, Zhang L, Song Y, Wei J, Li C et al (2014) Engineering an enhanced, thermostable, monomeric bacterial luciferase gene as a reporter in plant protoplasts. PLoS ONE 9:e107885. https://doi.org/10.1371/journal.pone.0107885

Dunn AK, Rader BA, Stabb EV, Mandel MJ (2015) Regulation of bioluminescence in *Photobacterium leiognathi* strain KNH6. J Bacteriol 197:3676–3685. https://doi.org/10.1128/JB.00524-15

Fatima Q, Zahin M, Khan MSA, Ahemad I (2010) Modulation of quorum sensing controlled behaviour of bacteria by growing seedling, seed and seedling extracts of leguminous plants. Indian J Microbiol 50:238. https://doi.org/10.1007/s12088-010-0025-x

Hauk P, Stephens K, Mckay R, Virgile CR, Ueda H, Ostermeier M, Ryu KS, Sintim HO, Bentley WE (2016) Insightful directed evolution of *Escherichia coli* quorum sensing promoter region of the lsrACDBFG operon: a tool for synthetic biology systems and protein expression. Nucleic Acids Res 44:10515–10525. https://doi.org/10.1093/nar/gkw981

Hooshangi S, Bentley WE (2011) LsrR Quorum sensing "Switch" is revealed by a bottom-up approach. PLoS Comput Biol 7:e1002172. https://doi.org/10.1371/journal.pcbi.1002172

Jung J, Park W (2015) *Acinetobacter* species as model microorganisms in environmental microbiology: current state and perspectives. Appl Microbiol Biotechnol 99:25–33. https://doi.org/10.1007/s00253-015-6439-y

Jung I, HB S, Lee J, Kim BC, Gu MB (2014) A dip-stick type biosensor using bioluminescent bacteria encapsulated in color-coded alginate microbeads for detection of water toxicity. Analyst 139:4696. https://doi.org/10.1039/c4an00308j

Kalia VC (2014) Microbes, antimicrobials and resistance: the battle goes on. Indian J Microbiol 54:1–2. https://doi.org/10.1007/s12088-013-0443-7

Kalia VC (2015) Microbes: the most friendly beings? In: Kalia VC (ed) Quorum sensing vs Quorum Quenching: a battle with no end in sight. Springer, India, pp 1–5. https://doi.org/10.1007/978-81-322-1982-8_1

Kalia VC, Kumar P (2015) Genome wide search for biomarkers to diagnose *Yersinia* infections. Indian J Microbiol 55:366. https://doi.org/10.1007/s12088-015-0552-6

Kalia VC, Prakash J, Koul S, Ray S (2017) Simple and rapid method for detecting biofilm forming bacteria. Indian J Microbiol 57:109–111. https://doi.org/10.1007/s12088-016-0616-2

Kaur G, Rajesh S, Princy SA (2015) Plausible drug targets in the *Streptococcus* mutans quorum sensing pathways to combat dental biofilms and associated risks. Indian J Microbiol 55:349. https://doi.org/10.1007/s12088-015-0534-8

Koul S, Kalia VC (2017) Multiplicity of quorum quenching enzymes: A potential mechanism to limit quorum sensing bacterial population. Indian J Microbiol 57:100–108. https://doi.org/10.1007/s12088-016-0633-1

Koul S, Prakash J, Mishra A, Kalia VC (2016) Potential emergence of multi-quorum sensing inhibitor resistant (MQSIR) bacteria. Indian J Microbiol 56:1–18. https://doi.org/10.1007/s12088-015-0558-0

Koutsoudis MD, Tsaltas D, Minogue TD, Von Bodman SB (2006) Quorum-sensing regulation governs bacterial adhesion, biofilm development, and host colonization in *Pantoea stewartii* subspecies *stewartii*. PNAS 103:5983–5988. https://doi.org/10.1073/pnas.0509860103

McConnell MJ, Actis L, Pachón J (2013) *Acinetobacter baumannii*: human infections, factors contributing to pathogenesis and animal models. FEMS Microbiol Rev 37:130–155. https://doi.org/10.1111/j.1574-6976.2012.00344.x

Mitchell RJ, Ahn JM, Gu MB (2005) Comparison of *Photorhabdus luminescens* and *Vibrio fischeri lux* fusions to study gene expression patterns. J Microbiol Biotechnol 15:48–54

Molina AJ, Abisado RG, Calugay RJ (2016) Bioluminescent *Vibrio* spp. with antibacterial activity against the nosocomial pathogens *Staphylococcus aureus* and *Klebsiella pneumonia*. AACL Bioflux 9:185–194

Pooja S, Pushpananthan M, Jayashree S, Gunasekaran P, Rajendhran J (2015) Identification of periplasmic a-amlyase from cow dung metagenome by product induced gene expression profiling (Pigex). Indian J Microbiol 55:57–65. https://doi.org/10.1007/s12088-014-0487-3

Ray S, Kalia VC (2017) Biomedical applications of polyhydroxyalkanaotes. Indian J Microbiol 57:39–47. https://doi.org/10.1007/s12088-016-0651-7

Read HM, Mills G, Johnson S, Tsai P, Dalton J, Barquist L, Print CG, Patrick WM, Wiles S (2016) The in vitro and in vivo effects of constitutive light expression on a bioluminescent strain of the mouse enteropathogen *Citrobacter rodentium*. PeerJ 4:e2130. https://doi.org/10.7717/peerj.2130

Rekadwad BN, Khobragade CN (2017a) Bacterial quorum sensing (QS) in rhizosphere (paddy soil): understanding soil signaling and N- recycling for increased crop production. In: Kalia V c (ed) Microbial applications vol.2 – biomedicine, agriculture and industry. Springer International Publishing AG, Switzerland., Springer Nature, pp 119–131. https://doi.org/10.1007/978-3-319-52669-0_6

Rekadwad BN, Khobragade CN (2017b) Microbial biofilm: role in crop productivity. In: Kalia V c (ed) Microbialapplications vol.2 – biomedicine, agriculture and industry. Springer Nature, Cham, pp 107–118. https://doi.org/10.1007/978-3-319-52669-0_5

Rekadwad BN, Khobragade CN (2017b) Microbial biofilm: role in crop productivity. In: Kalia V c (ed) Microbial applications vol.2 – biomedicine, agriculture and industry. Springer International Publishing AG, Switzerland., Springer Nature, pp 107–118. https://doi.org/10.1007/978-3-319-52669-0_5

Renoz F, Champagne A, Degand H, Faber AM, Morsomme P, Foray V, Hance T (2017) Toward a better understanding of the mechanisms of symbiosis: a comprehensive proteome map of a nascent insect symbiont. PeerJ 5:e3291. https://doi.org/10.7717/peerj.3291

Santala V, Karp M, Santala S (2016) Bioluminescence-based system for rapid detection of natural transformation. FEMS Microbiol Lett 363:125. https://doi.org/10.1093/femsle/fnw125

Scott SR, Hasty J (2016) Quorum sensing communication modules for microbial consortia. ACS Synth Biol 5:969–977. https://doi.org/10.1021/acssynbio.5b00286

Shivak DJ, MacKenzie KD, Watson NL, Pasternak JA, Jones BD, Wang Y, DeVinney R, Wilson H, Surette MG, White AP (2016) A modular, Tn7-based system for making bioluminescent or fluorescent Salmonella and *E. coli* strains. Appl Environ Microbiol 82:4931–4943. https://doi.org/10.1128/AEM.01346-16

Shukla SK, Mangwani N, Rao TS, Das S (2014) Biofilm-mediated bioremediation of polycyclic aromatic hydrocarbons. In: Das S (ed) Microbial biodegradation and bioremediation, Elsevier, p 203–232. https://doi.org/10.1016/B978-0-12-800021-2.00008-X

Siddiqui MF, Rzechowicz M, Harvey W, Zularisam AW, Anthony GF (2015) Quorum sensing based membrane biofouling control for water treatment: a review. J Water Proc Eng 30:112–122. https://doi.org/10.1016/j.jwpe.2015.06.003

Soma Y, Hanai T (2015) Self-induced metabolic state switching by a tunable cell density sensor for microbial isopropanol production. Metab Eng 30:7–15. https://doi.org/10.1016/j.ymben.2015.04.005

Tien S-M, Hsu C-Y, Chen B-S (2016) Engineering bacteria to search for specific concentrations of molecules by a systematic synthetic biology design method. PLoS ONE 11:e0152146. https://doi.org/10.1371/journal.pone.0152146

Urbanczyk H, Ast JC, Dunlap PV (2010) Phylogeny, genomics, and symbiosis of photobacterium. FEMS Microbiol Rev 35(2011):324–342. https://doi.org/10.1111/j.1574-6976.2010.00250.x

Vidal JE, Ludewick HP, Kunkel RM, Zähner D, Klugman KP (2011) The LuxS-dependent quorum-sensing system regulates early biofilm formation by *Streptococcus pneumoniae* strain D39. Infect Immun 79:4050–4060. https://doi.org/10.1128/IAI.05186-11

Yagur-Kroll S, Belkin S (2014) Molecular manipulations for enhancing luminescent bioreporters performance in the detection of toxic chemicals. In: Thouand G, Marks R (eds) Bioluminescence: fundamentals and applications in biotechnology–volume 2, advances in biochemical engineering/biotechnology. Springer-Verlag, Berlin, pp 137–149. https://doi.org/10.1007/978-3-662-43619-6_4

Strategies Behind Biosensors for Food and Waterborne Pathogens

8

Bhaskar Das, P. Balasubramanian, R. Jayabalan, N. Lekshmi, and Sabu Thomas

Abstract

Slackness in the quality control of food and water consumed by human and other animals has become a significant issue which enhances the possibilities of cross-contamination with harmful pathogenic microbes. Intake of the contaminated food and water are the causes for the over abundance of infectious diseases in both animals and humans, and this has thus emerged as a global health concern. Detection of microbial contamination in food and water has relied on conventional methods which demand intensified pre-enrichment steps followed by laborious biochemical identification techniques. Recently, most promising and advanced techniques in biological sensor development have dragged all the scientist's attention which primarily deals with rapid real-time sensing applications due to its selectivity, sensitivity and specificity. In this book chapter, the possible routes of pathogenic infections have been outlined along with its various detection mechanisms. Additionally, strategies for the biosensor development have also been elaborated based on their transducing properties.

B. Das · P. Balasubramanian
Bioenergy and Environmental Laboratory, Department of Biotechnology and Medical Engineering, National Institute of Technology, Rourkela, Odisha, India

R. Jayabalan
Food Microbiology and Bioprocess Laboratory, Department of Life Science, National Institute of Technology, Rourkela, Odisha, India
e-mail: jayabalanr@nitrkl.ac.in

N. Lekshmi · S. Thomas (✉)
Cholera and Biofilm Research Laboratory, Rajiv Gandhi Centre for Biotechnology, Thiruvananthapuram, Kerala, India
e-mail: sabu@rgcb.res.in

V. C. Kalia (ed.), *Quorum Sensing and its Biotechnological Applications*,
https://doi.org/10.1007/978-981-13-0848-2_8

Keywords
Food borne pathogens · Biosensor · Contamination · Detection methods · Health concerns

8.1 Introduction

Microorganisms are present in our environment from the beginning of life and have become one of the essential parts of the nature for maintaining the eco-system. The microorganism, such as viruses and bacteria are found in every facet of the environment, due to their high adaptable nature. Among all the microorganisms, bacteria play an important beneficial role towards animals and human, but certain potentially harmful bacteria can have a profound negative impact on people due to their pathogenicity. Infringement of bacterial contamination leads to disastrous infectious diseases worldwide and can affect human health in two possible ways. Firstly, due to the lack of quality control in food processing sectors, contamination of food by bacterial pathogens (such as *Escherichia coli, Salmonella typhimurium, Campylobacter jejuni, Legionella pneumophila, Staphylococcus aureus, Streptococci*, etc.) results in numerous food borne diseases (Doyle and Buchanan 2012). It is estimated that infectious diseases cause about 40% of approximately 50 million total annual deaths worldwide (World Health Organization 2008). Secondly, inadequate access to safe and portable clean water along with poor hygiene and sanitation facilities can lead to contamination with pathogenic bacteria such as *E. coli* O157:H7, *Vibrio cholerae, Salmonella enterica, Pseudomonas aeruginosa*, etc.) (Connelly and Baeumner 2012). Waterborne pathogens are capable of causing 10–20 million deaths and non-fatal infection of more than 200 million people each year (Berry et al. 2006). These food and waterborne pathogenic bacteria are resistant to environmental conditions, and most of the human population is susceptible to these pathogens which cause high fatality rate (World Health Organization 2008). Examples of these are the incidents that took place in 1997 Hudson ground beef recall and the 1996 incident where more than 9000 fell ill, and 313 died due to *E.coli* O157:H7 contamination (Ivnitski et al. 2000).

The current general practices for controlling the outspread of microbial diseases include careful control of various kinds of pathogenic bacteria by food safety, water quality control and environmental monitoring. Conventional techniques for detection and identification of pathogenic bacteria mainly depend upon accurate microbiological and biochemical identifications (Ferreira et al. 2011). Most of these methods can be sensitive and inexpensive and give both qualitative and quantitative analysis of the tested bacteria, but still have some ambiguities due to its low efficacy for detection of pathogens in samples with a less initial load of microbes. For example, standard methods like NF EN ISO 11290-1 method for the detection of *Listeria monocytogenes* needs nearly about 7 days to produce results as they rely on the ability of these microorganisms to produce visible colonies. Compared to this technique, some newer microbiological based test like ALOA® method (AES laboratories) uses chromogenic medium in combination with *Listeria* monodisk for the detection of *L. monocytogenes* which can decrease the detection time down to 3 days. This still presents difficulties in the quality control of semi-perishable foods (Kumar 2013). Additionally, the transformation of bacterial cells into a dormant state which are viable, but non culturable (VBNC) makes the detection of such pathogens more difficult. Biosensors are recently emerging as a rapid method of detection of microbes in food and water (Mehrotra 2016). This chapter covers a different aspect of the development of biosensors for the detection of food and waterborne pathogens. Various types of biosensors with their mechanisms are described. Additionally, a major part of the chapter is devoted to describing the pathogenicity of water and foodborne pathogens and the primary mechanistic principle for the detection of these pathogens using biosensors. Furthermore, possible drawbacks of the existing biosensor technologies with the comparison of commercial technologies are also discussed.

8.2 Importance of Food and Waterborne Pathogens

Numerous sufferings and deaths are caused by foodborne pathogens worldwide. Around five million deaths are calculated in a year under the age of 5 in Latin America, Asia and Africa due to gastroenteritis (Lanata et al. 2013). *Campylobacter* induced enteritis is a significant illness in children aged 0–4 years in Mexico and Thailand. Statistics on foodborne pathogens showed decrease in occurrences from 1996–1998 to 2005 for *Shigella, Yersinia, L. monocytogens, Campylobacter* species, *E.coli* O157:H7, and *S. typhimurium* but upsurges for *S. enteritidis, S. heidelbeg,* and *S. javiana*. Many of the foodborne pathogens are spread through reservoirs like animals and poultry. Milk, meat and egg products may act as vehicles for *E. coli* O157:H7, *C. jejuni, L. monocytogenes, S. enterica*, and *Yersinia enterocolitica* (Ferens and Hovde 2011). Novel approaches to control the pathogens at farm level help to decrease the pathogen load in processing industries. However, ready-to-eat (RTE) food products are in serious concern since RTE products do not receive any treatment before consumption (Martinović et al. 2016). Several foodborne outbreaks have happened recently as the result of consumption of minimally processed fruits and vegetables, undercooked or processed dairy products, and RTE meats (Centers for Disease Control and Prevention (CDC) 2006).

Apart from gastroenteritis, the food and water borne pathogens cause autoimmune polyneuritis, autoimmune disease (allergic encephalitis), atherosclerosis, hemolytic-uremic syndrome (Shiga like toxin from *E. coli* O157:H7), chronic rheumatoid conditions, and Guillain-Barre syndrome (*Campylobacter* infections). Based on the eating habits, some foodborne infections exist in specific countries (Martinović et al. 2016). Consumption of raw fish in Japan and meat and vegetables in Scandinavian and middle/eastern countries are the reasons for *Vibrio parahaemolyticus* and botulism cases, respectively (Brandl 2006).

Around seven decades ago, the primary pathogens transmitted through food and water were *Clostridium botulinum, Clostridium perfringens, Salmonella, and S. aureus*. During these periods, the food borne illnesses were viewed as a trouble only for a day or two rather than a danger to life (Velusamy et al. 2010). Most of the countries did not have a systematic reporting program, except UK and USA. In most of the situations, the outbreak was due to the improper handling and poor storage conditions of food, especially poultry and meat. *Campylobacter, E. coli* O157:H7, *L. monocytogenes* and *Y. enterocolitica* were emerging when the food service establishments were educated to handle the problems with *Clostridium, Salmonella* and *Staphylococcus* (Mor-Mur and Yuste 2010). Though the new pathogens were emerging, it took several years for the health service providers to understand the seriousness of these new pathogens. Large outbreaks during 1985 and 1993 from *L. monocytogenes* and *E. coli* O157:H7, respectively, in the US, has paved way for changes in food safety policies in the US and several other countries (Kramer et al. 2006).

Through some unexplored reasons, there is always the emergence of new pathogens or re-emergence of old pathogens those are responsible for the increased occurrences of foodborne diseases (Martinović et al. 2016). Food and water borne outbreaks are not only causing human sufferings and fatalities but also distressing financial losses to food processors and producers. The reasons like (1) increased surveillance and reporting, (2) changes in the food manufacturing and agricultural practices, (3) changes in eating habits, (4) increased vulnerable populations, (5) improved detection methods, and (6) emerging pathogens with tolerance to stressed conditions are reported to be the possible reasons for the greater numbers of outbreaks in recent years (Bhunia 2008).

8.3 General Routes of Infection and Spreading Diseases

Three forms of diseases caused by foodborne pathogens are foodborne infection, intoxication, and toxicoinfection. Because of water and food are major reservoirs for foodborne illness, oral route and intestine are the primary route and site of infection, respectively. To make a successful infection, microorganism must have to pass several hurdles, and several factors have to work in a host cooperatively. The food and water borne pathogens can be transmitted even through direct contact with an infected animal or human, through soil or an arthropod vector (Conner and Schmid 2003). Following are the factors which determine the success of an infection by a foodborne pathogen (Cossart and Sansonetti 2004):

(i) Pathogens must be present in adequate numbers to initiate the infection process.
(ii) Pathogens are able to endure the changing environment of the host and must be able to multiply (presence of capsules or not).
(iii) Pathogens should find a place for their colonization through adhesion and invasion factors, and chemotaxis.
(iv) Pathogens must have some mechanisms (toxins and enzymes) to escape from the host immune system.
(v) Pathogens must damage the host tissues and cells by their component or metabolites (exotoxins, endotoxins, enzymes, etc.) that cause cell death by necrosis or apoptosis and encourage bacterial survival and multiplication.

Intact living microorganisms are necessary to initiate the foodborne infection. Upon intake with food or water, pathogens reach the intestine after passing the stomach environment. Colonization starts in the intestine followed by crossing the intestinal barrier through the invasion process or through translocation by phagocytic M cells (Ribet and Cossart 2015). Few pathogens cause localized damage, and some may spread to the liver, spleen, lymph nodes, brain or to other extraintestinal sites. Acute food borne infections are quick and last only for short duration due to fast clearance of microorganisms by immunological mechanisms. Chronic food borne infections are long and immunological removal is not effective against pathogens. Patients recovering from a foodborne infection releases pathogens to the environment for a while. The infectious dose of the pathogens varies from 50 to 10^9 CFU (Colony Forming Units) per gram of food for live bacteria or 10^3–10^5 numbers for spores or 10–100 particles of virus or 10–100 cysts for protozoa (Kent et al. 2015). This difference in numbers depends on the infective potential of the organism, immunological nature of the host, type of food consumed, and presence and absence of the antibiotics in the host body. The human body is gifted with several mechanisms to protect itself from the invasion of the pathogens. Mucus and fluids constantly wash the epithelial cell surface of the gastrointestinal tract. Human body constantly attempts to expel the pathogens by mucus production, peristaltic movements, and by epithelial ciliary sweeping action. The presence of bile salts, proteolytic enzymes, and resident microflora also prevent colonization of the pathogens in the gastrointestinal tract (Martinović et al. 2016). After reaching the intestine, the pathogens must be able to attach themselves to the intestine and cross the intestinal barrier by the mechanisms listed in Table 8.1. Ingestion of preformed toxins (botulinum toxin, *Bacillus cereus* toxin, staphylococcal enterotoxin, and seafood toxins) results in the food borne intoxication. Actively growing pathogen releases toxins in the food which are ingested. Toxins must be ingested and absorbed in the epithelial lining of the gastrointestinal tract to cause the inflammation which evokes diarrhoea or vomiting. In case of foodborne toxiconfection, ingested bacteria along with food colonize the mucosal surface and produce exotoxins in the intestine (Iwamoto et al. 2010). Exotoxins either damage the local cells or tissues or enter the blood stream to induce the disease. Toxins of enterotoxigenic *E. coli* (heat-labile and heat-stable), *V. cholerae* (cholera toxin) and enterotoxins of *C. perfringens* are examples of exotoxins (Sibley 2004; Bhunia 2007; Ray and Bhunia 2007).

8.4 Detection of Pathogens in Food and Water

The food industry increasingly adopts several measures to improve the safety and quality of food (Scognamiglio et al. 2014). Hazard Analysis Critical Control Point (HACCP), a management tool highly promoted by various regulatory agencies in many countries is to attain a safer food supply and coordination of trading values. Rapid methods of monitoring are desirable to make the HACCP efficient (Aung and Chang 2014). Testing for the presence of specific pathogens like *C. jejuni*, *E. coli* O157:H7, *L. monocytogenes* and others is required to validate the HACCP. Hence, it is very essential to have rapid methods for the detection of the pathogens. Food and water borne pathogens are detected by several techniques (Fig. 8.1) which include culture based methods, electrical methods, methods based on ATP

Table 8.1 Characteristics of some food and water borne pathogens and toxins and their mechanism of infection (Bhunia 2007)

Pathogens	Infectious dose	Incubation period	Types of pili	Adhesion factor	Receptor	Toxins	Mode of action	Target
E. coli O157:H7	50–100 CFU	3–9 days	Type I and P pili – *E. coli*, Type IV and BFP (bundle forming pili) – EPEC	Intimin (94 kDa) – EPEC and EIEC	Translocated intimin receptor (TIR) – EPEC and EIEC	1. Hemolysin (*E. coli*)	1. Pore formation	1. Plasma membrane
						2. Heat-labile toxin (LT) (*E. coli*)	2. ADP ribosyltrans-ferase	2. G-proteins
						3. Heat-stable toxin (ST) (*E. coli*)	3.Stimulates Guanylate cyclase	3. Guanylatecyclase
L. monocytogenes	10^2–10^3 CFU	7–14 days or even longer	NA	Internalin A (88 kDa), Internalin B (65 kDa), Vip (virulence protein) (43 kDa), LAP (listeria adhesion protein) (104 kDa)	E-cadherin, cMet, gC1q-R/p32, Gp96	Listeriolysin O	Pore formation and Apoptosis	Cholesterol
Shigella spp.	10–100 CFU	12 h to 7 days, but generally in 1–3 days	NA	NA	NA	1. Shiga toxin or Shiga like toxin	1. N-glycosidase	1. 28S rRNA
						2. IpaB	2. Apoptosis	
V. cholerae/V. haemolyticus/V. vulnificus	10^4–10^{10} CFU/g	6 h to 5 days	Type IV	Toxin-coregulated pili (TCP), Mannose-fucose resistant cell-associated hemagglutinin (MSHA), mannose-sensitive hemagglutinin (MSHA)	Glycoprotein	Cholera toxin (*V. cholerae*)	ADP ribosyltransferase	G-protein

S. aureus cells	10^5–10^8 CFU/g	–	NA	Fibronectin-binding protein (FnBP)	Fibronectin	1. Alpha-toxin 2. Enterotoxins, toxin shock toxins	1. Pore formation on 2. Superantigens	1. Plasma membrane 2. TCR and MHC II
Staphylococcal enterotoxin	1 ng/g of food	1–6 h	NA	NA	NA	NA	1. Pore formation 2. Super antigens	1. Plasma membrane 2. Major Histocompatibility Complex (MHC)
B. cereus	10^5–10^8 CFU or spores/g	1–6 h (vomitting), 8–12 h (diarrhea)	NA	NA	NA	Heat and acid resistant nonhemolytic enterotoxin, hemolysin BL, cytoxin K, Cereulide	Pore formation	Plasma membrane
B. anthracis (inhalation anthrax)	8×10^3–10^4 spores	2–5 days	NA	NA	NA	1. Edema factor 2. Lethal factor	1. Adenylatecyclase 2. Metalloprotease	1. ATP 2. MAPKK1/MAPKK2
C. botulinum neurotoxin	0.9–0.15 µg (i.v. or i.m. route) and 70 µg (oral route)	12–36 h; 2 h when large quantities are ingested	NA	NA	NA	Neurotoxin	Zinc metalloprotease	Synaptobrevin, SNAP-25, Syntaxin
C. perfringens	10^7–10^9 CFU	8–12 h	NA	NA	NA	Perfringolsin O	Pore formation	Cholesterol
C. jejuni	5×10^2–10^4 CFU	1–7 days (24–48 h)	NA	CadF (37 kDa)	Fibronectin	Enterotoxin and Cytotoxin	Pore formation	Gut enterocytes

NA Not Applicable

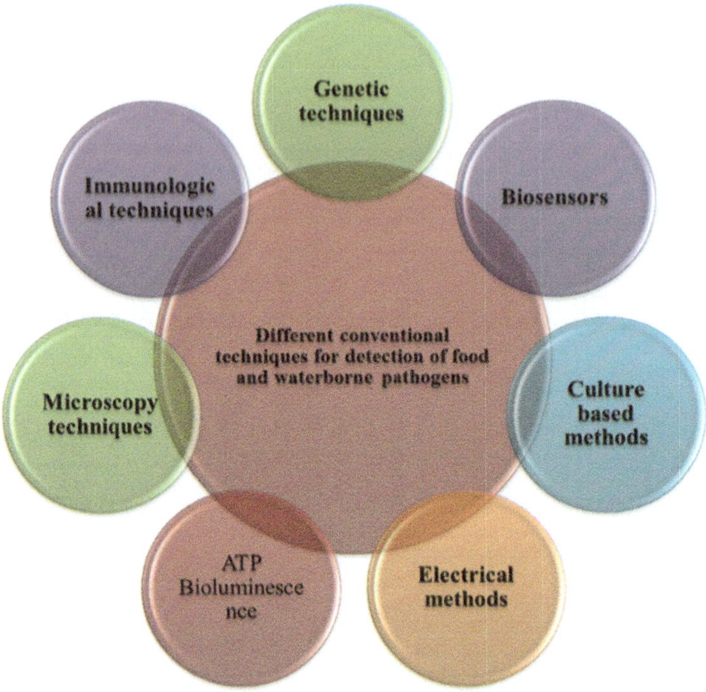

Fig. 8.1 Different detection methods of food and water borne pathogens

bioluminescence, microscopy techniques, immunological techniques, genetic techniques, and techniques using biosensor (Leonard et al. 2003). Comparison of the detection methods established for food borne pathogens is given in Table 8.2.

Culture methods detect the pathogens by growing or keeping them alive in a nutrient medium. Culture medium mainly comprises six components like amino-nitrogen compounds (peptones, other protein hydrolysates, infusions or extracts), energy sources (e.g. glucose), buffer salts (e.g. soluble phosphate salts, acetates and citrates), mineral salts and metals (phosphates, sulfates, calcium, magnesium, iron, manganese and trace metals), growth promoting factors (blood, serum, vitamins, NADH, etc.), and gelling agents (agar, gelatin, alginates, gums, etc). The basic culture medium can be made selective to allow the growth of the specific pathogens from food samples by adding selective compounds to which the pathogen of interest should be resistant (Blommel et al. 2007). Inorganic salts (sodium azide, lithium chloride, potassium tellurite added to control Gram negative bacteria; tetrathionate and sodium selenite are added to control Gram positive bacteria and coliforms), dyes (acriflavine, crystal violet, brilliant green, and malachite green), surface active agents (bile salts, cetrimide, lauryl sulphate and tergitol), and antibiotics are added to basic medium to give selective features. Apart from selective agents, components which provide differential features to the basic medium can also be added to make the basic medium a differential culture medium. Indicator dyes to indicate the pH

Table 8.2 Detection methods of food and water borne pathogens (Bhunia 2007)

Detection methods	*S. aureus*	*B. cereus*	*B. anthracis*	*C. botulinum*	*C. perfringens*	*S.enterica*
Culture methods	Baird-Parker agar; black, shiny, circular, convex, smooth with the entire margin forming a clear zone with an opaque zone (lecithinase halo) around the colonies	Determination of lecithinase activity in polymyxin-egg yolk mannitol-bromothymol agar (PEMBA)	Determination of non-hemolytic colonies on blood agar, lecithinase activity on egg yolk agar, visualization of capsules using India ink, confirmation of hydrolysis of casein, starch, and gelatin, acid production from salicin, inulin, and mannitol	Microscopic observation of cells with "tennis racket" appearance, observation of raised or flat, smooth or rough with some spreading, and irregular edged colonies with luster zone (pearly zone) when observed under oblique light on egg yolk agar	Isolation of *C. perfringens* using thioglycolate medium, tryptose-sulfite cycloserin agar containing egg yolk or brain heart infusion agar with 10% sheep blood under anaerobic incubation.	Pre-enrichment (lactose broth, tryptic soy broth, nutrient broth, skim milk or buffered peptone water), selective enrichment (Rappaport-Vassiliadis semisolid medium, selenite cysteine broth, or Muller Kauffmann tetrathionate broth) and isolation of pure cultures (Hektoen enteric agar, xylose lysine deoxycholateagar and brilliant green agar)
Cytotoxicity-based assays	Detection of mammalian cell damage by superantigens (Staphylococcal enterotoxins)	Detection of mammalian cell damage by emetic toxin	NA	NA	NA	NA
Nucleic acid-based methods	PCR based detection of enterotoxin genes (egc), exfoliative toxins A and B (etaA and etaB), methicillin- resistant (mecA) gene, and 16S rRNA	PCR based detection of toxin genes (ces, hblA, hblD, hblC, nheA, nheB, nheC, bceT, entFM, cytK, hlyII, hlyIII)	NA	Multiplex PCR for detection of *C. botulinum* serotypes A,B, E, and F	Multiplex PCR for detection of toxin genes: enterotoxin (*cpe*), alpha (*cpa*), beta (*cpb*), epsilon (*etx*), and iota (*iap*)	TaqMan based Q-PCR for *inv*A genes, NASBA method for detection of viable cells

(continued)

Table 8.2 (continued)

Detection methods	*S. aureus*	*B. cereus*	*B. anthracis*	*C. botulinum*	*C. perfringens*	*S.enterica*
Immunoassays	ELISA for superantigens	Antibody based assay for the detection of components of hemolysin and nonhemolytic enterotoxin	NA	Detection of toxins using polyclonal antibody	NA	Enzyme-linked immunosorbent assay (ELISA), surface adhesion immunofluorescent technique, dot-blot immunoassay, surface plasmon resonance (SRP) biosensor, piezoelectric biosensor, time-resolved immunofluorescence assay (TRF), fibre optic sensor
Other methods	Direct detection of whole cell or other metabolites, direct epifluorescence technique (DEFT), flow cytometry, impedimetry, ATP-bioluminescence	NA	NA	Mouse bioassay: intraperitoneal injection of toxin preparation to mice and observation for the symptoms include laboured breathing, pinching of the waist and paralysis, which develops in 1–4 days	NA	NA

Culture methods	Campylosel medium under microaerophilic conditions for *Campylobacter*	Observation of red colonies on modified cefsulodin-irgasan-novobiocin (CIN) agar with esculin (nonhydrolyzing pattern)	Observation of opaque, smooth, and round shaped with irregular edged colonies on blood agar, brain heart infusion agar, MacConkey agar	Observation of growth on selective thiosulfate citrate bile salt agar (TCBS), cellobiosepolymyxin B colistin (CPC) and mannitol-maltose agar	Usage of MacConkey, *Salmonella-Shigella*, Xylose-Lysine Deoxycholate, and Hektoen Enteric agars
	Modified charcoal cefoperazonedeoxycholate (mCCDA)for *Arcobacter* isolation				
Cytotoxicity-based assays	NA	NA	NA	NA	Plaque assay on cultured cell monolayer
Nucleic acid-based methods	PCR based detection of genes including flagellin (*fla*A), 16S rRNA, and 16S/23S intergenic spacer region for *Campylobacter*	PCR to detect yadA, virF, 16S rRNA genes and Q-PCR for *ail* gene	Real-time 5′ nuclease PCR for *pla* gene	Single or multigene-specific PCR that targets specific sequences in 16S rRNA, *tdh, trh, gyrb, toxR, ctxB, ctxAB*, and *tcpA*	Conventional and nested PCR for various species of *Shigella* which targets *ipaH, virA, ial* (invasion associated locus), LPS, and plasmid DNA
	Multiplex PCR targeting 16S and 23S rRNA genes for *Arcobacter*				

(continued)

Table 8.2 (continued)

Detection methods	*Campylobacter and Arcobacter*	*Y.enterocolitica*	*Y. pestis*	*Vibrio species (V. cholerae, V. parahaemolyticus, V. vulnificus)*	*Shigella species*
Immunoassays	NA	Indirect immunofluorescence with biopsy specimens	ELISA, direct immunofluorescence assay, and dipstick assays targeting F1 antigen or Pla protein	ELISA to detect intracellular and thermostable direct hemolysin (TDH) antigens	Wellcolex Color *Shigella* test (Latex Agglutination), EIA kits for *Shigella dysenteriae* (Shigel-Dot A), *Shigella flexneri* (Shigel-Dot B), *Shigella boydii* (Shigel-Dot C), *Shigella sonnei* (Shigel-Dot D)
Other methods	NA	Mouse bioassay for enterotoxin (Yst) determination or rabbit ileal loop assay	**NA**	NA	Rabbit ileal loop assay (RIL) to test the diarrheagenic action of toxins, newborn mouse model to study inflammation and tissue damage in the large intestine

NA Not Applicable

change in the medium (e.g. phenol red, neutral red, bromocresol purple), chromogens which can act as substrates for specific enzymes and changes the color of the medium, fluorogens which can produce fluorescence that can be detected when observed under UV illumination) (Sharma and Mutharasan 2013). Virulence factors of the pathogens can also be detected by adding blood (hemolysin), egg yolk emulsion and lecithin (phospholipase), and rabbit plasma fibrinogen (coagulase) (Jia et al. 2010). Altering the incubation conditions (pH, temperature and gaseous atmosphere) of the culture medium will also be helpful in selectively growing the targeted pathogen by suppressing the growth of others. Resuscitation media were developed to recover the injured cells by adding components which will reduce the damage to the targeted pathogen (e.g. blood, pyruvate, catalase, and cysteine to protect cells from reactive oxygen species). Several commercial kits are in the market (PetrifilmTM, CLONdiscTM, BD Biosciences, Oxoid Salmonella Rapid Test, Salmosyst®, Colilert®, ColiTrak®, and Quanti-Tray®) working based on culture methods for qualitative and quantitative analysis of food borne pathogens. Culture methods form the basis for all the detection techniques for pathogen in food. Culture based methods of detection are widely accepted due to their reliability, lower cost, ease of use, and universal acceptance (Stephens 2003; Alahi and Mukhopadhyay 2017).

An electrical method determines the response of microbial cultures to alternating current (AC) at specific frequencies. Electrolysis of growth medium and killing of microbes occurs at high currents, whereas the mediators (lipoic acid) present at the cell surface save the cells at lower current. Electrodes are required to be immersed in the growth medium or food homogenate (Luo et al. 2015). Impedance and conductance are the two parameters required to be determined either alone or in combination. Impedance is the measure of the AC equivalent of resistance to a direct current (DC) current. Capacitance and resistance are the two components of impedance at any frequency (cycles per second, Hz). Properties of the electrode, changes and conductance occurring between electrodes are related to capacitance (Singh et al. 2014). Reciprocal of the conductance is resistance. Impedance of the pure water and salt solutions is solely their conductance due to zero capacitance, whereas microbiological media have capacitance due to the presence of macromolecular content and microorganisms (Puttaswamy and Sengupta 2010). Changes in biomass results in changes in impedance but changes in conductance are the results of the effect of microbial metabolism on the charge-carrying capacity of the medium. Low molecular weight products produced through metabolism of proteins and carbohydrates are good charge carriers than the large molecular weight products. Hence, every growth medium has an impedance value based on the composition of the chemicals and inoculum (Sharma and Mutharasan 2013). Among the components of media, salts have major effect on impedance and hence the salty foods need dilution. Electrical methods are not useful in detecting pathogens in selective media with high salt levels. The assay temperature needs to be controlled due to the high-temperature coefficient of impedance which is 2% per degree (Gibson 2003; Alahi and Mukhopadhyay 2017).

ATP bioluminescence assay makes use of the availability of adenosine triphosphate (ATP) in all living cells as a energy donor. The intracellular concentration of ATP is measured to detect viable cells (Falzoni et al. 2013). Chemical energy associated with ATP is converted into light by luciferase-luciferin complex. Light detecting devices measures the emitted light and it is directly proportional to the ATP concentration. The theoretical sensitivity of this assay is high due to the availability of instruments which can detect every single photon emitted during the reaction. However, the presence of ATP in food (somatic cells) must be taken care before the assay. Somatic cells must be selectively lysed with non-ionic detergents (Triton X-100) and the released ATP of somatic cells must be enzymatically or chemically destructed. Cationic detergents are subsequently used to extract the ATP from microbes and are then measured by luciferase-luciferin. Microbial cell number is derived from the standard curves of ATP concentration or from emitted light in RLU (relative light units) to cell counting units (CFU/ml). ATP bioluminescence assay is made specific to particular pathogens by recognizing the target bacteria using antibodies or by specific bacteriophages followed by ATP assay (Griffiths and Brovko 2003; Noble and Weisberg 2005).

$$\text{Luciferin} + \text{ATP} + O_2 \rightarrow \text{Oxyluciferin} + \text{AMP} + CO_2 + \text{PPi} + \text{Light}$$

Interactions between specific antibodies and selective antigens on the pathogens are the basic mechanisms of the immunological techniques. Antibodies are targeted against the components on the outer cell wall, a protein on the flagella, or a metabolite or toxin produced by the pathogens during growth (Tlaskalová-Hogenová et al. 2011). Antibodies targeted against antigens on flagella are highly specific whereas the antibodies for somatic antigens are weak due to the sharing of antigens among a wide range of bacteria. Immunological methods are considered to be presumptive methods due to the sharing of common antigens across the microbial genus. The results from immunological methods are usually confirmed by culture based methods. Among the immunological methods ELISA (enzyme-linked immunosorbent assay) is the most widely used methods. ELFA (enzyme-linked fluorescent assays), a variant of ELISA, uses fluorescence based detection (Yeni et al. 2014). Magnetic beads coated with antibodies are used in IMS (immunomagnetic separation) to separate the target pathogen from the food components or other microbes. Different commercial kits based on immunological principles are available in the market for the detection of *Salmonella, Listeria, Campylobacter,* and *E. coli* O157:H7. Lateral flow devices are membrane based devices which rely on the immune-chromatography principle and provides rapid end-point testing. Apart from these techniques, simple agglutination tests are also widely accepted for the detection of food borne pathogens (Baylis 2003; McCarthy 2003).

Very precise detection of pathogens is possible by targeting specific sequences in DNA and RNA. Several methods are available for the amplification of specific sequences on the target pathogen. Polymerase Chain Reaction (PCR) technique relies on a thermostable polymerase, set of primers and nucleotide bases to amplify the specific sequence recognized by the primers. The amplified sequences are visualized by staining with ethidium bromide after agarose gel electrophoresis for the presence of a band or bands of the expected size. There are several types of PCR developed by improving the sensitivity of the traditional PCR (Girones et al. 2010). Nested PCR uses two sets of primers which flank the target sequence to be amplified. Reverse transcriptase (RT-PCR) amplifies the sequences in RNA. Simultaneous detection and quantification of a nucleotide signal in Real-time PCR is made possible by continuously measuring a fluorescent reporter during the reaction. NASBA (nucleic acid sequence-based amplification) is an isothermal amplification technique which involves the simultaneous activity of avian myoblastosis virus-reverse transcriptase (AMV-RT), ribonuclease H, and T7 RNA polymerase. Target sequences in the pathogens can also be detected by hybridization technique in which the single stranded nucleotides are annealed together based on the complementarity. The techniques mentioned above do not identify several pathogens simultaneously (Singh et al. 2014). Microarray technologies facilitate the simultaneous detection of several pathogens at a time. Advances in sequencing technologies and bioinformatics have led to a tremendous increase in the use of molecular subtyping protocols for the identification of pathogens. The chemotaxonomic method of identification is based on the analysis of antigenic characteristics, whole cell protein analysis and composition of fatty acids (Fizgerland and Swaminathan 2003; Sanderson and Nicholas 2003).

Detection of pathogenic bacteria using DNA amplification method has shown promising outcomes in the field of pathogen detection where PCR is used to enhance the sensitivity of the nucleic acid based assay. Target nucleic acid segment of defined length and sequence are amplified by following three steps of PCR such as denaturation, annealing, and extension of oligonucleotide primers by using thermostable DNA polymerase (Mandal et al. 2011). PCR technique has different advantages over culture and other conventional methods due to its specificity, sensitivity and rapid accuracy. However, there are still some difficulties due to its polymerase enzyme specificity towards environmental contaminants which leads to difficulties in quantifying the generation of false positives through the detection of naked nucleic acids, non-viable micro-organisms, or contamination of samples in the laboratory (Lampel et al. 2000). From industrial application point of view, regular detection of bacterial contamination (food and water borne) using PCR technique can be expensive and complicated which requires highly skilled personnel with accuracy (Singh et al. 2014).

Among all the conventional techniques for pathogen detection, immunological detection with the use of antibodies has shown some positive result for the detection of the bacterial cells, spores and viruses (Iqbal et al. 2000). Polyclonal antibodies can be raised rapidly and cheaply as compared to monoclonal antibodies. However, the limitation of polyclonal antibodies regarding specificity encourages the development of hybridoma techniques along with recombinant antibody phage display technology (Harris et al. 2004). Since last few years, immunological methods for the detection of bacterial pathogens have become more accurate, sensitive and reproducible with many commercial immunoassays available in the market (Uematsu et al. 2006). Even though both nucleic acid-based and antibody-based detection has been able to decrease the time consumed for the assay, they are still deficient in the ability to detect pathogens in "real-time". The requirement for a more sensitive, reliable and less time consuming and specific method of detecting a target analyte, at low cost, is the focus of many research, in particular for applications in environmental samples (Leonard et al. 2003).

The biosensor technology is one of the newly emerging techniques which offers the potential for detecting pathogens in real time. However, it still requires time-consuming pre-enrichment to detect small numbers of pathogenic bacteria in food and water. Advancement in the antibody dependent sensing techniques along with the emergence of phage displayed peptide biosensors show increased possibilities for the detection of water and foodborne bacteria (Benhar et al. 2001; Goodridge and Griffiths 2002).

8.5 Mechanism of Biosensors

A biosensor is an analytical device that consists of a bio-recognition element coupled to a signal transducer to detect an analyte of interest by converting a biological response to an electrical signal (Turner 2013). In most cases, "real-time" observation of a particular biological event (such as antigen-antibody reaction) can be done successfully by using this technology. Biosensors can enable the detection of analytes with a broad spectrum present in complex sample matrices and have shown promising outcome in areas such as food analysis, clinical diagnostics and environmental monitoring (Fitzpatrick et al. 2000). Common bio-recognition elements such as oligonucleotide probes, antibodies, enzymes, aptamers, cell-surface molecules and phages are called as bio-receptors which can recognize the target analyte molecules. Similarly, another major part of the biosensor is a transducer,which can be further classified into different aspects such as optical, electrochemical, thermometric, piezoelectric, magnetic and micromechanical or a combination of one or more than one of these techniques mentioned above (Arya et al. 2011). Biosensor devevloped to detect food and water borne pathogens must be having higher efficacy and real time validation. They should also have some basic and essential idealistic characteristics such as accuracy towards the pathogen detection with low or preferably zero probability of false positive and falst negative results (Rider et al. 2003). It has to be quick enough to produce a "real-time" response in case of perishable food analysis. The sensitivity of the biosensor has to be high to determine the pathogenic bacteria in food or water sample with a lower concentration, and it should detect false positive results efficiently. Additionally, along with higher sensitivity, the biosensor has to be highly specific for the target analyte; it should discriminate between target pathogen, toxin and other microorganisms. Along with these main features, a biosensor has to be reproducible, robust, and user-friendly (Zhao et al. 2014).

Nanotechnology has emerged as an elementary division of material science receiving global attention, owing to its wide array of applications. Nanoparticles are of great interest due to its small size, large surface to volume ratio and other novel characteristics (Dasgupta et al. 2015). Due to the wide range of applicability of this

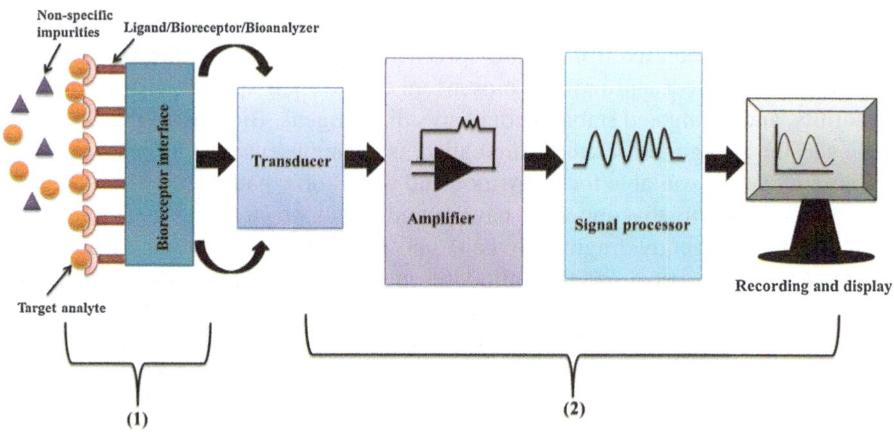

Fig. 8.2 Basic key components of a biosensor

nanoscience and technology, it has a simulated group of scientists to fabricate a nanomaterial based device for bio-analysis (combining nanomaterials with biological molecules). These bio recognition devices are capable of rapid and sensitive detection which can detect even a single cell of food and water borne pathogens within a few minutes (Arora et al. 2011). Development of nano-based materials such as nanoparticles, nanobelt, nanowire, nanofiber and nano-flakes have trans figured clinical and molecular biology by their significant use as bioanalyzer and biodetector. Yang et al. (2007) have described the application of polymeric nano-particles conjugated with biomolecules such as antibiotics, antibodies, adhesion molecules and particular DNA sequence for specific pathogen detection.

In general biosensors consist of two basic key components (Fig. 8.2) (1) Recognition element which can be categorized in to different substances such as, biological, enzymatic or cellular components (oligonucleotides, peptides, DNA sequence, and aptamers, etc.) basically termed as bio-receptors or bio-analyzers and (2) transducer (signal conversion unit). In addition, all biosensors consist of input/output interface (an electronic component which interacts with the instrument). The recognition element (bio-receptor/bioanalyzer), a ligand which binds directly or indirectly to the target molecule or component (analyte) is mainly responsible for producing a primary signal (Perumal and Hashim 2014). The transducer is the component that responds to the main signal from the recognition element and converts it into a form that can be amplified, stored, manipulated, displayed and analyzed. The signal produced by the recognition element can be generated by the direct interaction of ligand and analyte molecule which further can be analyzed directly with the help of a detector. This kind of biosensors is known as a direct (label free) biosensor (Ronkainen et al. 2010). Similarly, in some biosensors, generation of primary signal relies on the presence of any secondary molecules such as fluorescence labelled marker molecules which lead to the indirect detection of the target analyte. This type of biosensors is known as an indirect biosensor. Transducers work based on many physical principles including fluorescence, electrochemistry, optics, mass detection, etc. For the selection of a biosensor ligand (bio-receptor), two main criteria are followed such as affinity and specificity. Antibodies are broadly used as ligands in both direct and indirect biosensors due to their specific characteristics, versatility, and strong and stable binding specific antigens. Biosensors that use antibodies as the recognition element are called as immunosensors. Commercial antibodies are readily available for many food and waterborne pathogens. Most suitable recognition elements that appear to have potential outcomes for biosensor applications includes antibody fragments (Fab) and recombinant variation of antibody fragments (Emanuel et al. 2000). Antibodies in immunosensors can be produced by genetic immunization which involves the transfer of DNA specific for the antigen to stimulate antibody production and peptides by phage display techniques (Goldman et al. 2000). In the following chapters, different strategies behind the biosensors, their mechanisms and functionality will be discussed briefly.

8.6 Different Strategies for Pathogenic Bacteria Detection

Biosensors can be classified on the basis of their basic elementary part of the sensing system such as type of bio-receptors and transducer (Fig. 8.3). Additionally, for the detection of food and waterborne pathogens, several techniques have been described that allows the direct measurement of contamination from different liquid

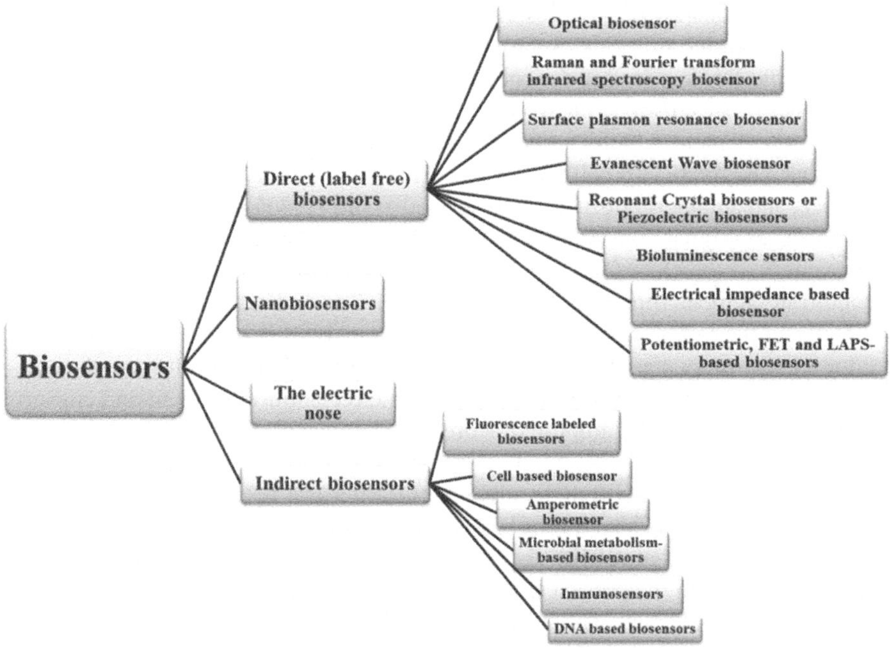

Fig. 8.3 Classification of biosensors

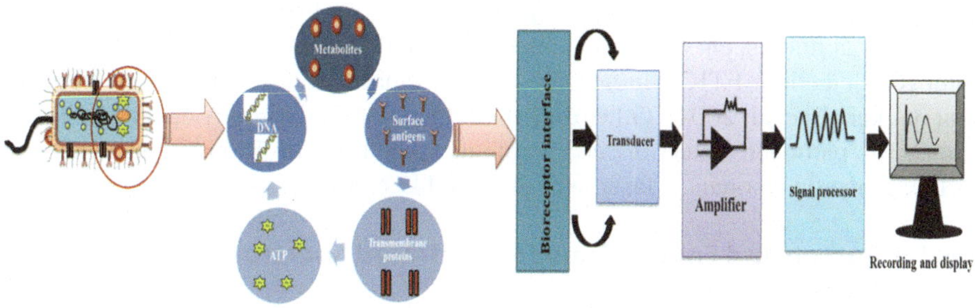

Fig. 8.4 Different strategies for detection of food and water borne pathogens

or solid interfaces (Mascini and Tombelli 2008). Several strategies have been developed to detect either the component (surface antigens, nucleic acids, and transmembrane proteins) or the metabolites produced by the pathogens (Fig. 8.4). In general, biosensors can be classified in two broad catagories (1) Direct biosensors and (2) Indirect biosensors.

8.6.1 Direct (Label Free) Biosensors

Direct (label free) biosensors are based on the direct measurement of a physical phenomenon occurring during a biochemical reaction on a transducer surface. Signal parameters such as a change in oxygen concentration, pH gradient and ion consumption, potential difference and current, resistance and optical properties can be monitored by using different electrochemical or optical transducers and amplified for the data storage and analysis. This direct technique can be further classified according to the process used for signal transduction (Perumal and Hashim 2014).

8.6.1.1 Optical Biosensors

Optical biosensors have gained popularity as the direct (label free) method for the detection of food and waterborne pathogenic bacteria amongst all of the other techniques due to their high selectivity and sensitivity. These sensors are capable of detecting tiny changes in refractive index or thickness, which occurs during the attachment of target cells or analyte component on the transducer surface. Amidst all other optical biosensors, very first commercially available sensor is fiber-optic which was marketed by Research International (Monroe, WA). The basic principle behind the detection mechanism of the fibre optic sensor relies upon the fluorescently labelled pathogen (target analyte) or toxins which when bound to the receptor molecules on the transducer surface gets excited by the laser wave at 635 nm (Tait et al. 2005). Fluorescent signals generated by continuous laser excitation were detected by the fluorescence detector in the real time system (Bhunia 2008). Various optical biosensors for the detection of pathogenic bacteria (Baeumner et al. 2003), toxins (Bae et al. 2004) and other contaminants from water and food samples have been developed so far, and out of them, the fluorescent biosensor has shown promising outcome due to its outstanding sensitivity and specificity. In fluorescent biosensors, a fluorescent compound attached to antibodies enhances the efficacy of pathogen detection. FTIC (fluorescein isothiocyanate) is a regularly used fluorescent marker (Li et al. 2004). Other than this technique, few recently developed techniques for pathogen detection use techniques such as the Raman spectroscopy, surface plasmon resonance, laser, etc. (Yoo and Lee 2016).

8.6.1.2 Raman and Fourier Transform Infrared Spectroscopy Biosensor

Spectroscopies based on the vibrational energy such as Raman spectroscopy and Fourier transform infrared (FT-IR) are the more recurrently reported whole organism fingerprinting techniques (Kloß et al. 2013). There is always a need of culturing the microorganisms for the detection of the specific analyte from the mixture of samples to get the highest amount of biomass. Raman spectroscopy is an optical technique based on the principle of light scattering and it has been utilized by many researchers as a mean of rapid detection of bacterial pathogens from food and water samples. Schmilovitch et al. (2005) employed a disperse system of the spectrophotometer with a 785 nm diode laser to detect both Gram-positive and Gram-negative bacteria which showed clear dissimilarity between samples containing bacteria and control (without bacteria). In this approach, a pathogen of interest along with target analytes are separated from the sample with the help of capture biomolecules, which add an extra layer of specificity there by synergistically enhancing the efficacy of the sensors.

8.6.1.3 Surface Plasmon resonance Biosensor

Surface plasmon resonance (SPR) works based on the principle of optical illumination can be utilized for biomolecular analysis (Scarano et al. 2010). Additionally, plasmon represents the excited free electron present in the outer surface of the metal layer. Compatible light energy photons are the source of this resonant excitement. The amplitude of this resulting plasmon electromagnetic wave is the maximum at the interface (Anker et al. 2008). Direct label free detection of food and waterborne pathogens is also possible with this method. Application of SPR based biosensors have been described by several researchers for the detection of food and water borne pathogen such as *L. monocytogenes, C. jejuni* (Koubova et al. 2001; Taylor et al. 2006; Ray and Bhunia 2007), *Salmonella* (Oh et al. 2004), and *E. coli* O157:H7 (Subramanian et al. 2006). Additionally, many commercial SPR based biosensors have been employed by many researchers for the detection of food and waterborne pathogens in recent years. Spreeta™ biosensors was used for the detection of *E. coli* O157:H7 (Waswa et al. 2007). Biacore 3000 was utilized by Leonard et al. (2004) for the detection of *L. monocytogenes.*

In Spreeta™ biosensors, the incident light from LED reflects off a gold surface, the angle and intensity analogous to the SPR minimum is measured representing the changes in the refractive index which corresponds to the coupling of antigen-antibody complex at the sensor surface (Waswa et al. 2007). This real-time assay was conducted, and the results were obtained after 30 min. The sensitivity of the assay was nearly 10^2–10^3 CFU/mL. Specificity of the assay was confirmed from the sensorgram as other pathogens (such as *E. coli* K12 and *Shigella*) did not show any changes during the analysis. Balasubramanian et al. (2007) also described the label free detection of *S. aureus* with the help of lytic phage as immensely precise and selective bio-recognition element and SPR based Spreeta™ sensor was chosen as a detection platform.

8.6.1.4 Evanescent Wave Biosensor

Generation of an evanescent wave can be defined as the total reflection of incident light at a particular angle. Under such conditions, an energy field is generated which penetrates a short distance past the reflecting surface. The circulation of the generated evanescent wave relies on the optical properties of the thin layer of the medium adjacent to the reflecting surface. These phenomena can be applied to the detection of pathogenic bacterial contamination in different food and water samples. Many fibre optics sensors use evanescent wave physics (Marazuela and Moreno-Bondi 2002). Evanescent based immunosensors are having a limit of detection (LOD) range between 1 and 10 ng/mL for large (>30 kDa) proteins (Nedelkov et al. 2000) depending on the molecular weight of the antigen and its affinity towards the antibody. Direct measurement of low molecular weight molecules such as mycotoxins with the molecular weight of 750 Da yield higher LODs as compared to the larger analytes. For the detection, at low range (LOD) various indirect methods such as a sandwich method or competitive assays are often used (Rasooly 2001).

8.6.1.5 Resonant Crystal Biosensors or Piezoelectric Biosensors

Resonant crystal biosensors are one of the widely used biosensors due to their simple mechanism of sensing technology and low cost. It is also known as quartz crystal microbalance (QCM) and piezoelectric biosensor (Marrazza 2014). The primary mechanism of this sensing technology relies on the changes of the acoustic resonant frequency of a quartz crystal during the attachment of a target analyte to the crystal surface. Quartz disk attached to electrodes works as a transducer in this sensor which can amplify the signal produced as an acoustic wave generated by application of an external oscillating electric potential across the device. The acoustic wave is generated through piezoelectric (PZ) effect resonating on the crystal at a particular frequency which is dependent on the mass of the analyte bound to the sensor surface. Resonant crystal biosensor allows real-time, direct and label free analysis of larger antigens which leads to the successful detection of pathogenic bacteria present in water and food samples (Law et al. 2015). Piezoelectric immunosensors were developed for *V. cholerae* (Chen et al. 2010) and *S. typhimurium* (Arora et al. 2011). Specificity of sensors are based on the receptor-analyte reaction, for example, in the immunogravimetric microbial assay (Kazemi-Darsanaki et al. 2012), for the detection of *Candida albicans* an anti-*C. albicans* antibody coated with PZ crystal was used and the sensitivity of the assay was in the range of 10^6–10^8 CFU/mL. Therewas no attachment of other species except *C. albicans* in the reaction, and no significant shifts in the frequencies were observed due to non-specific adsorption.

8.6.1.6 Bioluminescence Sensors

The emergence of bioanalytical sensing tools by using the ability of certain enzymes to produce a photon as a byproduct of their biochemical reactions have shown promising outcome for pathogen detection. This phenomenon is known as bioluminescence and can be used for the detection of physiological conditions of cells. Ulitzur and Kuhn (1987) first described the application of this bioluminescence for bacterial detection by luciferase reporter phages. In their work, they have introduced the luciferase gene in the genome of a bacteriophage, which can infect the bacterial cell. Bioluminescence sensors have been used for the detection of a wide range of bacteria (VanDorst et al. 2010). Blasco et al. (1998) have demonstrated the development of an accurate and sensitive method for the detection of *Salmonella spp.* and *E. coli.* They have used bacteriophages for the specific lysis of bacteria and the cell content released was measured by ATP bioluminescence. An increment in the sensitivity was obtained by focusing on bacterial adenylate kinase as the cell marker instead of using ATP. Emission of light was measured as proportional to the cell numbers over three orders of magnitude, and 10^3 cells were easily detectable in a 0.1 ml sample volume (Tallury et al. 2010).

8.6.1.7 Electrical Impedance Based Biosensor

Impedance is termed as the total resistance of a conductive system in AC supply and consists of two major basic part (1) capacitance which depends on the characteristics of the electrode and (2) conductance, which depends on the conductivity of the medium. In the case of impedance based biosensors, microbial metabolism results in an increase in capacitance and conductance due to the conversion of larger macromolecules and another component to its monomeric smaller form which led to increasing the charge carrying capacity (Varshney and Li 2009). This increment in the capacitance and conductance resulted in the decrease in the impedance. Therefore, the alteration of impedance, capacitance and conductance are only different ways of monitoring the test system and are all inter-related (Shimazaki et al. 2015). The relationship between impedance (Z), resistance (R), capacitance (C), and frequency (*f*) of a resistor and a capacitor series is articulated as follows (Miller et al. 2010):

$$Z^2 = R^2 + 1/(2pfC)^2$$

Bridge circuit measures this impedance phenomenon. There is always a need for a reference module to measure and exclude the nonspecific changes in the test module. Due to this requirement, one pair of electrode is used in these sensors. The reference module serves as a controller for evaporation, changes of dissolved gases, temperature and most importantly degradation of culture medium (Singh et al. 2014).

This impedance method is widely used by most of the researchers, and the Association of Official Analytical Chemists International (AOAC) accepted it as a first action method (Monaci and Visconti 2010). This approach is well established for the detection of specific food pathogens. A significant parameter for the pathogen in food and water sample is cell viability. Viable cells can be measured by using a microscope after suspending the cells in a dye such as Trypan Blue. A new emerging biosensor for real time monitoring of the concentration of growth and physiological state of cells was proposed by Monaci and Visconti (2010). This biosensor is based on the measurement of alteration in impedance during the growth of adherent cells on integrated electrode structures (Zeng et al. 2016). The impedance of the biosensor changes according to the cell density, growth and long-term behaviour of the cells grown on the electrode at different time interval. Most impedance analysis is completed within 20–25 h. Gracias and McKillip (2004) investigated the detection of *Salmonella* using this method in 250 food samples. Food samples for the analysis were pre-enriched 14–16 h at 37 °C in peptone water.

8.6.1.8 Potentiometric, FET and LAPS-Based Biosensors

These are amongst the least common directly measurable biosensors used for pathogen detection. The working principle is based on the detection of ions present in the test solution. There is always need of one inert reference electrode and one working electrode in contact with the sample. Detection of pathogenic bacteria is possible with these biosensors by continuous monitoring of the pH changes or fluctuation in ionic concentration during the in-situ analysis conditions. Bergveld (2003) employed the use of ion-selective field effect transistor (ISFETs) for biological detection events. They have also demonstrated the fabrication of this type of biosensor using p-type silicon substrate with two n-doped regions, one acting as a source and another as a drain with a gate in between which acts as an insulator due to the presence of SiO_2 covered region, which is further over encrusted by ion selective membrane. Another new technique has been evolved for the detection of food borne pathogen by combining ISFET and potentiometry with optical detection. It is known as a light-addressable potentiometric sensor (LAPS) (Perumal and Hashim 2014; Wu et al. 2015).

8.6.2 Indirect Biosensors

8.6.2.1 Fluorescence Labelled Biosensors

Microorganisms are composed of various biological entities like proteins and polysaccharides which act as antigens in immunogenic reaction. This phenomenon permits the development of immunoassay techniques for pathogenic bacterial detection. In fluorescent immunoassay (FIA), immunoglobulins are labelled with fluorochromes which absorb short-wavelength light and then emit light at higher wavelengths which can be detected using fluorescent microscopy. Fluorescein isothiocyanate and rhodamine isothiocyanate-bovine serum albumin are most commonly used fluorochromes to tag antibodies for the detection of bacteria contaminated samples in both direct and indirect methods (Parkinson and Pejcic 2005). For detection of food borne bacteria by using FIA, food samples are pre-enriched with culture mediums because of less number of viable bacterial cells in food sample and also to reduce the interference caused by the background fluorescence producing food particulates (Perumal and Hashim 2014).

Similarly, detection of waterborne pathogenic bacteria in water samples can be done by concentrating the bacterial cells through membrane filtration. Polycarbonate filters are commonly used in this procedure (Mandal et al. 2011). Using this technique, detection of waterborne pathogenic bacteria *E.coli* O157:H7 was possible in the range of 10^5–10^9 CFU/ml within 4 h of assay time periods. Additionally, it has also been utilized for the detection of *S. typhimurium* and *Klebsiella pneumoniae.* Chowdhury et al. (1995) used a similar kind of technique for detecting *V. cholerae* O1 and O139. Bacterial cells were incubated with yeast extract in the presence of nalidixic acid which leads to the growth of substrate responsive viable bacterial cells with elongated and enlarged shape and was readily detectable using fluorescent antibody.

8.6.2.2 Cell Based Biosensor

These biosensors also serve as a dependable tool for the detection of pathogens in food and water samples. Working principle of cell based assays depends upon the electrical properties of cells to figure out the changes in the cell's vicinity (Singh et al. 2013). Due to the presence of various biological molecules at the cell surface, it functions like a capacitor where fluid acts as a resistance element. Electrical impedance detects the minute alteration in cell growth, density as well as the differences in the regular activities of the cell due to the influence of the external environment. Detection of the pathogenic bacteria in food has been done with the help of mammalian cells (Gray 2004). The detection of enzymes and cofactors from the cell system has become easier due to the massive release of metabolites (chemical compounds) into the culture medium during the assay (Fratamico and Bhunia 2005).

8.6.2.3 Amperometric Biosensor

Amperometric biosensors can detect the electrochemically active analyte which can be oxidized or reduced on the electrode. Amperometric biosensors consist of thin film construction made of gold (Au), carbon or platinum. For screen printing, the electrodes, substrates (glass, plastic or ceramic) are coated with thin films of ink in a particular pattern. Different inks can be applied to get various dimensions and shapes of the biosensor. It includes a series of basic steps, selection of the screen, selection and preparation of inks, selection of substrate and finally drying and curing stages (Arora et al. 2011). The main advantages of this technology are design flexibility, process automation, good reproducibility and a wide choice of materials. In recent years, screen printed electrochemical cells, which are cheap, affordable and can be produced in large scale, are widely used for developing the amperometric biosensors for the detection of various foods and waterborne pathogens (Lazcka et al. 2007; Velusamy et al. 2010). Disposable nature of these sensors lead to decrease in chance of contamination during electrode fouling due to the over growth of microorganism which results in loss of sensitivity and accuracy of the biosensor. Amperometric biosensor depends upon the enzyme functionality which helps to convert electrochemically inactive analyte to electrochemically active analyte through a catalytic process. Horse radish peroxidase and alkaline phosphatase have been used commonly as the functional enzymes. These biosensors are used to develop immunosensors and genosensors for pathogen detection (Palchetti and Mascini 2008; Arora et al. 2011).

8.6.2.4 Microbial Metabolism-Based Biosensors

These types of biosensors have been developed for the detection of bacterial pathogens by their metabolic pathways. The amperometric transducers are used for tracking the biochemical reactions in the bacterial cell metabolism. In general, Clark type oxygen electrode is used for the measurement of oxygen consumed by the bacterial cells (Patel 2002; Timur et al. 2003). The detection of pathogens was carried out by the measurement of the cathodic peak current of oxygen during the proliferation period of bacterial cells (Ruan et al. 2002). One more detection mechanism using electrochemical transducer relies on the detection of specific marker enzymes after culturing the test sample in a suitable medium. By using this strategy, detection of coliforms in water and food samples have become readily possible due to the mere presence of enzymes such as β-D-glucuronide glucuronosohydrolase (GUS) and β-D-galactosidase (β-GAL). Detection of *E.coli* using conventional approaches to detect GUS enzymes or β-GAL is a much more lengthy process which relies on spectrometric detection of the bacterium. Conventionally, the bacterial cells are first treated with chromogenic substrates such as p-nitrophenyl-β-D-glucuronide (PNPG) and then monitored spectrometrically, until the release of chromophore indicators (p-nitrophenyl (PNP) and d-glucuronic acid) confirming the presence of *E. coli* in the test sample. To overcome this time-consuming protocol Mulchandani et al. (2005) developed an efficient electro-oxidative method for GUS detection using bacteria-based biosensor. They have immobilized food borne pathogenic bacteria on a carbon paste electrode that can degrade PNP and produces hydroquinone (intermediate) for oxidation at a lower potential. A rapid detection method for viable *E. coli* cells was developed by Pérez et al. (2001) using enzyme β-D-galactosidase that can convert 4-aminophenyl-β-D-galactopyranoside (4-APGal) to 4-aminophenol (A-AP) after hydrolysis.

8.6.2.5 Immunosensors

These types of biosensors work based on the principle of immunology where antigen-antibody interaction takes place. Specific antibodies are used for the detection of specific antigen or toxin of microorganisms. Due to the specific binding phenomenon, it has a high level of sensitivity towards the detection of particular water and food borne pathogens. Antibodies for this sake can be immobilized on the surface of the electrodes or magnetic beads that lead to differentiation of two different kinds of immunosensors (1) immunosensors based on antibody immobilized on the electrode surface and (2) immunosensors based on antibody attached on the surface of the magnetic beads (Arora et al. 2011). In this biosensors, enzyme-substrate catalysis happen in the presence of antibody and produces products such as ions, pH variation, or oxygen consumption, which further lead to the generation of electrical signal with the help of a transducer. Numerous approaches have been used in the immune-module operations of biosensors, which includes an antibody-based system for the detection of food and waterborne pathogens such as *E.coli*

O157:H7 and *Salmonella spp*. Immunomagnetic beads have also been utilized to enhance the selectivity of amperometric biosensors (Liu et al. 2001; Abbaspour et al. 2015). In this technique, *S. typhimurium* is sandwiched between antibody coated magnetic beads, and alkaline phosphatase (enzyme) labelled antibody. After that, by using a magnet, beads are localized onto the surface of a disposable graphite ink based electrode in a multiwell plate format. The presence of bacterial cells is detected by the oxidation of the electroactive enzyme product. This technique offers a LOD of 8×10^3 cells/ml in buffer sample within 80 min (Gehring et al. 1996).

8.6.2.6 DNA Based Biosensors

In recent years, newly emerged DNA based biosensorshave shown promising outcome in the field of pathogen detection. These biosensors consists of short nucleic acid sequences also known as probes with the specificity towards a particular bacterium conjugated on the surface of a transducer. The probe DNA sequence binds to the complementary DNA sequence of the target bacterium and leads to the detection of the pathogenic bacteria; this event is also known as hybridization. The level of hybridization indicates the presence of a complementary sequence in the sample which finally leads to the detection of the target pathogen (Nordin et al. 2016). Some reviewers (Kerman et al. 2003) have demonstrated the application of electrical transducers with the combination of DNA based detection. Different pathogenic bacteria can be detected easily by using disposable low-density genosensor arrays. This can be fabricated by using a screen-printed array of Au electrodes having immobilized thiol-tethered oligonucleotide and biotinylated signalling probes (Farabullini et al. 2007) for complimentary sequence detection. Analysis strategy depends on the identification of a toxin produced by the specific bacteria responsible for the production of toxin. This can be termed as one of the most crucial steps for the detection technique as the encoded gene of the target bacteria can frequently express toxin in food samples (Singh et al. 2014). Wang (2002) have successfully developed one novel detection method for the detection of *Cryptosporidium*, *E.coli*, and *Giardia* by using genosensor technology. Bacterial immobilization of specific oligonucleotides was done by using carbon pasted electrode, and chronopotentiometric techniques. Further simultaneous monitoring of hybridization outcomes were measured by real-time sensors for pathogen detection.

8.6.3 Nano-biosensors

The interplay between nanomaterials and biological system creates an emerging research field of vast importance. Unique features of nanoparticles such as small size, large surface to volume ratio and other novel characteristics make them tremendously applicable. In particular, application of nanomaterials for the development of nano-based biosensors for sensing applications has received considerable attention (McFarland and Van Duyne 2003). Integration of one dimensional (1D) nanomaterials, such as nanowires, or two dimensional (2D) nanotubes in electric devices offer substantial advantages for the detection of pathogenic bacteria and have more advantages over conventional optical bio-detection methods (Gruner 2006). Optical-affinity biosensors based on SPR do the qualitative and quantitative measurements of biomolecular interactions between immobilized biomolecule on a metal surface and target analyte in the test solution. Application of nanotechnology in the field of optical bio-detection has emerged significantly in recent years (Kumar et al. 2015). To enhance the sensitivity of an SPR based biosensor, Au nanoparticles (NPs) have been utilized to amplify their detection level. Joung et al. (2008) employed Au NPs in a signal amplification system to enhance the sensitivity for the detection of *E. coli* 16s rRNA by using peptide nucleic-acid probes with an SPR biosensor system. DNA based biosensor for the detection of particular pathogens has become a widely used technique. Recent advancements in nanotechnology have also resulted in the development of bio-barcode assay which can provide amplification without the use of PCR and can detect many target pathogens in one sample (Li et al. 2005; Stoeva et al. 2006). This method is based on the bio-functionalization of Au NPs with a ds-DNA or ss-DNA bio-barcode, bio-receptor (such as antibody or oligonucleotide) or single component modified microparticles (MMPs) containing another bio-receptor capable of binding with the target analyte. Complexes formed by sandwiching an analyte molecule between two bio-functionalized particles are then separated using a magnet (Rowland et al. 2016). The DNA bio-barcode is released and detected by using a chip-based detection method consisting of silver (Ag) enhanced Au NPs or a fluorophore bound to the bio-barcode (Oh et al. 2006). Magnetic nanoparticles have also been utilized by many researchers for the development of a sensing technology with higher specificity and sensitivity at a lower cost. One example is the work done by Koets et al. (2009) who have employed the development of magneto-resistant sensor using supermagnetic particles as detection labels for *E. coli* and *Salmonella*. Additionally, there are many examples in the literature for novel NPs-based materials for electrochemical bio-sensing which enhances the specificity and efficacy of real-time analysis (Kumar et al. 2015). For example, screen printed carbon electrode modified with Au NPs display a 13.1-fold increase in detection of *E. coli* O157:H7 compared to the traditional screen printed carbon electrode with a working range of 10^2–10^7 CFU/mL (Lin et al. 2008).

8.6.4 The Electronic Nose

Electronic nose system has gained popularity rapidly during the last few years, the majority of its applications concentrate in the field of pathogenic bacterial detection in different food and water samples. Electronic nose system comprises of sophisticated software, data preprocessing, and statistical analysis of collected data by pattern recognition (PR) software (Loutfi et al. 2015). This technology has been used extensively in the field of sensors due to its potential for detecting target samples based on acoustic waves, conducting polymers or semiconducting materials. Several reports can be found in recent years on the use of electronic nose to detect or identify bacteria. Schiffman et al. (2001) have investigated the efficacy of the electronic nose using 15 different kinds of metal-oxide sensors to classify bacteria and fungi.

Due to the microbial metabolism, they produce some volatile organic compounds along with gases during the growth period. These metabolic products can be monitored by these sensors because they have tremendous information potential and respond to both odorous and odourless volatile compounds. Magan et al. (2001) have demonstrated the application of electronic nose biosensor using 14 conducting polymeric sensors to detect volatile profiles produced by non-inoculated skimmed milk media and media inoculated with *B. cereus* and *Candida pseudotropicalis*. Bacteria present in the test sample were detected by the automated headspace analyzer which consists of conducting polymer sensors (Osmetech Microbial Analyzer, OMA). Each of the sensors is having different sensing abilities towards various volatile organic compounds. This system was also used for screening urine specimens by sampling urine headspace and analyzing the output of the multi-detector response (Aathithan et al. 2001).

8.7 Conclusion

The upsurge of infectious diseases caused by water and foodborne pathogens are perilous global health issues. Technologies which can rapidly, sensitively, and correctly detect their presence in accordance with water and food safety regulations at clinically significant levels are essential for the upgradation of health and quality of life for millions of people. In this chapter, note on water and foodborne pathogens, infectious diseases caused by them and their specific route of infections have been discussed briefly. Additionally, numerous strategies for the development of advanced sensing techniques (biosensors) for the detection of food and waterborne pathogens have been described along with their advantages over the conventional approaches. Biosensors have great potential in detection of food and waterborne bacterial pathogens. The sensitivity of different biosensors discussed in this chapter vary relying on transducer properties and specific biological recognition elements (bio-receptors). Though several sensing strategies are developed through research, only a few approaches have shown their potential to reach the commercial market. Many of the newly developed biosensors can detect single or few analytes, but a future aspect of biosensing will be the multiple-sensing element instruments. For example, immunosensors based on the use of different antibodies are placed in an orderly arrangement in the 2-dimensional format of the chip. Using this technique, various antigens of water and foodborne pathogenic bacteria can be detected by their binding to specific antibodies at unique positions. Similarly, DNA based biosensors have also demonstrated their efficacy at low concentrations, but they require a purification process in upstream which is time-consuming.

Current and future research need to be concerned in two important cases (1) detection of pathogens in their actual environment matrixes and (2) pre-processing/pre-enrichment steps required for the analysis. Additionally, it involves miniaturisation strategies, material research and emphasis on multiplexing so that various pathogenic target analytes can be detected at once in "real-time" scale. Taken together, this technology can provide novel approaches capable of providing high sensitivity, specificity and speed to replace the current and conventional strategies. This would hopefully, improve access to safe drinking water and safe food for consumption thereby reducing the global health issues due to water and foodborne diseases.

Acknowledgements Authors from NIT Rourkela acknowledge the PhD scholar institute fellowship given by NIT Rourkela, Odisha, India.

References

Aathithan S, Plant JC, Chaudry AN, French GL (2001) Diagnosis of bacteriuria by detection of volatile organic compounds in urine using an automated headspace analyzer with multiple conducting polymer sensors. J Clin Microbiol 39:2590–2593. https://doi.org/10.1128/JCM.39.7.2590-2593.2001

Abbaspour A, Norouz-Sarvestani F, Noori A, Soltani N (2015) Aptamer-conjugated silver nanoparticles for electrochemical dual-aptamer-based sandwich detection of *Staphylococcus aureus*. Biosens Bioelectron 68:149–155. https://doi.org/10.1016/j.bios.2014.12.040

Alahi ME, Mukhopadhyay SC (2017) Detection methodologies for pathogen and toxins: a review. Sensors 17:1885 (1–20). https://doi.org/10.3390/s17081885

Anker JN, Hall WP, Lyandres O, Shah NC, Zhao J, Van Duyne RP (2008) Biosensing with plasmonic nanosensors. Nat Mater 7:442–453. https://doi.org/10.1038/nmat2162

Arora P, Sindhu A, Dilbaghi N, Chaudhury A (2011) Biosensors as innovative tools for the detection of food borne pathogens. Biosens Bioelectron 15:1–2. https://doi.org/10.1016/j.bios.2011.06.002

Arya SK, Singh A, Naidoo R, Wu P, McDermott MT, Evoy S (2011) Chemically immobilized T4-bacteriophage for specific *Escherichia coli* detection using surface plasmon resonance. Analyst 136:486–492. https://doi.org/10.1039/C0AN00697A

Aung MM, Chang YS (2014) Traceability in a food supply chain: safety and quality perspectives. Food Control 39:172–184. https://doi.org/10.1016/j.foodcont.2013.11.007

Bae YM, Oh BK, Lee W, Lee WH, Choi JW (2004) Detection of insulin–antibody binding on a solid surface using imaging ellipsometry. Biosens Bioelectron 20:895–902. https://doi.org/10.1016/j.bios.2004.03.032

Baeumner AJ, Cohen RN, Miksic V, Min J (2003) RNA biosensor for the rapid detection of viable *Escherichia coli* in drinking water. Biosens Bioelectron 18:405–413. https://doi.org/10.1016/S0956-5663(02)00162-8

Balasubramanian S, Sorokulova IB, Vodyanoy VJ, Simonian AL (2007) Lytic phage as a specific and selective probe for detection of *Staphylococcus aureus*—a surface plasmon resonance spectroscopic study. Biosens Bioelectron 22:948–955. https://doi.org/10.1016/j.bios.2006.04.003

Baylis CL (2003) Immunological techniques: immunochromatography, enzyme linked immunofluorescent assays and agglutination techniques. In: McMeekin TA (ed) Detecting pathogens in food. Woodhead Publishing Limited, Cambridge, pp 217–240 ISBN 1855736705

Benhar I, Eshkenazi I, Neufeld T, Opatowsky J, Shaky S, Rishpon J (2001) Recombinant single chain antibodies in bioelectrochemical sensors. Talanta 13:899–907. https://doi.org/10.1016/S0039-9140(01)00497-0

Bergveld P (2003) Thirty years of ISFETOLOGY: what happened in the past 30 years and what may happen in the next 30 years. Sensors Actuators B Chem 88:1–20. https://doi.org/10.1016/S0925-4005(02)00301-5

Berry D, Xi C, Raskin L (2006) Microbial ecology of drinking water distribution systems. Curr Opin Biotechnol 17:297–302. https://doi.org/10.1016/j.copbio.2006.05.007

Bhunia AK (2007) Foodborne microbial pathogens: mechanisms and pathogenesis. Springer Science and Business Media, New York ISBN:0387745378

Bhunia AK (2008) Biosensors and bio-based methods for the separation and detection of foodborne pathogens. Adv Food Nutr Res 54:1–44. https://doi.org/10.1016/S1043-4526(07)00001-0

Blasco R, Murphy MJ, Sanders MF, Squirrell DJ (1998) Specific assays for bacteria using phage mediated release of adenylate kinase. J Appl Microbiol 84:661–666. https://doi.org/10.1046/j.1365-2672.1998.00393.x

Blommel PG, Becker KJ, Duvnjak P, Fox BG (2007) Enhanced bacterial protein expression during auto-induction obtained by alteration of lac repressor dosage and medium composition. Biotechnol Prog 23:585–598. https://doi.org/10.1021/bp070011x

Brandl MT (2006) Fitness of human enteric pathogens on plants and implications for food safety. Annu Rev Phytopathol 44:367–392. https://doi.org/10.1146/annurev.phyto.44.070505.143359

Centers for Disease Control and Prevention (CDC) (2006) Preliminary FoodNet data on the incidence of infection with pathogens transmitted commonly through food – 10 States, United States, 2005. MMWR Morb Mortal Wkly Rep 55:392 https://www.ncbi.nlm.nih.gov/pubmed/16617286

Chen H, H QY, Jiang JH, Shen GL, Yu RQ (2010) Construction of supported lipid membrane modified piezoelectric biosensor for sensitive assay of cholera toxin based on surface-agglutination of ganglioside-bearing liposomes. Anal Chim Acta 657:204–209. https://doi.org/10.1016/j.aca.2009.10.036

Chowdhury MAR, Xu B, Montilla R, Hasan JAK, Huq A, Colwell RR (1995) A simplified immune fluorescence technique for detection of viable cells of *Vibrio cholerae* O1 and O139. J Microbiol Methods 24:165–170. https://doi.org/10.1016/0167-7012(95)00066-6

Connelly JT, Baeumner AJ (2012) Biosensors for the detection of waterborne pathogens. Anal Bioanal Chem 402:117–127. https://doi.org/10.1007/s00216-011-5407-3

Conner SD, Schmid SL (2003) Regulated portals of entry into the cell. Nature 422:37–44. https://doi.org/10.1038/nature01451

Cossart P, Sansonetti PJ (2004) Bacterial invasion: the paradigms of entero invasive pathogens. Science 304:242–248. https://doi.org/10.1126/science.1090124

Dasgupta N, Ranjan S, Mundekkad D, Ramalingam C, Shanker R, Kumar A (2015) Nanotechnology in agro-food: from field to plate. Food Res Int 69:381–400. https://doi.org/10.1016/j.foodres.2015.01.005

Doyle MP, Buchanan RL (eds) (2012) Food microbiology: fundamentals and frontiers. ASM Press, Washington, DC. https://doi.org/10.1128/9781555818463

Emanuel PA, Dang J, Gebhardt JS, Aldrich J, Garber EA, Kulaga H, Dion-Schultz A (2000) Recombinant antibodies: a new reagent for biological agent detection. Biosens Bioelectron 14:751–759. https://doi.org/10.1016/S0956-5663(99)00058-5

Falzoni S, Donvito G, Di Virgilio F (2013) Detecting adenosine triphosphate in the pericellular space. Interface Focus 3:20120101. https://doi.org/10.1098/rsfs.2012.0101

Farabullini F, Lucarelli F, Palchetti I, Marrazza G, Mascini M (2007) Disposable electrochemical genosensor for the simultaneous analysis of different bacterial food contaminants. Biosens Bioelectron 22:1544–1549. https://doi.org/10.1016/j.bios.2006.06.001

Ferens WA, Hovde CJ (2011) *Escherichia coli* O157:H7: animal reservoir and sources of human infection. Foodborne Pathog Dis 8:465–487. https://doi.org/10.1089/fpd.2010.0673

Ferreira L, Sánchez-Juanes F, Muñoz-Bellido JL, González-Buitrago JM (2011) Rapid method for direct identification of bacteria in urine and blood culture samples by matrix-assisted laser desorption ionization time-of-flight mass spectrometry: intact cell vs. extraction method. Clin Microbiol Infect 17:1007–1012. https://doi.org/10.1111/j.1469-0691.2010.03339.x

Fitzpatrick J, Fanning L, Hearty S, Leonard P, Manning BM, Quinn JG, O'Kennedy R (2000) Applications and recent developments in the use of antibodies for analysis. Anal Lett:2563–2609. https://doi.org/10.1080/00032710008543210

Fizgerland C, Swaminathan B (2003) Genetic techniques: molecular subtyping methods. In: McMeekin TA (ed) Detecting pathogens in food. Woodhead Publishing Limited, Cambridge, pp 271–292 ISBN 1855736705

Fratamico PM, Bhunia AK (2005) Foodborne pathogens: microbiology and molecular biology. Caister Academic Press, Wymondham ISBN: 1-904455-00-X

Gehring AG, Crawford CG, Mazenko RS, Van Houten LJ, Brewster JD (1996) Enzyme-linked immonomagnetic electrochemical detection of *Salmonella typhimurium*. J Immunol Methods 195:15–25. https://doi.org/10.1016/0022-1759(96)00076-2

Gibson DM (2003) Electrical methods. In: McMeekin TA (ed) Detecting pathogens in food. Woodhead Publishing Limited, Cambridge, pp 147–164 ISBN 1855736705

Girones R, Ferrus MA, Alonso JL, Rodriguez-Manzano J, Calgua B, de Abreu CA, Hundesa A, Carratala A, Bofill-Mas S (2010) Molecular detection of pathogens in water–the pros and cons of molecular techniques. Water Res 44:4325–4339. https://doi.org/10.1016/j.watres.2010.06.030

Goldman ER, Pazirandeh MP, Mauro JM, King KD, Frey JC, Anderson GP (2000) Phage-displayed peptides as biosensor reagents. J Mol Recognit 13:382–387. https://doi.org/10.1002/1099-1352(200011/12)13:6<382::AID-JMR511>3.0.CO;2-W

Goodridge L, Griffiths M (2002) Reporter bacteriophage assays as a means to detect foodborne pathogenic bacteria. Food Res Int 35:863–870. https://doi.org/10.1016/S0963-9969(02)00094-7

Gracias KS, McKillip JL (2004) A review of conventional detection and enumeration methods for pathogenic bacteria in food. Can J Microbiol 50:883–890. https://doi.org/10.1139/w04-080

Gray KMN (2004) Cytotoxicity and cell-based sensors for detection of *Listeria monocytogenes* and *Bacillus cereus*. Ph.D. Thesis (AAI3150767), Purdue University, West Lafayette. http://docs.lib.purdue.edu/dissertations/AAI3150767/

Griffiths M, Brovko L (2003) ATP bioluminescence. In: McMeekin TA (ed) Detecting pathogens in food. Woodhead Publishing Limited, Cambridge, pp 165–185 ISBN 1855736705

Gruner G (2006) Carbon nanotube transistors for biosensing applications. Anal Bioanal Chem 384:322–335. https://doi.org/10.1007/s00216-005-3400-4

Harris RJ, Shire SJ, Winter C (2004) Commercial manufacturing scale formulation and analytical characterization of therapeutic recombinant antibodies. Drug Dev Res 61:137–154. https://doi.org/10.1002/ddr.10344

Iqbal SS, Mayo MW, Bruno JG, Bronk BV, Batt CA, Chambers JP (2000) A review of molecular recognition technologies for detection of biological threat agents. Biosens Bioelectron 15:549–578. https://doi.org/10.1016/S0956-5663(00)00108-1

Ivnitski D, Abdel-Hamid I, Atanasov P, Wilkins E, Stricker S (2000) Application of electrochemical biosensors for detection of food pathogenic bacteria. Electroanalysis 12:317–325. https://doi.org/10.1002/(SICI)1521-4109(20000301)12:5<317::AID-ELAN317>3.0.CO;2-A

Iwamoto M, Ayers T, Mahon BE, Swerdlow DL (2010) Epidemiology of seafood-associated infections in the United States. Clin Microbiol Rev 23:399–411. https://doi.org/10.1128/CMR.00059-09

Jia A, Woo NY, Zhang XH (2010) Expression, purification, and characterization of thermolabile hemolysin (TLH) from *Vibrio alginolyticus*. Dis Aaquat Org 90:121–127. https://doi.org/10.3354/dao02225

Joung HA, Lee NR, Lee SK, Ahn J, Shin YB, Choi HS, Kim MG (2008) High sensitivity detection of 16s rRNA using peptide nucleic acid probes and a surface plasmon resonance biosensor. Anal Chim Acta 630:168–173. https://doi.org/10.1016/j.aca.2008.10.001

Kazemi-Darsanaki R, Azizzadeh A, Nourbakhsh M, Raeisi G, AzizollahiAliabadi M (2012) Biosensors: functions and applications. J Biol Today's World 2:20–23. https://doi.org/10.15412/J.JBTW.01020105

Kent RM, Fitzgerald GF, Hill C, Stanton C, Ross RP (2015) Novel approaches to improve the intrinsic microbiological safety of powdered infant milk formula. Nutrients 7:1217–1244. https://doi.org/10.3390/nu7021217

Kerman K, Kobayashi M, Tamiya E (2003) Recent trends in electrochemical DNA biosensor technology. Meas Sci Technol 15:R1. https://doi.org/10.1088/0957-0233/15/2/R01

Kloß S, Kampe B, Sachse S, Rösch P, Straube E, Pfister W, Kiehntopf M, Popp J (2013) Culture independent Raman spectroscopic identification of urinary tract infection pathogens: a proof of principle study. Anal Chem 85:9610–9616. https://doi.org/10.1021/ac401806f

Koets M, Van der Wijk T, Van Eemeren JTWM, Van Amerongen A, Prins MWJ (2009) Rapid DNA multi-analyte immunoassay on a magneto-resistance biosensor. Biosens Bioelectron 24:1893–1898. https://doi.org/10.1016/j.bios.2008.09.023

Koubova V, Brynda E, Karasova L, Škvor J, Homola J, Dostalek J, Tobiška P, Rošický J (2001) Detection of foodborne pathogens using surface plasmon resonance biosensors. Sensors Actuators B Chem 74:100–105. https://doi.org/10.1016/S0925-4005(00)00717-6

Kramer A, Schwebke I, Kampf G (2006) How long do nosocomial pathogens persist on inanimate surfaces? A systematic review. BMC Infect Dis 6:130. https://doi.org/10.1186/1471-2334-6-130

Kumar N (2013) Development of enzyme substrate based assay for detection of listeria monopcytogenes in milk. Ph.D. Thesis NDRI, Karnal. http://krishikosh.egranth.ac.in/handle/1/83937

Kumar S, Ahlawat W, Kumar R, Dilbaghi N (2015) Graphene, carbon nanotubes, zinc oxide and gold as elite nanomaterials for fabrication of biosensors for healthcare. Biosens Bioelectron 70:498–503. https://doi.org/10.1016/j.bios.2015.03.062

Lampel KA, Orlandi PA, Kornegay L (2000) Improved template preparation for PCR-based assays for detection of food-borne bacterial pathogens. Appl Environ Microbiol 66:4539–4542. https://doi.org/10.1128/AEM.66.10.4539-4542.2000

Lanata CF, Fischer-Walker CL, Olascoaga AC, Torres CX, Aryee MJ, Black RE (2013) Global causes of diarrheal disease mortality in children <5 years of age: a systematic review. PLoS One 8:e72788. https://doi.org/10.1371/journal.pone.0072788

Law JWF, AbMutalib NS, Chan KG, Lee LH (2015) Rapid methods for the detection of foodborne bacterial pathogens: principles, applications, advantages and limitations. Front Microbiol 5:770. https://doi.org/10.3389/fmicb.2014.00770

Lazcka O, Del Campo FJ, Munoz FX (2007) Pathogen detection: a perspective of traditional methods and biosensors. Biosens Bioelectron 22:1205–1217. https://doi.org/10.1016/j.bios.2006.06.036

Leonard P, Hearty S, Brennan J, Dunne L, Quinn J, Chakraborty T, O'Kennedy R (2003) Advances in biosensors for detection of pathogens in food and water. Enzym Microb Technol 32:3–13. https://doi.org/10.1016/S0141-0229(02)00232-6

Leonard P, Hearty S, Quinn J, O'Kennedy R (2004) A generic approach for the detection of whole *Listeria monocytogenes* cells in contaminated samples using surface plasmon resonance. Biosens Bioelectron 19:1331–1335. https://doi.org/10.1016/j.bios.2003.11.009

Li Y, Dick WA, Tuovinen OH (2004) Fluorescence microscopy for visualization of soil microorganisms- a review. Biol Fertil Soils 39:301–311. https://doi.org/10.1007/s00374-004-0722-x

Li Y, Cu YTH, Luo D (2005) Multiplexed detection of pathogen DNA with DNA-based fluorescence nanobarcodes. Nat Biotechnol 23:885. https://doi.org/10.1038/nbt1106

Lin YH, Chen SH, Chuang YC, Lu YC, Shen TY, Chang CA, Lin CS (2008) Disposable amperometricimmunosensing strips fabricated by Au nanoparticles-modified screen-printed carbon electrodes for the detection of foodborne pathogen *Escherichia coli* O157: H7. Biosens Bioelectron 23:1832–1837. https://doi.org/10.1016/j.bios.2008.02.030

Liu Y, Che Y, Li Y (2001) Rapid detection of *Salmonella typhimurium* using immunomagnetic separation and immune-optical sensing method. Sensors Actuators B Chem 72:214–218. https://doi.org/10.1016/S0925-4005(00)00663-8

Loutfi A, Coradeschi S, Mani GK, Shankar P, Rayappan JBB (2015) Electronic noses for food quality: a review. J Food Eng 144:103–111. https://doi.org/10.1016/j.jfoodeng.2014.07.019

Luo Y, Liu T, Zhu J, Kong L, Wang W, Tan L (2015) Label-free and sensitive detection of thrombomodulin, a marker of endothelial cell injury, using quartz crystal microbalance. Anal Chem 87:11277–11284. https://doi.org/10.1021/acs.analchem.5b02447

Magan N, Pavlou A, Chrysanthakis I (2001) Milk-sense: a volatile sensing system recognizes spoilage bacteria and yeasts in milk. Sensors Actuators B Chem 72:28–34. https://doi.org/10.1016/S0925-4005(00)00621-3

Mandal PK, Biswas AK, Choi K, Pal UK (2011) Methods for rapid detection of foodborne pathogens: an overview. Am J Food Technol 6:87–102. https://doi.org/10.3923/ajft.2011.87.102

Marazuela M, Moreno-Bondi M (2002) Fiber-optic biosensors-an overview. Anal Bioanal Chem 372:664–682. https://doi.org/10.1007/s00216-002-1235-9

Marrazza G (2014) Piezoelectric biosensors for organophosphate and carbamate pesticides: a review. Biosensors 4:301–317. https://doi.org/10.3390/bios4030301

Martinović T, Andjelković U, Gajdošik MŠ, Rešetar D, Josić D (2016) Foodborne pathogens and their toxins. J Proteome 147:226–235. https://doi.org/10.1016/j.jprot.2016.04.029

Mascini M, Tombelli S (2008) Biosensors for biomarkers in medical diagnostics. Biomarkers 13:637–657. https://doi.org/10.1080/13547500802645905

McCarthy (2003) Immunological techniques: ELISA. In: Mcmeekin TA (ed) Detecting pathogens in food. Woodhead Publishing Limited, Cambridge, pp 147–162 ISBN 1855736705

McFarland AD, Van Duyne RP (2003) Single silver nanoparticles as real-time optical sensors with zeptomole sensitivity. Nano Lett 3:1057–1062. https://doi.org/10.1021/nl034372s

Mehrotra P (2016) Biosensors and their applications. J Oral Biol Craniofacial Res 6:153–159. https://doi.org/10.1016/j.jobcr.2015.12.002

Miller JR, Outlaw RA, Holloway BC (2010) Graphene double-layer capacitor with ac line-filtering performance. Science 329:1637–1639. https://doi.org/10.1126/science.1194372

Monaci L, Visconti A (2010) Immunochemical and DNA-based methods in food allergen analysis and quality assurance perspectives. Trends Food Sci Technol 21:272–283. https://doi.org/10.1016/j.tifs.2010.02.003

Mor-Mur M, Yuste J (2010) Emerging bacterial pathogens in meat and poultry: an overview. Food Bioprocess Technol 3:24–35. https://doi.org/10.1007/s11947-009-0189-8

Mulchandani P, Hangarter CM, Lei Y, Chen W, Mulchandani A (2005) Amperometric microbial biosensor for p-nitrophenol using Moraxella sp.-modified carbon paste electrode. Biosens Bioelectron 21:523–527. https://doi.org/10.1016/j.bios.2004.11.011

Nedelkov D, Rasooly A, Nelson RW (2000) Multitoxin biosensor–mass spectrometry analysis: a new approach for rapid, real-time, sensitive analysis of *Staphylococcal* toxins in food. Int J Food Microbiol 60:1–13. https://doi.org/10.1016/S0168-1605(00)00328-7

Noble RT, Weisberg SB (2005) A review of technologies for rapid detection of bacteria in recreational waters. J Water Health 3:381–392. https://doi.org/10.2166/wh.2005.051

Nordin N, Yusof NA, Abdullah J, Radu S, Hushiarian R (2016) Sensitive detection of multiple pathogens using a single DNA probe. Biosens Bioelectron 86:398–405. https://doi.org/10.1016/j.bios.2016.06.077

Oh BK, Lee W, Kim YK, Lee WH, Choi JW (2004) Surface plasmon resonance immunosensor using self-assembled protein G for the detection of *Salmonella paratyphi*. J Biotechnol 111:1–8. https://doi.org/10.1016/j.jbiotec.2004.02.010

Oh BK, Nam JM, Lee SW, Mirkin CA (2006) A fluorophore-based bio-barcode amplification assay for proteins. Small 2:103–108. https://doi.org/10.1002/smll.200500260

Palchetti I, Mascini M (2008) Electroanalytical biosensors and their potential for food pathogen and toxin detection. Anal Bioanal Chem 391:455–471. https://doi.org/10.1007/s00216-008-1876-4

Parkinson G, Pejcic B (2005) Using biosensors to detect emerging infectious diseases. Prepared for The Australian Biosecurity Cooperative Research Centre, pp 1–80. https://doi.org/10.1.1.475.7819&rep=rep1&type=pdf

Patel PD (2002) (Bio) sensors for measurement of analytes implicated in food safety: a review. TrAC Trends Anal Chem 21:96–115. https://doi.org/10.1016/S0165-9936(01)00136-4

Pérez F, Tryland I, Mascini M, Fiksdal L (2001) Rapid detection of *Escherichia coli* in water by a culture-based amperometric method. Anal Chim Acta 427:149–154. https://doi.org/10.1016/S0003-2670(00)00984-3

Perumal V, Hashim U (2014) Advances in biosensors: principle, architecture and applications. J Appl Biomed 12:1–5. https://doi.org/10.1016/j.jab.2013.02.001

Puttaswamy S, Sengupta S (2010) Rapid detection of bacterial proliferation in food samples using microchannel impedance measurements at multiple frequencies. Sens Instrumen Food Qual 4:108–118. https://doi.org/10.1007/s11694-010-9101-5

Rasooly A (2001) Surface plasmon resonance analysis of staphylococcal enterotoxin B in food. J Food Prot 64:37–43. https://doi.org/10.4315/0362-028X-64.1.37

Ray B, Bhunia A (eds) (2007) Fundamental food microbiology, 4th edn. CRC Press, Boca Raton, pp 43–55 ISBN: 1420007742, 9781420007749

Ribet D, Cossart P (2015) How bacterial pathogens colonize their hosts and invade deeper tissues. Microbiol Infect 17:173–183. https://doi.org/10.1016/j.micinf.2015.01.004

Rider TH, Petrovick MS, Nargi FE, Harper JD, Schwoebel ED, Mathews RH, Blanchard DJ, Bortolin LT, Young AM, Chen J, Hollis MA (2003) AB cell-based sensor for rapid identification of pathogens. Science 301:213–215. https://doi.org/10.1126/science.1084920

Ronkainen NJ, Halsall HB, Heineman WR (2010) Electrochemical biosensors. Chem Soc Rev 39:1747–1763. https://doi.org/10.1039/B714449K

Rowland CE, Brown CW, Delehanty JB, Medintz IL (2016) Nanomaterial-based sensors for the detection of biological threat agents. Mater Today 19:464–477. https://doi.org/10.1016/j.mattod.2016.02.018

Ruan C, Yang L, Li Y (2002) Rapid detection of viable *Salmonella typhimurium* in a selective medium by monitoring oxygen consumption with electrochemical cyclic voltammetry. J Electroanal Chem 519:33–38. https://doi.org/10.1016/S0022-0728(01)00711-2

Sanderson K, Nicholas D (2003) Genetic techniques: PCR, NASBA, hybridization and microarrays. In: McMeekin TA (ed) Detecting pathogens in food. Woodhead Publishing Limited, Cambridge, pp 259–270 ISBN 1855736705

Scarano S, Mascini M, Turner AP, Minunni M (2010) Surface plasmon resonance imaging for affinity-based biosensors. Biosens Bioelectron 25:957–966. https://doi.org/10.1016/j.bios.2009.08.039

Schiffman SS, Wyrick DW, Gutierrez-Osuna R, Nagle HT (2001) Effectiveness of an electronic nose for monitoring bacterial and fungal growth. In: Gardner JW, Persaud KC (eds) Electronic noses and olfaction 2000: proceedings of the 7th international symposium on olfaction and electronic noses, Brighton, UK, July 2000, Series: series in sensors, pp 173–180. CRC Press. ISBN 9780750307642

Schmilovitch Z, Mizrach A, Alchanatis V, Kritzman G, Korotic R, Irudayaraj J, Debroy C (2005) Detection of bacteria with low-resolution Raman spectroscopy. Trans ASAE 48:1843–1850. https://doi.org/10.13031/2013.19982

Scognamiglio V, Arduini F, Palleschi G, Rea G (2014) Biosensing technology for sustainable food safety. TrAC Trends Anal Chem 62:1–0. https://doi.org/10.1016/j.trac.2014.07.007

Sharma H, Mutharasan R (2013) Review of biosensors for foodborne pathogens and toxins. Sensors Actuators B Chem 183:535–549. https://doi.org/10.1016/j.snb.2013.03.137

Shimazaki Y, Yamamoto M, Borzenets IV, Watanabe K, Taniguchi T, Tarucha S (2015) Generation and detection of pure valley current by electrically induced Berry curvature in bilayer graphene. arXiv preprint arXiv:1501.04776. https://doi.org/10.1038/nphys3551

Sibley LD (2004) Intracellular parasite invasion strategies. Science 304:248–253. https://doi.org/10.1126/science.1094717

Singh A, Poshtiban S, Evoy S (2013) Recent advances in bacteriophage based biosensors for foodborne pathogen detection. Sensors 13:1763–1786. https://doi.org/10.3390/s130201763

Singh R, Mukherjee MD, Sumana G, Gupta RK, Sood S, Malhotra BD (2014) Biosensors for pathogen detection: a smart approach towards clinical diagnosis. Sensors Actuators B Chem 197:385–404. https://doi.org/10.1016/j.snb.2014.03.005

Stephens P (2003) Culture methods. In: McMeekin TA (ed) Detecting pathogens in food. Woodhead Publishing Limited, Cambridge, pp 123–146 ISBN 1855736705

Stoeva SI, Lee JS, Smith JE, Rosen ST, Mirkin CA (2006) Multiplexed detection of protein cancer markers with biobarcoded nanoparticle probes. J Am Chem Soc 128:8378–8379. https://doi.org/10.1021/ja0613106

Subramanian A, Irudayaraj J, Ryan TA (2006) Mixed self-assembled monolayer-based surface plasmonimmunosensor for detection of *E-coli* O157:H7. Biosens Bioelectron 21:998–1006. https://doi.org/10.1016/j.bios.2005.03.007

Tait CR, Anderson GP, Ligler FS (2005) Evanescent wave fluorescence biosensors. Biosens Bioelectron 20:2470–2487. https://doi.org/10.1016/j.bios.2004.10.026

Tallury P, Malhotra A, Byrne LM, Santra S (2010) Nanobioimaging and sensing of infectious diseases. Adv Drug Deliv Rev 62:424–437. https://doi.org/10.1016/j.addr.2009.11.014

Taylor AD, Ladd J, Yu Q, Chen S, Homola J, Jiang S (2006) Quantitative and simultaneous detection of four foodborne bacterial pathogens with a multi-channel SPR sensor. Biosens Bioelectron 22:752–758. https://doi.org/10.1016/j.bios.2006.03.012

Timur S, Pazarlioğlu N, Pilloto R, Telefoncu A (2003) Detection of phenolic compounds by thick film sensors based on *Pseudomonas putida*. Talanta 61:87–93. https://doi.org/10.1016/S0039-9140(03)00237-6

Tlaskalová-Hogenová H, Štěpánková R, Kozáková H, Hudcovic T, Vannucci L, Tučková L, Rossmann P, Hrnčíř T, Kverka M, Zákostelská Z, Klimešová K (2011) The role of gut microbiota (commensal bacteria) and the mucosal barrier in the pathogenesis of inflammatory and autoimmune diseases and cancer: contribution of germ-free and gnotobiotic animal models of human diseases. Cell Mol Immunol 8:110–120. https://doi.org/10.1038/cmi.2010.67

Turner AP (2013) Biosensors: sense and sensibility. Chem Soc Rev 42:3184–3196. https://doi.org/10.1039/C3CS35528D

Uematsu S, Jang MH, Chevrier N, Guo Z, Kumagai Y, Yamamoto M, Kato H, Sougawa N, Matsui H, Kuwata H, Hemmi H (2006) Detection of pathogenic intestinal bacteria by Toll-like receptor 5 on intestinal CD11c+ lamina propria cells. Nat Immunol 7:868–874. https://doi.org/10.1038/ni1362

Ulitzur S, Kuhn J (1987) Introduction of lux genes into bacteria, a new approach for specific determination of bacteria and their antibiotic susceptibility. In: Schlomerich J, Andreesen R, Kapp A, Ernst M, Woods WG (eds) Bioluminescence and chemiluminescence new perspectives. Wiley, New York, pp 463–472 ISBN 0-471-091470-3

VanDorst B, Mehta J, Bekaert K, Rouah-Martin E, De Coen W, Dubruel P, Robbens J (2010) Recent advances in recognition elements of food and environmental biosensors: a review. Biosens Bioelectron 26:1178–1194. https://doi.org/10.1016/j.bios.2010.07.033

Varshney M, Li Y (2009) Interdigitated array microelectrodes based impedance biosensors for detection of bacterial cells. Biosens Bioelectron 24:2951–2960. https://doi.org/10.1016/j.bios.2008.10.001

Velusamy V, Arshak K, Korostynska O, Oliwa K, Adley C (2010) An overview of foodborne pathogen detection: in the perspective of biosensors. Biotechnol Adv 28:232–254. https://doi.org/10.1016/j.biotechadv.2009.12.004

Wang J (2002) Electrochemical nucleic acid biosensors. Anal Chim Acta 469:63–71. https://doi.org/10.1016/S0003-2670(01)01399-X

Waswa J, Irudayaraj J, Roy DC (2007) Direct detection of *Escherichia coli* O157: H7 in selected food systems by a surface plasmon resonance biosensor. LWT Food Sci Technol 40:187–192. https://doi.org/10.1016/j.lwt.2005.11.001

World Health Organization (2008) The global burden of disease: 2004 update. World Health Organization, Geneva ISBN 978 92 4 156371 0

Wu C, Bronder T, Poghossian A, Werner CF, Schöning MJ (2015) Label-free detection of DNA using a light-addressable potentiometric sensor modified with a positively charged polyelectrolyte layer. Nanoscale 7:6143–6150. https://doi.org/10.1039/C4NR07225A

Yang H, Qu L, Lin Y, Jiang X, Sun YP (2007) Detection of *Listeria monocytogenes* in biofilms using immunonanoparticles. J Biomed Nanotechnol 3:131–138. https://doi.org/10.1166/jbn.2007.019

Yeni F, Acar S, Polat ÖG, Soyer Y, Alpas H (2014) Rapid and standardized methods for detection of foodborne pathogens and mycotoxins on fresh produce. Food Control 40:359–367. https://doi.org/10.1016/j.foodcont.2013.12.020

Yoo SM, Lee SY (2016) Optical biosensors for the detection of pathogenic microorganisms. Trends Biotechnol 34:7–25. https://doi.org/10.1016/j.tibtech.2015.09.012

Zeng Y, Zhu Z, Du D, Lin Y (2016) Nanomaterial-based electrochemical biosensors for food safety. J Electroanal Chem 781:147–154. https://doi.org/10.1016/j.jelechem.2016.10.030

Zhao X, Lin CW, Wang J, Oh DH (2014) Advances in rapid detection methods for foodborne pathogens. J Microbiol Biotechnol 24:297–312. https://doi.org/10.4014/jmb.1310.10013

Transcriptome: A Tool for Biotechnological Applications of Quorum Sensing Using Single Cell and Viruses

9

Bhagwan Rekadwad

Abstract

Cutting edge sequencing (NGS) technology is a sound methodologically for exploration of information of to carry out advanced research in the various field of biology such as microbiology, biotechnology, agricultural microbiology, microbial ecology and community analyses for determination of cellular activities and gene expression under adverse environmental conditions. For the transcriptome analyses and its quantification, RNA-Seq has provided unlimited access to modern bio-analysis. This chapter presents an awful description of quorum sensing, quorum quenching, transcriptome analyses, NGS and correlation as well as an association of microorganism with other organisms such as human, plants, animal, microorganisms (eukaryotes and prokaryotes) and viruses are explained as well. Thus, transcriptome analysis widens the possibilities to get more in-depth/to get more top to bottom information about the modern RNA world in genetically similar cells or in single cell and viruses.

Keywords

Quorum sensing · Quorum quenching · Plant-microbe interactions · Next generation sequencing · Microbiome · Single cell genomics

B. Rekadwad (✉)
National Centre for Microbial Resource, National Centre for Cell Science, Pune, India

V. C. Kalia (ed.), *Quorum Sensing and its Biotechnological Applications*,
https://doi.org/10.1007/978-981-13-0848-2_9

Abbreviations

Dtr machinery	DNA transfer and replication machinery
HOHL	N-(3-hydroxy-octanoyl)-homoserine lactone (3OH,C8-HSL)
T4SS	Type-4 secretion system

9.1 Introduction

Microorganisms are inescapable. They are discovered wherever on the planet in different environmental specialities. They could be useful, pathogenic, saprophytic or advantageous in nature. Microorganisms have particular resistance. They are equipped for creating assortments of little or vast biomolecules either intracellular oozed outside or extracellular fit for infiltrating inside theirselves or others incorporate proteins, catalysts, chitin, exopolysaccharides, peptidoglycan which have the part in the start of resistance or immunity. Higher living beings (Human, plants and creatures or animals) and microorganisms are related with endophytic microscopic organisms in plants and microorganisms or human and creatures as the microbiome, particularly in the gut. Microorganisms capable of triggering persistent and apparent defence mechanism which do not harm to the host or producers (Fig. 9.1). Such microorganisms have better adjustment and adaptability with quickly changing unfriendly conditions. Microbial cooperation i.e. relationship with human, creature (animals- native and genetically modified animals), plants (native and genetically modified animals) and microorganisms (both eukaryotes and prokaryotes) are controlled systemically and managed or auto-directed by independent assortment of genes worked by the data put away in genes particularly in RNA. In this way, RNA sequencing is speedy and favored strategies for the top to the bottom portrayal of parts of transcriptome and their action in various ecological unfavorable conditions. The "transcriptomology" is defined as "utilization of cutting edge sequencing technology to assure total mRNAs from the genes of organisms expressed in hostile conditions in same cell or organisms in same time and vice versa". There are some possibilities we won't deny the involvement of virus genome (DNA/RNA) or virus gene involved in may help/modify/disrupt functions of transcriptome while expression of genes in quorum sensing and quorum quenching. Evaluation, monitoring and expression of thousands of genes albeit provide an unexplained view of working adaptive responses of all together total genes in response to extrinsic and intrinsic stimuli. This part depicts few of these uses of advances for viable applications spreading over biotechnology field.

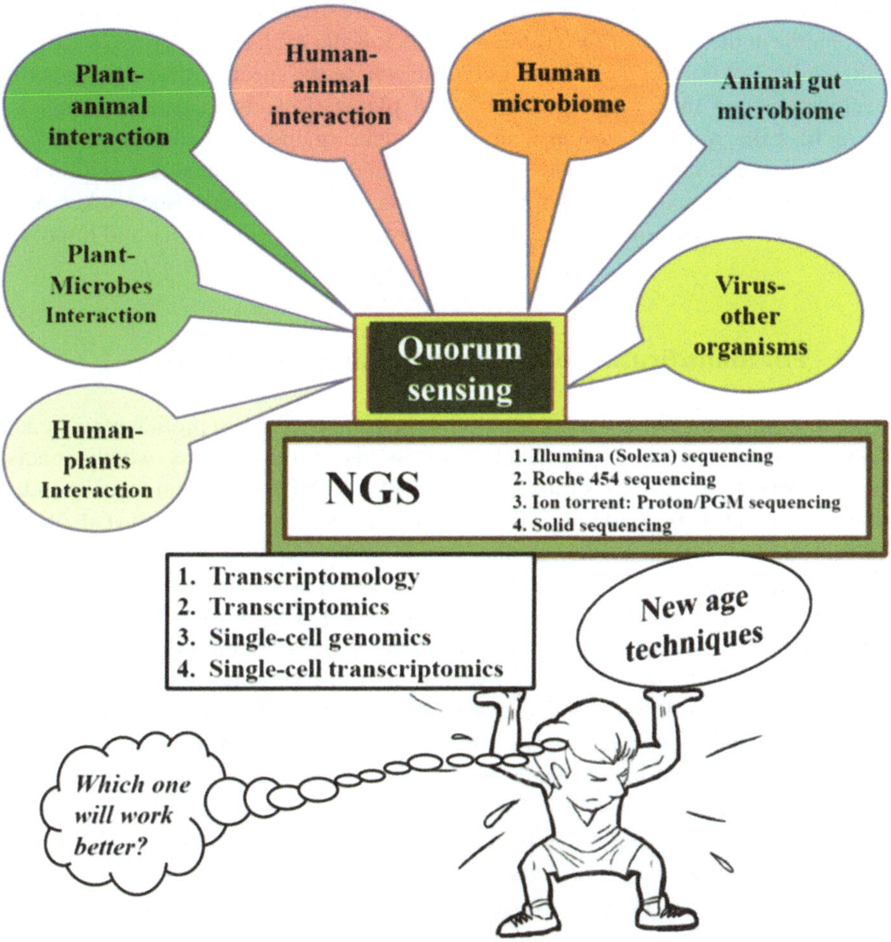

Fig. 9.1 Genealogy of new generation sequencing for quorum sensing and applications in biotechnology

9.2 Role of Transcriptome Analysis

9.2.1 In Revealing Plant-Microbe-Human and Human-Microbe-Animal Association

Microorganisms especially bacteria are associated with rhizosphere and in the rhizospheric soil, with phyllosphere and endosphere specifically as the microbiome. These entire have beneficial, detrimental or many times neutral associations influence always on health and development of plants (Chaparro et al. 2014; Kalia 2014, 2015; Mendes et al. 2014; Kalia et al. 2014; Kalia and Kumar 2015; Kothari et al. 2016). Next generation tolls such as meta transcriptomics and metabolomics are powerful tools which can generate great and comprehensive frontline of actively involved microbiome at the specific time (Kaur et al. 2015; Kalia et al. 2017; Kim et al. 2017).

Regardless of interaction with indirectly, they have the specific role in the growth, development and increasing productivity of plants (Rekadwad and Khobragade 2017a, b). Like, quorum sensing with plants microorganisms inhabiting gut (especially human gut, carnivores and grazing animals) possesses its core microbiome occupied from the external environment. These have beneficial effects takes part in localization or disposal or degradation of unwanted large molecules and improve health.

9.2.2 For Identification of Quorum Sensing Regulation

Agrobacterium tumefaciens strain P4 (ATP4) is the natural strain proficient to create AHls (Hao et al. 2012) as a product of its metabolic process which enacts *Agrobacterium tumefaciens* strain NT1 (ATNT1). ATNT1 is quorum sensing indicator strain reacts (sense) to long chain AHLs (Kim et al. 2015; Schikora et al. 2016; Prateeksha et al. 2017). While, another strain *Chromobacterium violaceum* strain CVO26 is quorum sensing marker (indicator strain) which sense short chain AHLs (Gonzalez and Keshwan 2006; Zhu et al. 2011). The TLC spot shape and Rf esteem demonstrate that quorum sensing molecule secreted by ATP4 might be HOHL. The genome of *Agrobacterium tumefaciens* strain P4 (ATP4) contains *luxI* like quality which displays 783 bp ORF and encodes a protein (29.6 kDa; 260 amino acids). This obscure quality was named as *cinI* quality. The inclusion mutagenesis in the *cinI* quality and transcriptome examinations allowed the recognizable proof of 32 *cinI*-controlled qualities in ATP4. Proteins encoded by these genes are responsible for the conjugative transfer (Mondy et al. 2013; Cabezón et al. 2015) of ATP4. The targeted in this study were *cinR, cinI, cinX, avhB5, avhb11, AGROTU_05920, traA, traG, gyrB, blcR* etc. These genes perform distinctive functions. The *avhB* gene encodes a T4SS which forms conjugation apparatus. *tra* gene encodes *Dtr* machinery, *cinI* and two (2) *luxR* orthologs. Conjugation experiments confirm the conjugative transfer of pAtP4 is regulated by HOHL. Root colonization showed that quorum sensing regulation of the conjugation of the pAtP4 does not present a pickup or lost wellness to the bacterial host in the tomato plant rhizosphere (Hartmann et al. 2014; Mhedbi-Hajri et al. 2016).

9.2.3 For Detection of Cell-Cell (Bacteria-Other Prokaryotes/ Eukaryotes) Communication

Microscopic organisms are much wiser than we can consider. They embrace diverse survival procedures to make their life agreeable (Seymour et al. 2017). Investigates on bacterial correspondence to date recommend that microorganisms can speak with each other utilizing compound flagging atoms and additionally utilizing particle channel intervened electrical flagging. Despite the fact that in recent decades the extents of compound flagging have been explored widely, those of electrical flagging have gotten less consideration. Bacteria timely conduct and share information and develop biofilm through quorum sensing and electrical signalling under low nutrient supplementation (Kalia 2014; Koul et al. 2016; Koul and Kalia 2017; Majumdar and Pal 2017; Ray and Kalia 2017). This brings up the issue of whether individual bacterial ancestries react to the nearness of their nearby relatives by changing their gene expression or, rather, regardless of whether gatherings essentially go about as the arithmetic addition of their individual segments. The transcriptome sequencing helps to answer the raised question (Kumar et al. 2017). The transcriptomes of two firmly related strains of extreme halophile *Salini bacterruber* developed axenically and in dual culture/coculture. The strains utilized here co-occurred in the indigenous habitat and are 100% indistinguishable in their 16S rRNAgene, and each strain harbors an extra genome about 10% of its total genome. Generally, transcriptomic designs from unadulterated societies were fundamentally the same as for both strains. Expression was recognized along for all intents and purposes the entire genome but with a few qualities at low levels. A subset of qualities was profoundly communicated in both strains, including qualities coding for the light-driven proton (H^+) pump xanthorhodopsin, gene taking part in the anxiety reaction, and genes coding for transcriptional controllers(González-Torres et al. 2015; Krause et al. 2017). Expression contrasts between unadulterated societies influenced fundamentally qualities required in ecological detecting. At the point when the strains were developed in coculture, there was an unobtrusive however critical change in their individual interpretation designs contrasted with those in unadulterated culture. Each strain detected the nearness of the other and reacted in a particular way, which focuses to fine intraspecific transcriptomic tweak (Charbonneau et al. 2017; Long et al. 2017).

9.2.4 Role in Regulation of Nitrogen Oxide(s) Fluxes in Nitrification Process

Varieties of microorganisms especially bacteria produce/sense diffusible molecules in the form of chemical signals include acyl-homoserine lactones (AHLs). Bacteria participating in the conversion of ammonia to nitrate through nitrite during aerobic oxidation process produce AHLs. These nitrogen oxides (NO_2, N_2O, NO) takes part in global warming. These greenhouse gases produced during microbial processes have an impact in microbial culture system such production of important enzymes, protein or small biochemical or macromolecules. Hence, inhibition of nitrogen oxide production is necessary to avoid the issues created by inhibitory molecules (Mellbye et al. 2016). Thus, there should inhibition of quorum sensing mechanism (Basavaraju et al. 2016). This has important role in the control of biofouling like situations. Such type of issues and problem would be overcome by using modern techniques such as the quorum quenching transcriptomic technique for the in-depth analyses of quorum sensing bacteria participating in nitrogen cycle such as *Nitrobacter winogradskyi*. It was revealed that there is link between nitrogen oxides metabolism and quorum sensing in *N. winogradskyi* and *N. winogradskyi* like nitrifying microorganisms. Quorum quenching transcriptomics experiments helps to understand that *N. winogradskyi* and *N. winogradskyi* like nitrifying bacteria acts as nitrogen sinks or source of nitrogen oxides and nitrous oxides (Pérez et al. 2015; Sayavedra-Soto et al. 2015; Feltner et al. 2016; Shen et al. 2016; Welsh and Blackwell 2016).

9.3 Single-Cell Genomics (SSGs) and Single-Cell Transcriptomics (SSTs) for Assessing Biology of Single-Cell (SC)

Strategies, for example, microarrays and all the more as of late cutting edge sequencing are broadly used to comprehend the relationship amongst phenotype and genotype. Apparently, it has provided glimpses at the genome-wide scale (Tang et al. 2009; Picelli et al. 2013; Shapiro et al. 2013; Deng et al. 2014; Wu et al. 2014). NGS methods, for example, single-cell transcriptome (SST) sequencing system and single-cell genome (SSG) sequencing strategy are the promising genetic tools giving the measurement of genes and their inconsistent expression among individual or same cell. Be that as it may, these have a few constraints i.e. it will evaluate either genome or single-cell. Presently, there no any utilitarian innovative and functional technology is available to investigate and correlation amongst genes and their constant expression in the same cell (Grün et al. 2014; Junker and van Oudenaarden 2014). Here we portray a strategy to at evaluate both the genome and transcriptome of a similar cell in the same cell and at the same time. Since the bacterial mRNA has the half-life around a couple of minute shifts from not as much as the single moment to 20 min. In eukaryotes like human normal half-existence of mRNA is around 10-h changing between 30 min and 24 h (Sciencing; http://sciencing.com/degradation-mrna-6196816.html). This would be conceivable to some degree through sequencing both genomic DNA (gDNA) and mRNA from a similar cell permits to compare genome variability and transcriptome heterogeneity of the same cell. The quasilinear intensification gDNA and mRNA permitted to discover the genes with high cell-to-cell fluctuation in transcript numbers, by and large, have brought down genomic duplicate numbers, and the other way around, recommending that copy number may drive inconstancy in gene expression among individual cells (Dey et al. 2015; Wei et al. 2016; Shirota and Kinoshita 2016; Guan and Rosenecker 2017). Utilisation of modern sequencing methodology could extend from picking up bits of knowledge into, quorum sensing, bacterial evolution, and human diseases, especially in tumor biology.

9.4 Applications of QS in Decision Taking Cell Through Viral Genome for Defence or Resistance or Eradication of Disease

Signalling molecules such as AHLs in many Gram-negative bacteria are produced in quorum sensing. In vivo these molecules may trigger selective response by which bacteria can act in such a way that cell infected by lytic virus may be turned to lysogenic phase due to induction of alternative paradigm for prophage induction due to signal produced by high number of host cell either infected or healthy (Ghosh et al. 2009; Høyland-Kroghsbo et al. 2013). This may happen due to the presence of bacterial genes in incorporated into the viral genome. For instance, bacteriophage phiCDHM1 is a virus of *Clostridium difficile*. It is closely related to the mycoviruses which infect *C. difficile*. It was discovered that it has many genes (*ArcB, ArcC, ArcD*) which are ever not reported in any bacteriophage. The discovery of quorum sensing genes virus/phage homologous with their hot could influence their bacterial abundance/population. The transfer of genes/genetic material through horizontal gene transfer and distribution within the species may results in alteration of functions or production some specific molecules (Hargreaves et al. 2014). As stated earlier, viruses/phages belong to the SPbeta group can produce small molecules for communication by which they would capable of coordinating between them and take the decision on cell cycle to be carried out for infected cell i.e. lytic or lysogenic decisions on infected cell. Infecting viruses use a phage/virus specific peptide communication code for taking the lysogenic decision. This is called as the arbitrium system. Arbitrium system has three genes- *aimP* (for peptide production), *aimR* (production of intracellular peptide receptor) and *aimX* (for negative regulation of lysogeny). This system allows descendant bacteriophages to communicate and take decision its predecessors so that they decide to take either take up the lytic or lysogenic cycle (Erez et al. 2017).

References

Basavaraju M, Sisnity VS, Palaparthy R, Addanki PK (2016) Quorum quenching: signal jamming in dental plaque biofilms. J Dent Sci 11:349–352. https://doi.org/10.1016/j.jds.2016.02.002

Cabezón E, Ripoll-Rozada J, Peña A, de la Cruz F, Arechaga I (2015) Towards an integrated model of bacterial conjugation. FEMS Microbiol Rev 39:81–95. https://doi.org/10.1111/1574-6976.12085

Chaparro JM, Badri DV, nVivanco JM (2014) Rhizosphere microbiome assemblage is affected by plant development. ISME J 8:790–803. https://doi.org/10.1038/ismej.2013.196

Charbonneau ARL, Forman OP, Cain AK, Newland G, Robinson C, Boursnell M, Parkhill J, Leigh MDJ, Waller AS (2017) Defining the ABC of gene essentiality in streptococci. BMC Genomics 18:426. https://doi.org/10.1186/s12864-017-3794-3

Deng Q, Ramsköld D, Reinius B, Sandberg R (2014) Single-cell RNA-seq reveals dynamic, random monoallelic gene expression in mammalian cells. Science 343:193–196. https://doi.org/10.1126/science.1245316

Dey SS, Kester L, Spanjaard B, Bienko M, van Oudenaarden A (2015) Integrated genome and transcriptome sequencing of the same cell. Nat Biotechnol 33:295–292. https://doi.org/10.1038/nbt.3129

Erez Z, Steinberger-Levy I, Shamir M, Doron S, Stokar-Avihail A, Peleg Y, Melamed S, Leavitt A, Savidor A, Albeck S, Amitai G, Sorek R (2017) Communication between viruses guides lysis–lysogeny decisions. Nature 541:488–493. https://doi.org/10.1038/nature21049

Feltner JB, Wolter DJ, Pope CE, Groleau M, Smalley NE, Greenberg EP, Mayer-Hamblett N, Burns J, Déziel E, Hoffman LR, Dandekar AA (2016) LasR variant cystic fibrosis isolates reveal an adaptable quorum sensing hierarchy in *Pseudomonas aeruginosa*. MBio 7:e01513-16. https://doi.org/10.1128/mBio.01513-16

Ghosh D, Roy K, Williamson KE, Srinivasiah S, Wommack KE, Radosevich M (2009) Acyl-homoserine lactones can induce virus production in lysogenic bacteria: an alternative paradigm for prophage induction. Appl Environ Microbiol 75:7142–7152. https://doi.org/10.1128/AEM.00950-09

Gonzalez JE, Keshwan ND (2006) Messing with bacterial quorum sensing. Microbiol Mol Biol Rev 70:859–875. https://doi.org/10.1128/MMBR.00002-06

González-Torres P, Pryszcz LP, Santos F, Martínez-García M, Gabaldón T, Antón J (2015) Interactions between closely related bacterial strains are revealed by deep transcriptome sequencing. Appl Environ Microbiol 81:8445–8456. https://doi.org/10.1128/AEM.02690-15

Grün D, Kester L, van Oudenaarden A (2014) Validation of noise models for single-cell transcriptomics. Nat Methods 11:637–640. https://doi.org/10.1038/nmeth.2930

Guan S, Rosenecker J (2017) Nanotechnologies in delivery of mRNA therapeutics using nonviral vector-based delivery systems. Gene Ther. https://doi.org/10.1038/gt.2017.5

Hao X, Xie P, Johnstone L, Miller SJ, Rensing C, Wei G (2012) Genome sequence and mutational analysis of plant-growth-promoting bacterium *Agrobacterium tumefaciens* CCNWGS0286 isolated from a Zinc-Lead mine tailing. Appl Environ Microbiol 78:5384–5394. https://doi.org/10.1128/AEM.01200-12

Hargreaves KR, Kropinski AM, Clokie MRJ (2014) What does the talking? Quorum sensing signalling genes discovered in a bacteriophage genome. PLoS One 9:e85131. https://doi.org/10.1371/journal.pone.0085131

Hartmann A, Rothballer M, Hense BA, Schröder P (2014) Bacterial quorum sensing compounds are important modulators of microbe-plant interactions. Front Plant Sci 5:131. https://doi.org/10.3389/fpls.2014.00131

Høyland-Kroghsbo NM, Mærkedahl RB, Svenningsen SL (2013) A quorum-sensing-induced bacteriophage defense mechanism. MBio 4:e00362-12. https://doi.org/10.1128/mBio.00362-12

Junker JP, van Oudenaarden A (2014) Every cell is special: genome-wide studies add a new dimension to single-cell biology. Cell 157:8–11. https://doi.org/10.1016/j.cell.2014.02.010

Kalia VC (2014) Microbes, antimicrobials and resistance: the battle goes on. Indian J Microbiol 54:1–2. https://doi.org/10.1007/s12088-013-0443-7

Kalia VC (2015) Microbes: the most friendly beings? In: Kalia VC (ed) Quorum sensing vs quorum quenching: a battle with no end in sight. Springer, New Delhi, pp 1–5. https://doi.org/10.1007/978-81-322-1982-8_1

Kalia VC, Kumar P (2015) Genome wide search for biomarkers to diagnose *Yersinia* infections. Indian J Microbiol 55:366. https://doi.org/10.1007/s12088-015-0552-6

Kalia VC, Kumar P, Pandian SK, Sharma P (2014) Biofouling control by quorum quenching. Hb_25. In: Kim SK (ed) Springer handbook of marine biotechnology, vol 15. Springer, Berlin, pp 431–440

Kalia VC, Prakash J, Koul S, Ray S (2017) Simple and rapid method for detecting biofilm forming bacteria. Indian J Microbiol 57:109–111. https://doi.org/10.1007/s12088-016-0616-2

Kaur G, Rajesh S, Princy SA (2015) Plausible drug targets in the *Streptococcus* mutans quorum sensing pathways to combat dental biofilms and associated risks. Indian J Microbiol 55:349. https://doi.org/10.1007/s12088-015-0534-8

Kim HS, Lee S, Byun Y, Park HD (2015) 6-Gingerol reduces Pseudomonas aeruginosa biofilm formation and virulence via quorum sensing inhibition. Sci Rep 5:8656. https://doi.org/10.1038/srep08656

Kim MK, Zhao A, Wang A, Brown ZZ, Muir TW, Stone HA, Bassler BL (2017) Surface-attached molecules control *Staphylococcus aureus* quorum sensing and biofilm development. Sci Rep 2:17080. https://doi.org/10.1038/nmicrobiol.2017.80

Kothari R, Singh RP, Kothari V (2016) Application of next generation sequencing technologies in revealing plant-microbe interactions. Next Gener Seq Applic 3:1. https://doi.org/10.4172/2469-9853.1000e108

Koul S, Kalia VC (2017) Multiplicity of quorum quenching enzymes: a potential mechanism to limit quorum sensing bacterial population. Indian J Microbiol 57:100–108. https://doi.org/10.1007/s12088-016-0633-1

Koul S, Prakash J, Mishra A, Kalia VC (2016) Potential emergence of multi-quorum sensing inhibitor resistant (MQSIR) bacteria. Indian J Microbiol 56:1–18. https://doi.org/10.1007/s12088-015-0558-0

Krause S, Bremges A, Münch PC, McHardy AC, Gescher J (2017) Characterisation of a stable laboratory co-culture of acidophilic nanoorganisms. Sci Rep 7:3289. https://doi.org/10.1038/s41598-017-03315-6

Kumar R, Verma H, Haider S, Bajaj A, Sood U, Ponnusamy K, Nagar S, Shakarad MN, Negi RK, Singh Y, Khurana JP, Gilbert JA, Lal R (2017) Comparative genomic analysis reveals habitat-specific genes and regulatory hubs within the genus *Novosphingobium*. mSystems 2:e00020-17. https://doi.org/10.1128/mSystems.00020-17

Long SW, Olsen RJ, Eagar TN, Beres SB, Zhao P, Davis JJ, Brettin T, Xia F, Musser (2017) Population genomic analysis of 1,777 extended-spectrum beta-lactamase-producing *Klebsiella pneumoniae* isolates, Houston, Texas: unexpected abundance of clonal group 307. MBio 8:e00489-17. https://doi.org/10.1128/mBio.00489-17

Majumdar S, Pal S (2017) Bacterial intelligence: imitation games, time-sharing, and long-range quantum coherence. J Cell Commun Signal. https://doi.org/10.1007/s12079-017-0394-6

Mellbye BL, Giguere AT, Bottomley PJ, Sayavedra-Soto LA (2016) Quorum quenching of *Nitrobacter winogradskyi* suggests that quorum sensing regulates fluxes of nitrogen oxide(s) during nitrification. MBio 7:e01753-16. https://doi.org/10.1128/mBio.01753-16

Mendes LW, Kuramae EE, Navarrete AA, Van Veen JA, Tsai SM (2014) Taxonomical and functional microbial community selection in soybean rhizosphere. ISME J 8:1577–1587. https://doi.org/10.1038/ismej.2014.17

Mhedbi-Hajri N, Yahiaoui N, Mondy S, Hue N, Pélissier F, Faure D, Dessaux Y (2016) Transcriptome analysis revealed that a quorum sensing system regulates the transfer of the pAt megaplasmid in *Agrobacterium tumefaciens*. BMC Genomics 17:661. https://doi.org/10.1186/s12864-016-3007-5

Mondy S, Lalouche O, Dessaux Y, Faure D (2013) Genome sequence of the quorum-sensing-signal-producing nonpathogen *Agrobacterium tumefaciens* strain P4. Genome Announc 1:e00798–e00713. https://doi.org/10.1128/genomeA.00798-13

Pérez J, Buchanan A, Mellbye B, Ferrell R, Chang JH, Chaplen F, Bottomley PJ, Arp DJ, Sayavedra-Soto LA (2015) Interactions of *Nitrosomonas europaea* and *Nitrobacter winogradskyi* grown in co-culture. Arch Microbiol 197:79–89. https://doi.org/10.1007/s00203-014-1056-1

Picelli S, Björklund AK, Faridani OR, Sagasser S, Winberg G, Sandberg R (2013) Smart-seq2 for sensitive full-length transcriptome profiling in single cells. Nat Methods 10:1096–1098. https://doi.org/10.1038/nmeth.2639

Prateeksha SBR, Shoeb M, Sharma S, Naqvi AH, Gupta VK, Singh BN (2017) scaffold of selenium nanovectors and honey phytochemicals for inhibition of *Pseudomonas aeruginosa* quorum sensing and biofilm formation. Front Cell Infect Microbiol 7:93. https://doi.org/10.3389/fcimb.2017.00093

Ray S, Kalia VC (2017) Biomedical applications of polyhydroxyalkanaotes. Indian J Microbiol 57:39–47. https://doi.org/10.1007/s12088-016-0651-7

Rekadwad BN, Khobragade CN (2017a) Bacterial quorum sensing (QS) in rhizosphere (paddy soil): understanding soil signaling and N- recycling for increased crop production. In: Kalia VC (ed) Microbial applications vol.2 – biomedicine, agriculture and industry. Springer Nature, New Delhi, pp 119–131. https://doi.org/10.1007/978-3-319-52669-0_6

Rekadwad BN, Khobragade CN (2017b) Microbial biofilm: role in crop productivity. In: Kalia VC (ed) Microbial applications vol.2 – biomedicine, agriculture and industry. Springer International Publishing, Cham., AG, Springer Nature, pp 107–118. https://doi.org/10.1007/978-3-319-52669-0_5

Sayavedra-Soto L, Ferrell R, Dobie M, Mellbye B, Chaplen F, Buchanan A, Chang J, Bottomley P, Arp D (2015) *Nitrobacter winogradskyi* transcriptomic response to low and high ammonium concentrations. FEMS Microbiol Lett 362:1–7. https://doi.org/10.1093/femsle/fnu040

Schikora A, Schenk ST, Hartmann A (2016) Beneficial effects of bacteria-plant communication based on quorum sensing molecules of the N-acyl homoserine lactone group. Plant Mol Biol 90:605–612. https://doi.org/10.1007/s11103-016-0457-8

Seymour JR, Amin SA, Raina JB, Stocker R (2017) Zooming in on the phycosphere: the ecological interface for phytoplankton–bacteria relationships. Nat Microbiol 2:17065. https://doi.org/10.1038/nmicrobiol.2017.65

Shapiro E, Biezuner T, Linnarsson S (2013) Single-cell sequencing-based technologies will revolutionize whole-organism science. Nat Rev Genet 14:618–630. https://doi.org/10.1038/nrg3542

Shen Q, Gao J, Liu J, Liu S, Liu Z, Wang Y, Guo B, Zhuang X, Zhuang G (2016) A new acyl-homoserine lactone molecule generated by *Nitrobacter winogradskyi*. Sci Rep 6:22903. https://doi.org/10.1038/srep22903

Shirota M, Kinoshita K (2016) Discrepancies between human DNA, mRNA and protein reference sequences and their relation to single nucleotide variants in the human population. Database (Oxford) 2016:baw124. https://doi.org/10.1093/database/baw124

Tang F, Barbacioru C, Wang Y, Nordman E, Lee C, Xu N, Wang X, Bodeau J, Tuch BB, Siddiqui A, Lao K, Surani MA (2009) mRNA-Seq whole-transcriptome analysis of a single cell. Nat Methods 6:377–382. https://doi.org/10.1038/nmeth.1315

Wei G, Yu J, Wang J, Gu P, Birch DS, Chen Y (2016) Hairpin DNA-functionalized gold nanorods for mRNA detection in homogenous solution. J Biomed Opt 21:097001. https://doi.org/10.1117/1.JBO.21.9.097001

Welsh MA, Blackwell HE (2016) Chemical probes of quorum sensing: from compound development to biological discovery. FEMS Microbiol Rev 40:774–794. https://doi.org/10.1093/femsre/fuw009

Wu AR, Neff NF, Kalisky T, Dalerba P, Treutlein B, Rothenberg ME, Mburu FM, Mantalas GL, Sim S, Clarke MF, Quake SR (2014) Quantitative assessment of single-cell RNA-sequencing methods. Nat Methods 11:41–46. https://doi.org/10.1038/nmeth.2694

Zhu H, He CC, Chu QH (2011) Inhibition of quorum sensing in Chromobacterium violaceum by pigments extracted from Auricularia auricular. Lett Appl Microbiol 52:269–274. https://doi.org/10.1111/j.1472-765X.2010.02993.x

Part II

Energy

Application of Quorum Sensing Systems in Production of Green Fuels

10

Jyotsana Prakash and Vipin Chandra Kalia

Abstract

Microorganisms have been used in diverse areas of biotechnology. The focus in recent times has been on exploiting the microbial communication for biofuel production. This communication known as Quorum sensing (QS) helps bacteria to sense their environments and enable them to survive in diverse habitats. QS based communication works through signal molecules. Exploiting the communication signals for the production of energy can help overcome the increasing energy crisis. A number of areas in energy sector including bio-hydrogen, biodiesel, bio-ethanol and bio-electricity production have started using QS for the improving the efficiency of theses bioprocesses. Here, we present recent advances in improving the efficiency of bioenergy production process by exploiting bacterial cell-cell communication.

Keywords

Biodiesel · Bioenergy · Bioethanol · Fuel cells · Hydrogen · Quorum sensing

J. Prakash (✉)
Microbial Biotechnology and Genomics, CSIR – Institute of Genomics and Integrative Biology (IGIB), Delhi University Campus, New Delhi, India

Academy of Scientific & Innovative Research (AcSIR), New Delhi, India
e-mail: jyotsana.prakash@igib.in

V. C. Kalia
Molecular Biotechnology Lab, Department of Chemical Engineering, Konkuk University, Seoul, Republic of Korea
e-mail: vckaliaku@gmail.com

V. C. Kalia (ed.), *Quorum Sensing and its Biotechnological Applications*,
https://doi.org/10.1007/978-981-13-0848-2_10

10.1 Introduction

The interaction of microbes with plants, animals and among themselves has served numerous purposes. Such interactions have paved the ways to develop efficient bioremediation, bioenergy production, agricultural practices and in preventing bacterial infections (Deshmukh et al. 2016; Lun et al. 2016; Valim et al. 2016; Ahiwale et al. 2017; Das et al. 2017). In recent times, biological processes in food, agriculture, medical or energy sector, have gained immense interest (Begum et al. 2016; Hernández-Saldaña et al. 2016; Varsha et al. 2016; Saini and Keum 2017). The main reasons are rapid depletion of fossil fuels and increasing deterioration of the environment due to the use of chemical processes.

Talking about energy production, emissions from electricity generation, coal mining, oil refineries and gas extraction cause an immense pollution (Jaramillo and Muller 2016). This makes such processes worthy of condemnation when healthy environment is a priority. However, with increasing industrialization, the world's energy demand is supposed to increase up to 18 billion tonne of oil equivalent (Btoe) by 2035 (International Energy Agency 2011). The increase in energy production will be associated with increased greenhouse gas emissions, posing a huge risk to human survival. The requirement is thus of an alternative, clean and sustainable energy source. Among renewable energy sources, biological energy production holds promise to provide a rescue from such crisis.

Biological energy production uses biomass, which is converted by microbes into a range of products including ethanol, diesel, hydrogen (H_2) and methane (Cavinato et al. 2011; Patel and Kalia 2013; Yasin et al. 2013; Arasu et al. 2015; Bandyopadhyay et al. 2015; Gomma et al. 2015; Shiva Krishna et al. 2015). The brilliantly armed bacteria can utilize a whole range of substrates as food materials (Go et al. 2015; Spier et al. 2015; Kurm et al. 2017). These bacteria once thought to be mute organisms are now well known for their efficient social interactions. These bacteria can sense their environment and through cell-cell communication can modulate their gene expressions. Such a bacterial interaction is known as Quorum sensing (QS) (Montgomery et al. 2013; Kalia et al. 2014). Humans have evolved to exploit their surroundings for self-benefit and with such abilities they have not let bacterial world untouched. The amazing bacterial communication is well researched and is exploited for human benefits. Energy production is one such area that has attracted microbial support. Here, we deal with microbial communication as an aid to bioenergy production.

10.2 Applications of Quorum Sensing in Bioenergy

Microbes produce a host of bioactive molecules and undergo physiological and metabolic changes to withstand environmental pressures (Sanchart et al. 2017; Sharma and Lal 2017; Thakur et al. 2017). One of the most interesting strategies adopted by certain microbes is to change their behaviour from a single celled organism to a "multi-cellular" organism. This shift in behaviour is regulated through a cell density dependent system – QS system (QSS) (Montgomery et al. 2013). It has been observed to operate in various microbial activities such as bioluminescence, bio-film formation, virulence, antibiotic production, swarming, conjugation, toxin secretion, exo-enzyme production, bio-corrosion, and symbiosis (Kalia 2013). QS operates via diffusible signals, where these molecules allow microbes to sense their outer environment and communicate with other microbes by making appropriate responses. These signals and responses represent bacterial languages, which might be used for species specific communication or for inter-species correspondence. Microbes, especially pathogens have a repertoire of multiple QSS to communicate at personal/commercial levels (Koul et al. 2016).

Bacteria have developed QSS for their personal usage. However, it has been realized that by defining the physiological conditions under which QSS operates, the same can be exploited for human benefits. A few areas, where QS has been found to have the potential to generate bioproducts of high values include bioenergy, waste treatment, food preservatives, biosensors, health, and agricultural activities (Kalia and Purohit 2011; Pastorella et al. 2012; Kalia 2013; Kaur et al. 2015).

10.2.1 Bio-hydrogen Production

Bio-H_2 has been recognised as the cleanest fuel for the future. Fermentative H_2 production has made significant advancements using diverse microbes (Kalia et al. 2003, 2016; Porwal et al. 2008; Kumar (P) et al. 2013, 2015a; Patel et al. 2015). However, H_2 yields have been found to be quite low and virtually stagnant in a narrow range of 0.3–3.8 mole/mole hexose sugars like glucose (Patel et al. 2012). Efforts to retain large bacterial population within the bioreactors have proven effective in enhancing H_2 yields. An obviously effective approach has been to immobilize H_2-producers on different support materials (Patel et al. 2010). However, a more innovative strategy can be the use of self-flocculating or biofilm forming bacteria. These hold greater promise to retain large population of fermenting organisms. Within the bioreactor, exopolysaccharides (EPS) secreting microbes allow large population of bacteria to be entrapped within the mucilage, which are thus prevented from being washed away (Keskin et al. 2012; Abe et al. 2013; Ercan and Demirci 2015).

Among a number of potential H_2-producers, quite a few of them have an ability to express QS mediated biofilm formation, which include bacterial species belonging to *Bacillus, Clostridium, Streptococcus, Sinorhizobium, Enterobacter, Klebsiella, Caldicellulosiruptor* and *Escherichia* (Kalia and Purohit 2008). Biofilm formers can be easily identified using simple screening tests (Kalia et al. 2017). These biofilms can be exploited by using them as support for immobilizing microbes within the bioreactors for effective H_2 production. Co-cultures of thermophillic bacteria, *Caldicellulosiruptor* species have been used for biofilm formation to enhance H_2 production. The two bacteria together resulted in 2.5 times enhanced H_2 yield and 5 times higher H_2 productivity to the tune of 20 mmol/L/h at a dilution rate of >1.0 h^{-1} in Up-flow anaerobic reactors, in comparison to individually employed cultures. These enhancements in H_2 productivity and yield were reported to be due to an increased biofilm formation in co-cultures, which is dependent upon QS mediated high production of c-di-GMP (Table 10.1) (Brune and Bayer 2012; Pawar et al. 2015). In another recent study, biofilm forming *Bacillus amyloliquefaciens* strain CD16 was shown to produce 1.18 times more H_2 than the non-biofilm forming counterpart *Bacillus thuringenesis* strain EGU45 in a continuous culture system (Prakash et al. 2017). The two strains had similar H_2 producing efficiency in a batch system. However, when switched to continuous mode the biofilm forming bacteria proved to be more effective in retaining larger cell mass in the system and resulted in higher yields of H_2 (Prakash et al. 2017).

Biofilm formation also holds opportunity in increasing H_2 production through Microbial electrolytic cells (MECs). Biofilm in these fuel cells may prevent the microbial dispersion and hence increase the effective gas production. QS can be used as an effective strategy to form and control biofilm formation in such fuel cells for their efficient performance (Zhou et al. 2013). Addition of small chain length acyl-homoserine lactones (AHLs – QS signalling molecules) to regulate biofilm formation on bioelectrodes in MECs has shown to enhance H_2 yields. An increase of 81.82% in H_2 yield was recorded at an external voltage of 0.4 V. Overall reactor performance including energy efficiency, electron recovery efficiency and coulombic efficiency were also enhanced (Cai et al. 2016).

Apart from biofilm formation, genetic approaches may be used to redirect the metabolic pathways towards the synthesis of desired products. RNA*seq* of *Rhodopseudomonas palustris* CGA009 has revealed the presence of a transcript antisense to QS regulatory protein rpaR. Such intergenic regions could have regulatory roles in microbes, engineering of which can redirect the metabolic flux towards the H_2 production (McKinlay 2014).

10.2.2 Bio-electricity Production by Fuel Cells

An attractive way to generate electricity is through bio-electrochemical devices known as microbial fuel cells (MFCs) (Manogari and Daniel 2015). The electrons generated by the microbial oxidation of substrate are transferred to the electrodes for electricity generation. This transfer of electrons require electron shuttles either produced by the microbes or added exogenously (Kumar (R) et al. 2015b). To avoid the use of exogenous addition of electron shuttles, exoelectrogens can be used. Exoelectrogens are the microbes capable of producing molecules that can exogenously transfer the electrons to electrodes (Kumar (R) et al. 2015b). QS plays an important role in the development of biofilms in a number of exoelectrogens (Diggle et al. 2003; Allesen-Holm et al. 2006; Kumar (R) et al. 2015b). The formation of biofilm in MFCs has been reported to increase the current output substantially. Microbes able to produce thick biofilms have proved to be potent current producers than those producing monolayer biofilms (Wrighton et al. 2011; Zhou et al. 2013).

QS mediated production of pyocyanins and phenazines are known to act as redox mediators in MFCs. These have been reported to increase the energy output from 5% to 50% in biofuel cells (Table 10.1) (Rabaey et al. 2005). A wild type *Pseudomonas aeruginosa* PA14 used for bio-electricity generation showed a 28-fold

Table 10.1 Potential applications of quorum sensing in production of green fuels

Organism	QS mediated effect	Fuel production	References
Bio-hydrogen			
Caldicellulosiruptor saccharolyticus and *C. owensensis*	C-di-GMP production and increase in biofilm formation	2.5 times enhanced H_2 yield and 5 times higher productivity	Pawar et al. (2015)
Bacillus amyloliquefaciens strain CD16	Biofilm formation	1.18 times more H_2 production than non-biofilm former	Prakash et al. (2017)
Mixed microbial culture	Addition of AHLs for biofilm regulation on MEC's bioelectrodes	81.82% increase in yields	Cai et al. (2016)
Bio-electricity			
Geobacter sulfurreducens	Biofilm formation	Current increased from 1.4 to 5.2 mA with increase in biofilm pillar height from 10 to 40 μM	Kumar (R) et al. (2015b)
Pseudomonas aeruginosa	Pyocyanins and phenazines production	Increased energy output from 5% to 50%	Rabaey et al. (2005)
P. aeruginosa PA14	Lactones, phenazines and *retS* mutant	28 to 48-fold increase in current generation	Venkataraman et al. (2010)
P. aeruginosa	Overexpression of *phzM* – pyocyanins production	Enhanced power output up to 166.68 μW/cm^2 by MFC	Yong et al. (2014)
Shewanella oneidensis MR-1	AHLs regulated P*tac* controlled *LuxR*	Regulated current generation	Hu et al. (2015)
Halanaerobium praevalens	Biofilm formation by exogenous addition of 100 nM quinolone	30% increase in energy generation	Monzon et al. (2016)
Bio-ethanol			
Wild yeasts and Lactic acid bacteria (LAB)	Biocides production	Killing of wild yeasts and lactic acid bacteria increasing ethanol producing yeast population	Brexó and Sant'Ana (2017)
Saccharomyces cerevisiae	Upregulation of transcription factors Mig1p and Cat8p	Increased stress tolerance by yeast	Westman and Franzén (2015)
Zymomonas mobilis	Increased signalling by addition of AI-2	50% enhancement in fuel production	Yang (2011)
Bio-diesel			
Algal cells	QS containing *Escherichia coli* symbiotically associated with algal cells	Nitrogen stress induced biodiesel production	Wyss (2013)
Butanediol			
Aeromonas hydrophila AH-1N	Exogenous AHLs addition	Fivefold increase in productivity	Houdt et al. (2007)

(continued)

Table 10.1 (continued)

Organism	QS mediated effect	Fuel production	References
Acetic acid			
Gluconacetobacter intermedius	Blocking of QS system GinI/GinR	30% enhancement in productivity	Lida et al. (2008)
Isobutanol			
E. coli	Blocking QS genes *luxS* or *lsrA*	Threefold increase in yield	Huo et al. (2011)

increase in current output with exogenous addition of QS signalling molecules (lactones). The enhanced phenazines production by the strain resulted in 48-fold increase in current generation. A 45-fold increase in current generation by *retS* mutant was also observed clearly indicating the effect of QS on the production of current (Venkataraman et al. 2010). The study thus paves the way for the development of high energy generating biofuel cells. A number of other studies have also shown an increased energy output with the over-production of phenazines or pyocyanins in *Pseudomonas* (Yong et al. 2014, 2015). One of such study reported 1.6-fold increase in pyocyanins (electron shuttles) production by overexpression of *phzM* in *P. aeruginosa.* This caused an enhanced power output up to 166.68 $\mu W/cm^2$ by MFC (Yong et al. 2014).

Genetic engineering approaches have been used to design MFCs, the electricity generation of which is controlled by exogenous addition of QS signals. The engineered MFC contains IPTG controlled P*tac* promoter and P*tac* controlled *LuxR* (QS signal regulator) in *Shewanella oneidensis* MR-1 (a*mtrA* knockout mutant). This AND logic gate construction allows the current generation in the presence of AHLs (QS signals) and Isopropyl β-D-1-thiogalactopyranoside (IPTG), thus allowing a regulated current production (Hu et al. 2015).

Engineered MFC that can generate electricity from high salinity wastewaters has been designed. The fuel cell used extremophile *Halanaerobium praevalens* for electricity generation. An exogenous addition of QS signals (100 nM quinolone) was used to enhance biofilm formation by the bacterium resulting in 90% increase in cell mass and 30% increase in energy density generated (Monzon et al. 2016).

10.2.3 Bio-ethanol and Bio-diesel Production

Another important class of biofuels is bioethanol and biodiesel. The biofuels being cleaner alternatives to fossil fuels have gained immense attraction. A number of studies suggest the potential involvement of QS in ethanol fermentation (Kuipers et al. 1998; Di Cagno et al. 2011; Branco et al. 2014). Certain cyclic peptides and higher alcohols act as QS signalling molecules in yeasts. Such a signalling among wild yeasts and lactic acid bacteria (LAB) through QS tend to dominate ethanol producing yeasts population, causing loss to ethanol industry. Stimulation of biocides through QS in the fermentation might provide an economical alternative to prevent such losses (Table 10.1) (Brexó and Sant'Ana 2017). Another strategy to use QS for increasing ethanol yields is the regulation of QS responsive genes in *S. cerevisiae*. This might lead to stationary phase induction maintaining a lower biomass and higher ethanol production (Chen and Fink 2006; Dickinson 2008; Albuquerque and Casadevall 2012; Mas et al. 2014).

A number of stress tolerant genes are also reported to be activated by QS in *S. cerevisiae*. The aromatic alcohols, tryptophol and phenylethanol are the important QS signalling molecules in yeast (Hlavácek et al. 2009). These are known to upregulate transcription factors such as Mig1p and Cat8p, helping the high cell density yeast cells to tolerate the stress during ethanol production (Westman and Franzén 2015). QS mediated flocculation might also be involved in increased stress tolerance by ethanol producing yeast culture (Gasch et al. 2000; Zi et al. 2013; Conrad et al. 2014). Addition of QS signals AI-2 during bioethanol production by *Zymomonas mobilis* has shown a 50% enhancement in fuel production (Yang 2011). Genetic engineering approaches to create AI-2 synthesizing and ethanol producing strains may further stimulate biofuel production.

QS system when incorporated into *Escherichia coli*, has been used for biodiesel production by algal cells. The QS containing *E.coli* allowed the symbiotically associated algal cells to sense their high cell densities. This triggered nitrogen stress in algal cells inducing biodiesel production (Wyss 2013).

10.2.4 Others

QS has been exploited to increase the production efficiency of a number of other value added products. Exogenous addition of QS signalling molecules – AHLs has shown to increase butanediol production by fivefold in *Aeromonas hydrophila* (Houdt et al. 2007). Blocking QS system GinI/GinR in *Gluconacetobacter intermedius* resulted in enhanced bacterial growth leading to a 30% increase in acetic acid production (Lida et al. 2008). Deletion of QS genes *luxS* or *lsrA* has also shown to be associated with an increase in isobutanol production. An approximate increase of 66% in isobutanol yield was observed on blocking QS genes (Huo et al. 2011).

10.3 Conclusion

QS being an important means of communication in microbial world can help us exploit microbial metabolism. Inhibition of such microbial cross talk can help us fight infections. On the other hand, the same communication can lead to the production of various industrially important compounds. The increasing energy crisis demands energy production through cheap alternatives. QS has proved to be an effective strategy in the production of bioelectricity and biofuels.

Acknowledgments This work was supported by Brain Pool grant (NRF-2018H1D3A2001746) by National Research Foundation of Korea (NRF) to work at Konkuk University (VCK).

References

Abe A, Furukawa S, Watanabe S, Morinaga Y (2013) Yeasts and lactic acid bacteria mixed-species biofilm formation is a promising cell immobilization technology for ethanol fermentation. Appl Biochem Biotechnol 171:72–79. https://doi.org/10.1007/s12010-013-0360-6

Ahiwale SS, Bankar AV, Tagunde S, Kapadnis BP (2017) A bacteriophage mediated gold nanoparticle synthesis and their anti-biofilm activity. Indian J Microbiol 57:188–194. https://doi.org/10.1007/s12088-017-0640-x

Albuquerque P, Casadevall A (2012) Quorum sensing in fungi-a review. Med Mycol 50:337–345. https://doi.org/10.3109/13693786.2011.652201

Allesen-Holm M, Barken KB, Yang L, Klausen M, Webb JS, Kjelleberg S, Molin S, Givskov M, Tolker-Nielsen T (2006) A characterization of DNA release in *Pseudomonas aeruginosa* cultures and biofilms. Mol Microbiol 59:1114–1128. https://doi.org/10.1111/j.1365-2958.2005.05008.x

Arasu MV, Al-Dhabi NA, Rejiniemon TS, Lee KD, Huxley VAJ, Kim DH, Duraipandiyan V, Karuppiah P, Choi KC (2015) Identification and characterization of *Lactobacillus brevis* P68 with antifungal, antioxidant and probiotic functional properties. Indian J Microbiol 55:19–28. https://doi.org/10.1007/s12088-014-0495-3

Bandyopadhyay P, Mishra S, Sarkar B, Swain SK, Pal A, Tripathy PP, Ojha SK (2015) Dietary *Saccharomyces cerevisiae* boosts growth and immunity of IMC *Labeo rohita* (Ham.) juveniles. Indian J Microbiol 55:81–87. https://doi.org/10.1007/s12088-014-0500-x

Begum IF, Mohan Kumar R, Jeevan M, Ramani K (2016) GC–MS analysis of bioactive molecules derived from *Paracoccus pantotrophus* FMR19 and the antimicrobial activity against bacterial pathogens and MDROs. Indian J Microbiol 56:426–432. https://doi.org/10.1007/s12088-016-0609-1

Branco P, Francisco D, Chambon C, Hébraud M, Arneborg N, Almeida MG, Caldeira J, Albergaria H (2014) Identification of novel GAPDH-derived antimicrobial peptides secreted by *Saccharomyces cerevisiae* and involved in wine microbial interactions. Appl Microbiol Biotechnol 98:843–853. https://doi.org/10.1007/s00253-013-5411-y

Brexó RP, Sant'Ana AS (2017) Microbial interactions during sugar cane must fermentation for bioethanol production: does quorum sensing play a role? Crit Rev Biotechnol 38:1–14. https://doi.org/10.1080/07388551.2017.1332570

Brune KD, Bayer TS (2012) Engineering microbial consortia to enhance biomining and bioremediation. Front Microbiol 3:203. https://doi.org/10.3389/fmicb.2012.00203

Cai W, Zhang Z, Ren G, Shen Q, Hou Y, Ma A, Deng Y, Wang A, Liu W (2016) Quorum sensing alters the microbial community of electrode-respiring bacteria and hydrogen scavengers toward improving hydrogen yield in microbial electrolysis cells. Appl Energy 183:1133–1141. https://doi.org/10.1016/j.apenergy.2016.09.074

Cavinato C, Bolzonella D, Fatone F, Cecchi F, Pavan P (2011) Optimization of two-phase thermophilic anaerobic digestion of biowaste for hydrogen and methane production through reject water recirculation. Bioresour Technol 102:8605–8611. https://doi.org/10.1016/j.biortech.2011.03.084

Chen H, Fink GR (2006) Feedback control of morphogenesis in fungi by aromatic alcohols. Genes Dev 20:1150–1161. https://doi.org/10.1101/gad.1411806

Conrad M, Schothorst J, Kankipati HN, Van Zeebroeck G, Rubio-Texeira M, Thevelein JM (2014) Nutrient sensing and signaling in the yeast *Saccharomyces cerevisiae*. FEMS Microbiol Rev 38:254–299. https://doi.org/10.1111/1574-6976.12065

Das K, Rajawat MVS, Saxena AK, Prasanna R (2017) Development of *Mesorhizobium ciceri*-based biofilms and analyses of their antifungal and plant growth promoting activity in chickpea challenged by fusarium wilt. Indian J Microbiol 57:48–59. https://doi.org/10.1007/s12088-016-0610-8

Deshmukh R, Khardenavis AA, Purohit HJ (2016) Diverse metabolic capacities of fungi for bioremediation. Indian J Microbiol 56:247–264. https://doi.org/10.1007/s12088-016-0584-6

Di Cagno R, De Angelis M, Calasso M, Gobbetti M (2011) Proteomics of the bacterial cross-talk by quorum sensing. J Proteome 74:19–34. https://doi.org/10.1016/j.jprot.2010.09.003

Dickinson JR (2008) Filament formation in *Saccharomyces cerevisiae*-a review. Folia Microbiol (Praha) 53:3–14. https://doi.org/10.1007/s12223-008-0001-6

Diggle SP, Winzer K, Chhabra SR, Worrall KE, Camara M, Williams P (2003) The *Pseudomonas aeruginosa* quinolone signal molecule overcomes the cell density-dependency of the quorum sensing hierarchy, regulates *rhl*-dependent genes at the onset of stationary phase and can be produced in the absence of LasR. Mol Microbiol 50:29–43. https://doi.org/10.1046/j.1365-2958.2003.03672.x

Ercan D, Demirci A (2015) Current and future trends for biofilm reactors for fermentation processes. Crit Rev Biotechnol 35:1–14. https://doi.org/10.3109/07388551.2013.793170

Gasch AP, Spellman PT, Kao CM, Carmel-Harel O, Eisen MB, Storz G, Botstein D, Brown PO (2000) Genomic expression programs in the response of yeast cells to environmental changes. Mol Biol Cell 11:4241–4257 doi: NA

Go T-H, Cho K-S, Lee S-M, Lee O-M, Son H-J (2015) Simultaneous production of antifungal and keratinolytic activities by feather-degrading *Bacillus subtilis* S8. Indian J Microbiol 55:66–73. https://doi.org/10.1007/s12088-014-0502-8

Gomma AE, Lee SK, Sun SM, Yang SH, Chung G (2015) Improvement in oil production by increasing malonyl-CoA and glycerol-3-phosphate pools in *Scenedesmus quadricauda*. Indian J Microbiol 55:447–455. https://doi.org/10.1007/s12088-015-0546-4

Hernández-Saldaña OF, Valencia-Posadas M, de la Fuente-Salcido NM, Bideshi DK, Barboza-Corona JE (2016) Bacteriocinogenic bacteria isolated from raw goat milk and goat cheese produced in the Center of México. Indian J Microbiol 56:301–308. https://doi.org/10.1007/s12088-016-0587-3

Hlavácek O, Kucerová H, Harant K, Palková Z, Váchová L (2009) Putative role for ABC multidrug exporters in yeast quorum sensing. FEBS Lett 583:1107–1113. https://doi.org/10.1016/j.febslet.2009.02.030

Hu Y, Yang Y, Katz E, Song H (2015) Programming the quorum sensing-based AND gate in *Shewanella oneidensis* for logic gated-microbial fuel cells. Chem Commun 51:4184–4187. https://doi.org/10.1039/c5cc00026b

Huo YX, Cho KM, Rivera JGL, Monte E, Shen CR, Yan Y, Liao JC (2011) Conversion of proteins into biofuels by engineering nitrogen flux. Nat Biotechnol 29:346–351. https://doi.org/10.1038/nbt.1789

Jaramillo P, Muller NZ (2016) Air pollution emissions and damages from energy production in the US: 2002–2011. Energy Policy 90:202–211. https://doi.org/10.1016/j.enpol.2015.12.035

Kalia VC (2013) Quorum sensing inhibitors: an overview. Biotechnol Adv 31:224–245. https://doi.org/10.1016/j.biotechadv.2012.10.004

Kalia VC, Purohit HJ (2008) Microbial diversity and genomics in aid of bioenergy. J Ind Microbiol Biotechnol 35:403–419. https://doi.org/10.1007/s10295-007-0300-y

Kalia VC, Purohit HJ (2011) Quenching the quorum sensing system: potential antibacterial drug targets. Crit Rev Microbiol 37:121–140. https://doi.org/10.3109/1040841X.2010.532479

Kalia VC, Lal S, Ghai R, Mandal M, Chauhan A (2003) Mining genomic databases to identify novel hydrogen producers. Trends Biotechnol 21:152–156. https://doi.org/10.1016/S0167-7799(03)00028-3

Kalia VC, Wood TK, Kumar P (2014) Evolution of resistance to quorum-sensing inhibitors. Microb Ecol 68:13–23. https://doi.org/10.1007/s00248-013-0316-y

Kalia VC, Prakash J, Koul S (2016) Biorefinery for glycerol rich biodiesel industry waste. Indian J Microbiol 56:113–125. https://doi.org/10.1007/s12088-016-0583-7

Kalia VC, Prakash J, Koul S, Ray S (2017) Simple and rapid method for detecting biofilm forming bacteria. Indian J Microbiol 57:109–111. https://doi.org/10.1007/s12088-016-0616-2

Kaur G, Rajesh S, Princy SA (2015) Plausible drug targets in the *Streptococcus mutans* quorum sensing pathways to combat dental biofilms and associated risks. Indian J Microbiol 55:349–356. https://doi.org/10.1007/s12088-015-0534-8

Keskin T, Giusti L, Azbar N (2012) Continuous biohydrogen production in immobilized biofilm system versus suspended cell culture. Int J Hydrog Energy 37:1418–1424. https://doi.org/10.1016/j.ijhydene.2011.10.013

Koul S, Prakash J, Mishra A, Kalia VC (2016) Potential emergence of multi-quorum sensing inhibitor resistant (MQSIR) bacteria. Indian J Microbiol 56:1–18. https://doi.org/10.1007/s12088-015-0558-0

Kuipers OP, PGGA DR, Kleerebezem M, De Vos WM (1998) Quorum sensing-controlled gene expression in lactic acid bacteria. J Biotechnol 64:15–21. https://doi.org/10.1016/S0168-1656(98)00100-X

Kumar P, Patel SKS, Lee JK, Kalia VC (2013) Extending the limits of *Bacillus* for novel biotechnological applications. Biotechnol Adv 31:1543–1561. https://doi.org/10.1016/j.biotechadv.2013.08.007

Kumar P, Mehariya S, Ray S, Mishra A, Kalia VC (2015a) Biodiesel industry waste: a potential source of bioenergy and biopolymers. Indian J Microbiol 55:1–7. https://doi.org/10.1007/s12088-014-0509-1

Kumar R, Singh L, Wahid ZA, Din MF (2015b) Exoelectrogens in microbial fuel cells toward bioelectricity generation: a review. Int J Energ Res 39:1048–1067. https://doi.org/10.1002/er.3305

Kurm V, Putten WH, Boer W, Naus-Wiezer S, Hol WH (2017) Low abundant soil bacteria can be metabolically versatile and fast growing. Ecology 98:555–564. https://doi.org/10.1002/ecy.1670

Lida A, Yasuo O, Sueharu H (2008) Control of acetic acid fermentation by quorum sensing via N-acylhomoserine lactones in *Gluconacetobacter intermedius*. J Bacteriol 190:2546–2555. https://doi.org/10.1128/JB.01698-07

Lun L, Li D, Yin Y, Li D, Xu G, Zhao Z, Li S (2016) Characterization of chromium waste form based on biocementation by microbacterium. Indian J Microbiol 56:353–360. https://doi.org/10.1007/s12088-016-0579-3

Manogari R, Daniel DK (2015) Isolation, characterization and assessment of *Pseudomonas* sp. VITDM1 for electricity generation in a microbial fuel cell. Indian J Microbiol 55:8–12. https://doi.org/10.1007/s12088-014-0491-7

Mas A, Guillamon JM, Torija MJ, Beltran G, Cerezo AB, Troncoso AM, Garcia-Parrilla MC (2014) Bioactive compounds derived from the yeast metabolism of aromatic amino acids during alcoholic fermentation. Biomed Res Int 2014:1–7. https://doi.org/10.1155/2014/898045

McKinlay JB (2014) Systems biology of photobiological hydrogen production by purple nonsulfur bacteria. In: Zannoni D, De Philippis R (eds) Microbial bioenergy: hydrogen production, advances in photosynthesis and respiration. Springer, Dordrecht, pp 155–176. https://doi.org/10.1007/978-94-017-8554-9_7

Montgomery K, Charlesworth JC, LeBard R, Visscher PT, Burns BP (2013) Quorum sensing in extreme environments. Life 3:131–148. https://doi.org/10.3390/life3010131

Monzon O, Yang Y, Li Q, Alvarez PJ (2016) Quorum sensing autoinducers enhance biofilm formation and power production in a hypersaline microbial fuel cell. Biochem Eng J 109:222–227. https://doi.org/10.1016/j.bej.2016.01.023

Pastorella G, Gazzola G, Guadarrama S, Marsili E (2012) Biofilms: applications in bioremediation. In: Lear G, Lewis GD (eds) Microbial biofilms-current research and applications. Horizon Scientific Press, Norwich, pp 73–98

Patel SKS, Kalia VC (2013) Integrative biological hydrogen production: an overview. Indian J Microbiol 53:3–10. https://doi.org/10.1007/s12088-012-0287-6

Patel SKS, Purohit HJ, Kalia VC (2010) Dark fermentative hydrogen production by defined mixed microbial cultures immobilized on ligno-cellulosic waste materials. Int J Hydrog Energy 35:10674–10681. https://doi.org/10.1016/j.ijhydene.2010.03.025

Patel SKS, Kumar P, Kalia VC (2012) Enhancing biological hydrogen production through complementary microbial metabolisms. Int J Hydrog Energy 37:10590–10603. https://doi.org/10.1016/j.ijhydene.2012.04.045

Patel SKS, Kumar P, Singh M, Lee JK, Kalia VC (2015) Integrative approach to produce hydrogen and polyhydroxybutyrate from biowaste using defined bacterial cultures. Bioresour Technol 176:136–141. https://doi.org/10.1016/j.biortech.2014.11.029

Pawar SS, Vongkumpeang T, Grey C, van Niel Ed WJ (2015) Biofilm formation by designed co-cultures of *Caldicellulosiruptor* species as a means to improve hydrogen productivity. Biotechnol Biofuels 8:19. https://doi.org/10.1186/s13068-015-0201-7

Porwal S, Kumar T, Lal S, Rani A, Kumar S, Cheema S, Purohit HJ, Sharma R, Singh Patel SK, Kalia VC (2008) Hydrogen and polyhydroxybutyrate producing abilities of microbes from diverse habitats by dark fermentative process. Bioresour Technol 99:5444–5451. https://doi.org/10.1016/j.biortech.2007.11.011

Prakash J, Gupta R, Priyanka XX, Kalia VC (2017) Bioprocessing of biodiesel industry effluent by immobilized bacteria to produce value added products. Appl Biochem Biotechnol 185(1):179–190

Rabaey K, Boon N, Höfte M, Verstraete W (2005) Microbial phenazine production enhances electron transfer in biofuel cells. Environ Sci Technol 39:3401–3408. https://doi.org/10.1021/es048563o

Saini RK, Keum Y-S (2017) Progress in microbial carotenoids production. Indian J Microbiol 57:129–130. https://doi.org/10.1007/s12088-016-0637-x

Sanchart C, Rattanaporn O, Haltrich D, Phukpattaranont P, Maneerat S (2017) *Lactobacillus futsaii* CS3, a new GABA-producing strain isolated from Thai fermented shrimp (*Kung–Som*). Indian J Microbiol 57:211–217. https://doi.org/10.1007/s12088-016-0632-2

Sharma A, Lal R (2017) Survey of (Meta)genomic approaches for understanding microbial community dynamics. Indian J Microbiol 57:23–38. https://doi.org/10.1007/s12088-016-0629-x

Shiva Krishna P, Sudheer Kumar B, Raju P, Murty MSR, Prabhakar Rao T, Singara Charya MA, Prakasham RS (2015) Fermentative production of pyranone derivate from marine *Vibrio* sp. SKMARSP9: isolation, characterization and bioactivity evaluation. Indian J Microbiol 55:292–301. https://doi.org/10.1007/s12088-015-0521-0

Spier F, Buffon JG, Burkert CA (2015) Bioconversion of raw glycerol generated from the synthesis of biodiesel by different oleaginous yeasts: lipid content and fatty acid profile of biomass. Indian J Microbiol 55:415–422. https://doi.org/10.1007/s12088-015-0533-9

Thakur R, Sharma KC, Gulati A, Sud RK, Gulati A (2017) Stress-tolerant *Viridibacillus arenosi* strain IHB B 7171 from tea rhizosphere as a potential broad-spectrum microbial inoculant. Indian J Microbiol 57:195–200. https://doi.org/10.1007/s12088-017-0642-8

Valim FP, Aguiar-Oliveira E, Kamimura ES, Alves VD, Maldonado RR (2016) Production of star fruit alcoholic fermented beverage. Indian J Microbiol 56:476–481. https://doi.org/10.1007/s12088-016-0601-9

Van Houdt R, Aertsen A, Michiels CW (2007) Quorum-sensing-dependent switch to butanediol fermentation prevents lethal medium acidification in *Aeromonas hydrophila* AH-1N. Res Microbiol 158:379–385. https://doi.org/10.1016/j.resmic.2006.11.015

Varsha KK, Nishant G, Sneha SM, Shilpa G, Devendra L, Priya S, Nampoothiri KM (2016) Antifungal, anticancer and aminopeptidase inhibitory potential of a phenazine compound produced by *Lactococcus* BSN307. Indian J Microbiol 56:411–416. https://doi.org/10.1007/s12088-016-097-1

Venkataraman A, Rosenbaum M, Arends JB, Halitschke R, Angenent LT (2010) Quorum sensing regulates electric current generation of *Pseudomonas aeruginosa* PA14 in bioelectrochemical systems. Electrochem Commun 12:459–462. https://doi.org/10.1016/j.elecom.2010.01.019

Westman JO, Franzén CJ (2015) Current progress in high cell density yeast bioprocesses for bioethanol production. Biotechnol J 10:1185–1195. https://doi.org/10.1002/biot.201400581

Wrighton KC, Thrash JC, Melnyk RA, Bigi JP, Byrne-Bailey KG, Remis JP, Schichnes D, Auer M, Chang CJ, Coates JD (2011) Evidence for direct electron transfer by a Gram-positive bacterium isolated from a microbial fuel cell. Appl Environ Microbiol 77:7633–7639. https://doi.org/10.1128/AEM.05365-11

Wyss SC (2013) Design of a cross-domain quorum sensing pathway for algae biofuel applications. Doctoral dissertation, Ohio University. https://etd.ohiolink.edu/pg_10?0::NO:10:P10_ACCESSION_NUM:ouhonors1367239424

Yang JW (2011) Enhanced bioethanol production by *Zymomonas mobilis* in response to the quorum sensing molecules AI-2. http://etheses.dur.ac.uk/3231/1/jw_PhD.pdf?DDD1+

Yasin NH, Mumtaz T, Hassan MA, Abd Rahman N (2013) Food waste and food processing waste for biohydrogen production: a review. J Environ Manag 130:375–385. https://doi.org/10.1016/j.jenvman.2013.09.009

Yong XY, Shi DY, Chen YL, Jiao F, Lin X, Zhou J, Wang SY, Yong YC, Sun YM, OuYang PK, Zheng T (2014) Enhancement of bioelectricity generation by manipulation of the electron shuttles synthesis pathway in microbial fuel cells. Bioresour Technol 152:220–224. https://doi.org/10.1016/j.biortech.2013.10.086

Yong YC, Wu XY, Sun JZ, Cao YX, Song H (2015) Engineering quorum sensing signaling of *Pseudomonas* for enhanced wastewater treatment and electricity harvest: a review. Chemosphere 140:18–25. https://doi.org/10.1016/j.chemosphere.2014

Zhou M, Wang H, Hassett DJ, Gu T (2013) Recent advances in microbial fuel cells (MFCs) and microbial electrolysis cells (MECs) for wastewater treatment, bioenergy and bioproducts. J Chem Technol Biotechnol 88:508–518. https://doi.org/10.1002/jctb.4004

Zi L-H, Liu C-G, Xin C-B, Bai F-W (2013) Stillage backset and its impact on ethanol fermentation by the flocculating yeast. Process Biochem 48:753–758. https://doi.org/10.1016/j.procbio.2013.03.014

Applications of Quorum Sensing in Microbial Fuel Cell

11

Deepika Jothinathan, Nasrin Fathima AH, and Pachaiappan Raman

Abstract

Microbes in the biofilm coordinate themselves for a proper electron transfer inside the microbial fuel cell. Certain microbes use external mediators for the effective electron transfer. There are few exoelectrogens which can directly transfer the electrons to the anode via cytochromes and others through an indirect electron transfer, where the mechanism either takes place by bacteria's own mediators or by some chemical mediators added in the anode chamber. Bacteria in order to observe their population density, they use an autoinducer ligand and this process is so called quorum sensing.

Keywords

Biofilm · Acylhomoserine lactones · Exopolysaccharides · Electron acceptors · Redox

11.1 Introduction

Quorum sensing (QS) is used to designate an environmental sensing system that permits microbe to coordinate their genetic expression and physiological behavior in a cell-density dependent manner (Bassler and Losick 2006). Based on the

D. Jothinathan
Department of Life Sciences, Central University of Tamil Nadu, Thiruvarur, India

Nasrin Fathima AH
Department of Plant Biology and Plant Biotechnology, Presidency College, Chennai, India

P. Raman (✉)
Department of Biotechnology, School of Bioengineering, SRM University, Kattankulathur, Tamil Nadu, India
e-mail: pachaiappan.ra@ktr.srmuniv.ac.in

V. C. Kalia (ed.), *Quorum Sensing and its Biotechnological Applications*,
https://doi.org/10.1007/978-981-13-0848-2_11

signaling molecules, the Quorum sensing is classified into three categories namely acyl homoserine lactones (AHL), autoinducing peptides and autoinducer-2 which are used by gram positive, gram negative bacteria and both the species respectively for cell communication (Shrout and Nerenberg 2012; Lade et al. 2014). The anodic biofilm forms the basis for the bioelectricity production in wastewater treatment via microbial fuel cell. This biofilm is the major biological factor influencing the Microbial Fuel Cell (MFC) performance. Biofilms in general can be efficiently used for the bioenergy production from biological wastes and a wide range of microbes can contribute for the biodiesel and biohydrogen production etc. (Kalia et al. 2016). The electricity is generated by converting the organic biomass and carbohydrates with the help of catalytic activity of microorganism used by microbial fuel cell. In our search for clean and renewable sources of energy, the Microbial Fuel Cell is considered to have a favourable chance of success. Exoelectrogens is the name given to the bacteria that transfer electrons to the anode in Microbial Fuel Cells.

11.2 Biofilm Formation

Most of the bacteria form biofilm which includes gram positive bacteria such as *Staphylococcus* sp. and lactic acid bacteria, gram negative bacteria such as *Escherichia coli* and *Pseudomonas aeruginosa.* During the biofilm formation on the surface, the bacteria themselves communicate to each other by quorum sensing products such as AHL. The biofilm is usually composed of extracellular polymeric substances (EPS) which is a combination of exopolysaccharides, few proteins and nucleic acids (Branda et al. 2006). The EPS acts as scaffold that holds the cells together and the biofilm matrix accounts to 3–6% exopolysaccharides, 2–5% microbial cells and a majority portion of water (Branda et al. 2005). The composition of EPS alters according to the type of microbial group and the environmental conditions and also the thickness of the EPS increases with the biofilm maturity (Flemming and Wingender 2002; Kolter and Greenberg 2006). Quorum Quenching is a mechanism by which the molecules that are capable of quenching the QS system can be identified and this can be attained by disturbing the biofilm architecture (Kalia and Purohit 2011). To screen the microbes which produce more biofilm, a method employing medium 16 supplemented with Casein enzyme hydrolysate (CEH) was used (Kalia et al. 2017).

11.3 Energy Generation from MFC

The important function of MFC technology is to produce electricity from chemical energy present in a biomass catalyzed by microbes. According to the Carnot cycle, the limited thermal efficiency is ignored and theoretically a highest change competence can be achieved (>70%) similar to traditional chemical fuel cells. Chaudhuri and Lovley (2003) explored that *R. ferrireducens* could give a maximum electron yield of 80% of electricity. Similarly, there is also a report where 89% of energy recovery was achieved (Rabaey et al. 2003).

Tender et al. (2002) showed that even with high coulombic efficiency, the MFC power generation is still very low. One way to overcome this low rate of power generation is to store the electricity in batteries and then distribute it later to end users (Ieropoulos et al. 2003). MFC Technology thus shows great potential for sustainable long-term power applications. But first the technology must resolve the health and safety issues that can arise due to the usage of microoganisms.

11.3.1 Electron Transfer Mechanism

Direct electron transfer and indirect electron transfer are two electron transfers used in MFC's (Xi and Sun 2008). The Direct electron transfer mechanism usually needs an efficient binding between the microbial cell surface attachment and the surface of the electrode. The electrons are accepted from the cytochromes present in the external part of the cell. *Shewanella putrefaciens*, *Geobacter sulferreducens*, *Rhodoferax ferrireducens* etc. has the ability to switch over electrons from the cell to e^- acceptors via biofilms which are highly conductive appendages called pili and c-type cytochromes (Lovley 2008). In an indirect electron transfer, the mechanism either takes place by bacteria's own mediators or by some chemical mediators added in the anode chamber. Mediators assist the microbes to produce electrochemically effective output. The cell can be penetrated by the reduced form of mediator, which receives electrons from the e^- transporter to the anode surface (Lovley 2006).

Electron transfer by mediators: The electron transfer in this mechanism either takes place by bacteria's own mediators thereby promoting extracellular electron transfer or with the help of some chemical mediators added in the anode chamber. Mediators offer a dais for the microbes to produce electrochemically energetic products (Lovley 2006). Thionine, Methylene blue, Neutral red, Phenazines, anthraquinone-2, 6-disulfonate and iron chelates are some of the redox intermediaries that are commonly used in MFC (Du et al. 2007). *E. coli*, *Pseudomonas species*, *Proteus vulgaris* require a mediator as they are unable to transfer electrons outside the cell. An active mediator must enter the cell membrane and capture the electrons from the carrier in the Electron transport, must be stable even after long periods of redox cycling and should not harm the microbes (Du et al. 2007; Osman et al. 2010).

11.4 Role of Quorum Sensing in Microbial Fuel Cell

Among the gram negative microorganisms, *Pseudomonas* sp. has been predominantly reported for the bioelectricity production (Rabaey et al. 2004). There are few reports pertaining to the biofilm development in this microbe by pyocyanin pigment (Dietrich et al. 2008, 2013) which is regulated by Quorum sensing system (Yong et al. 2015). There are two AHL-based QS systems (LasI-LasR and RhlI-RhlR) (Duan and Surette 2007) and a Pseudomonas quinolone signal-based QS system (PQS) (de Kievit 2009) used by *P. aeruginosa*. These QS systems rely on self-generated signaling molecules to coordinate gene expression in response to population density. Along with the increase of the cell population density, the concentration of autoinducer increased significantly accordingly, this would then trigger the QS systems and positively regulate the pyocyanin (PYO) production (Recinos et al. 2012). The increase of PYO production would promote more and more *P. aeruginosa* cells adhered to the anode surface to form biofilm and facilitate the interfacial electron transfer (Qiao et al. 2015). Here, it is speculated that the integrated aerobic-anaerobic strategy allowed much fast growth of *P. aeruginosa* PAO1 which might quickly reach the threshold cell density to active the hierarchical cascade QS systems during the aerobic start-up stage, which would induce an increased PYO production and anode biofilm formation, thus might finally enhance the electricity output of MFC.

G. sulfurreducens is a Gram-negative bacterium, having the ability to oxidize acetate into electrons and protons and thus efficiently reduce Fe (III) oxides (Orellana et al. 2013). In its monolayer biofilms, the transfer of electron is aided by outer-membrane c-Cyts or through the riboflavin secretion that interacts with c-Cyts to transport the electrons out of the cell. Conductive proteinaceous pili made of PilA monomer units are present on the other side of the multilayer biofilms that are produced by *G. sulfurreducens* (Inoue et al. 2010; Malvankar and Lovley 2012). There are more reports supporting the electron transfer mechanism using pili in the microbial fuel cell system (Inoue et al. 2010; Kotloski and Gralnick 2013). There is an interesting study related to the interspecies electron transfer between *Pelotomaculum thermopropionicum*, and a methanogen, *Methanothermobacter thermautotrophicus* with the help of electrically conductive appendages was observed (Pham et al. 2008). The confirmation proposes that bacteria communicate with other species within the biofilm via the quorum sensing (QS) chemicals (e.g., p-coumaroyl-homoserine lactones and fatty acyl-homoserine lactones) (Schaefer et al. 2008).

11.4.1 Bacterial Role in MFC

MFC is purely dependent to oxidation reduction reaction (Redox). MFC contains two different types of compartments, anode and cathode. Both anode and cathode portion holds an electrode and that are alienated by cation permeable membrane. Acetate is an electron donor present in anode chamber is oxidized by respiratory bacteria in the deficiency of oxygen producing protons and electrons. Protons are transported across the cell membrane to adenosine triphosphate (ATP) while electrons are moved through an electron transport chain (ETC).

11.4.2 Recent Studies on Quorum Sensing in *P. aeruginosa*

Biofilm formation in *P. aeruginosa* requires the process of Quorum Sensing, as recent studies have discovered. LasI mutants do not build up into established biofilm, instead of that they prevent the biofilm formation at the micro colony stage (Davies et al. 1998). These mutants are supplemented to wild type biofilm secretion by exogenous addition of the LasI contingent HSL autoinducer

N-(-3-oxododecanoyl)-homoserine lactone. *P. aeruginosa* is one of the basic pathogens found in human lung affected with cystic fibrosis (CF). Analysis of CF sputum samples confirms the presence of *P. aeruginosa* biofilm *in vivo*. It was also detected the presence of LasI and Rh1I autoinducers in the sputum sample of the patients affected by CF (Parsek and Greenberg 2000; Singh et al. 2000). Thus it was concluded that biofilm formation by P. aeruginosa could prove to be lethal for human lungs and so antimicrobial treatments could be effective for CF treatment.

11.5 Waste Used in MFC Till Date

MFC is a recent technology which mostly uses waste to generate energy. Earlier, researchers have concentrated much on treating wastewater both industrial and domestic. In recent times, some have reported work on agrowastes such as starch-rich wastewater during the starch production from cassava (Kaewkannetra et al. 2009). The highest power density obtained was 22.19 W/m^3 with a single-chambered MFC utilizing cassava mill wastewater (Prasertsung et al. 2012). These results evidently show that cassava mill wastewater can be used as a biodegradable waste in MFC. An interesting research of using biodiesel waste for the simultaneous production of biohydrogen and polyhydroxyalkanoates has been reported (Patel 2015). These categories of study construct a path for the forthcoming scientists to use a single waste product to generate multiple valuable products.

Certain studies in MFC are pertained to cellulose for power production. Rismani et al. (2007) have carried out an experiment on cellulose for bioelectricity generation by means of rumen microbes from cattle (Rismani et al. 2007). Sedky et al. (2012), have resorted cellulose as substrate for generation of electricity using a dual-chambered MFC where *Streptomyces enissocaesilis* KNU and *Nocardiopsis sp.* KNU were employed for cellulose dilapidation in the anode chambers (Sedky et al. 2012). Vegetable waste was used for bioelectricity production in MFC by Clauwaert et al. (2008). There are few reports available which use slaughter house wastewater for electricity generation. A slaughterhouse wastewater with an anaerobic mixed sludge as a source of inoculum was fed in a dual-chambered microbial fuel cell was used (Katuri et al. 2012). Chaturvedi and Verma (2013), preformed a work on dilapidation of chicken feathers by *P. aeruginosa* with associated electricity production in MFC.

However, recently much work has been done on various waste products which is been utilized to generate power in a long term basis. An interesting approach of using human waste has received a wide publicity where Prof. Caitlyn Butler has developed a latrine that is efficient enough to purify the human waste simultaneously converting them to good compost for agriculture. This was installed in Ghana where the sanitation facilities are poor. The fuel is the waste organic matter which is filtered before use and nitrate is the oxidant used in the cathode. Similarly, scientists developed EcoBot-III, a robot which used human waste to produce energy (Ieropoulo et al. 2010). Table 11.1 shows the types of MFC's and the corresponding bioenergy production.

Table 11.1 Types of MFC's and bioenergy production

MFC type	Source/substrates	Power/current density	References
Single chamber	*Desulfovibrio vulgaris*	0.30 V	Lojou et al. (2002)
Single chamber	*Geobacter sulfurreducens*	2.4–12 A/m^3, 0.40 V	Dumas et al. (2008)
Dual Chamber-batch fed	*Acinetobacter calcoaceticus*	110 A/m^3 (7.8 W/m^3)	Rabaey et al. (2008)
Dual chamber – continuously fed	Enrichment culture	120 A/m^3, 0.50 V	Rozendal et al. (2008)
U-tube MFC	*Ochrobactrum anthropi*	89 mW/m^2	Rezaei et al. (2009)
U-tube MFC	*Enterobacter cloacae*/Cellulose	0.02 mA/cm^2	Justin Biffinger et al. (2009)
MFC array	*Shewanella sp.* and *Arthrobacter sp.*	2.69 mW/m^2 and 1.86 mW/m^2	Vega and Fernandez (1987)
Plant MFC	Paddy crop with compost	700 mV	Xi and Sun (2008)
Algal MFC in terrocotta pots	Mixed algal culture	44 μW	Lovley (2008)
Algal MFC	*Synechococcus leopoliensis*/BG-11 medium	42.5 W m^{-3}	Kristen and Brastad (2013)
Desalination MFC	Anaerobic and aerobic sludge/ Synthetic anode solution with acetate	13.16 mA	Coates (1996)

11.6 Exoelectrogens Involved in MFC

Both pure cultures and mixed cultures have been used in Microbial fuel cells and the power production has been monitored using a multimeter. The most studied microrganism in this field are *Geobacteria* and *Shewenella* which gave better results when inoculated as pure cultures.

The most well-known bacterial groups used in Microbial Fuel Cells are *Geobacteraceae* family and *Shewanella* genus. When compared with *Shewanella putrefaciens*, *Geobacter sulfurreducens* provides a 3000-fold increase in electron movement (Coates et al. 1996; Bond and Lovely 2003). *Geobacteraceae* family and *S. putrefaciens* are not the only microbial groups capable of Fe (III) reduction with surface-active cytochromes. *Clostridium butyricium, Clostridium beijerinckii, Desulfotomacum reducens, Thiobacillus ferroxidans Rhodobacter capsultatus* and Geovibrio genus are all competent of usage in a mediatorless fuel cell (Park et al. 2001). *Klebsiella pneumoniae, Saccharomyces cerevisiae, Staphylococcus aureus* and sewage sample used in MFC produced a higher voltage production (Aishwarya et al. 2011). Some Gram-positive bacteria including *Staphylococcus carnosus, Bacillus subtilis* and *Micrococcus luteus* were studied by cyclic voltammetry, an electrochemical method and also shown to achieve direct electron transfer,

Table 11.2 List of Microbial sources and their respective power output in MFC

Inoculum	Substrate	pH	Power	References
Hydrogen producing mixed culture	Composite/combined chemical wastewater	5.5	0.747 mA	Venkata Mohan et al. (2008)
Geobacter species	Dairy manure wastewater and potato water	8.3, 6.1	0.9 V	Patrick et al. (2011)
Anaerobes	Anaerobic sludge, Sodium Acetate as carbon source	7.0	377 ± 6 mWm^{-2}	Boris Tartakovsky et al. (2011)
Domestic wastewater consortium	Biodiesel waste, Glycerin media	7.0	470 ± 5 mV	Yujie Feng et al. (2008)
Anaerobic mixed consortia	Synthetic wastewater	5.5	274 mW/g	Venkata Mohan (2007)
Thermincola carboxydophila	Marine Marsh Sediment	6.8	209–254 mA/m^2	Mathis et al. (2008)
Domestic wastewater consortium	Cellulose	6.48–6.54	1070 mW/m^{-2}	Logan et al. (2006)
Lactic acid bacteria	Whey	6.46 ± 0.19	29.1 ± 4.9 W/m^2	Kassongo (2011)
γ-Proteobacteria	Acetate and 2-furfural	7.0	3490 mW/m^2	Abhijeet et al. (2009)
Enterobacter aerogenes	Glucose	7.0	0.41–0.42 mA,	Zhuang et al. (2011)
E. coli	Luria–Bertani (LB) medium	7.0	263.94 mW/m^2	Xi and Sun (2008)
Acetogenic bacteria	Formate	5.0	1 mA	Phuc et al. (2008)

presenting a potentially prevalent ability among bacteria. *Enterobacter aerogenes,* facultative anaerobes, are well known, powerful and resourceful producers of hydrogen. *Ochrobactrum anthropi* YZ-1, exoelectrogenic bacteria which was isolated from U-tube-shaped MFC produced power using acetate as the electron donor (Yi et al. 2008). Table 11.2 shows the list of microbial sources used in MFC. Combination of two bacteria such as *B. cereus* and *E. cloacae* has proved to produce much higher hydrogen when compared to the mixed or pure cultures (Patel 2014).

11.7 Opinion

The research area of Microbial fuel cell is very limited in terms of Quorum sensing. More reports are expected in future to expose the bacterial population response to the environment. The mediators which aid in the electron transfer in the microbial fuel cell should be studied for all the microbes in detail. This should be well correlated with the quorum sensing inducers or inhibitors in the forthcoming years. Till now most of the microbes which are studied in microbial fuel cell research are gram-negative bacteria. May be in the near future, gram positive bacteria and their QS systems should be focused.

References

Abhijeet PB, Jonathan RM, Tatiana AV, Choo Y, Hamilton (2009) Controlling accumulation of fermentation inhibitors in biorefinery recycle water using microbial fuel cells. Biotechnol Biofuels 2:7. https://doi.org/10.1186/1754-6834-2-7

Aishwarya D, Dalvi, Neha M, Shinde OA, Kininge PT (2011) Microbial fuel cell for production of bioelectricity from whey and biological waste treatment. Int J Adv Biotechnol Res 2:263–268

Bassler BL, Losick R (2006) Bacterially speaking. Cell 125:237–246. https://doi.org/10.1016/j.cell.2006.04.001

Bond DR, Lovley DR (2003) Electricity production by *Geobacter sulfurreducens* attached to electrodes. Appl Environ Microbiol 69:1548–1555. https://doi.org/10.1128/AEM.69.3.1548-1555.2003

Boris T, Poonam M, Bourque JS, Serge RG (2011) Electrolysis-enhanced anaerobic digestion of wastewater. Bioresour Technol 102:5685–5691. https://doi.org/10.1016/j.biortech.2011.02.097

Branda SS, Vik S, Friedman L, Kolter R (2005) Biofilms: the matrix revisited. Trends Microbiol 13:20–26. https://doi.org/10.1016/j.tim.2004.11.006

Branda SS, Chu F, Kearns DB, Losick R, Kolter R (2006) A major protein component of the *Bacillus subtilis* biofilm matrix. Mol Microbiol 59:1229–1238. https://doi.org/10.1111/j.1365-2958.2005.05020.x

Chaturvedi V, Verma P (2013) Metabolism of chicken feathers and concomitant electricity generation by pseudomonas aeruginosa by employing microbial fuel cell (MFC). J Waste Manag. https://doi.org/10.1155/2013/282798

Chaudhuri SK, Lovely DR (2003) Electricity generation by direct oxidation of glucose in mediatorless microbial fuel cells. Nat Biotechnol 21:1129–1232. https://doi.org/10.1038/nbt867

Clauwaert P, van der Ha D, Verstraete W (2008) Energy recovery from energy rich vegetable products with microbial fuel cells. Biotechnol Lett 30:1947–1951. https://doi.org/10.1007/s10529-008-9778-2

Coates JD, Phillips EJP, Lonergan DJ, Jenter H, Lovely DR (1996) Isolation of Geobacter species from diverse sedimentary environments. Appl Environ Microbiol 62:1531–1536 PMCID: PMC167928

Davies DG, Parsek MR, Pearson JP, Iglewski BH, Costerton JW, Greenberg EP (1998) The involvement of cell-to-cell signals in the development of a bacterial biofilm. Science 280:295–298 PMID: 9535661

de Kievit TR (2009) Quorum sensing in *Pseudomonas aeruginosa* biofilms. Environ Microbiol 11:279–288. https://doi.org/10.1111/j.1462-2920.2008.01792.x

Dietrich LEP, Teal TK, Price-Whelan A, Newman DK (2008) Redox-active antibiotics control gene expression and community behavior in divergent bacteria. Science 321:1203–1206. https://doi.org/10.1126/science.116061

Dietrich LEP, Okegbe C, Price-Whelan A, Sakhtah H, Hunter RC, Newman DK (2013) Bacterial community morphogenesis is intimately linked to the intracellular redox state. J Bacteriol 195:1371–1380. https://doi.org/10.1128/JB.02273-12

Du Z, Li H, Gu T (2007) A state of the art review on microbial fuel cells: a promising technology for wastewater treatment and bioenergy. Biotechnol Adv 25:464–482. https://doi.org/10.1016/j.biotechadv.2007.05.004

Duan K, Surette MG (2007) Environmental regulation of *Pseudomonas aeruginosa* PAO1 Las and Rhl quorum-sensing systems. J Bacteriol 189:4827–4836. https://doi.org/10.1128/JB.00043-07

Dumas C, Basseguy R, Bergel A (2008) Microbial electrocatalysis with Geobacter sulfurreducens biofilm on stainless steel cathodes. Electrochim Acta 53:2494–2500. https://doi.org/10.1016/j.electacta.2007.10.018

Flemming HC, Wingender J (2002) Extracellular polymeric substances (EPS): structural, ecological and technical aspects. In: Bitton G (ed) Encyclopedia of environmental microbiology. Wiley, New York, pp 1223–1231

Ieropoulo SI, Greenman J, Melhuish C, Horsfield I, EcoBot-III-A (2010) Robot with guts. ALIFE 733–740 at https://mitpweb2.mit.edu/sites/default/files/titles/alife/0262290758chap131.pdf

Ieropoulos I, Greenman J, Melhuish C (2003) Imitation metabolism: energy autonomy in biologically inspired robots. In: Proceedings of the 2nd international symposium on imitation of animals and artifacts, pp 191–194

Inoue K, Leang C, Franks AE, Woodard TL, Nevin KP, Lovley DR (2010) Specific localization of the c-type cytochrome Omc Z at the anode surface in current producing biofilms of *Geobacter sulfurreducens*. Environ Microbiol Rep 3:211–217. https://doi.org/10.1111/j.1758-2229.2010.00210.x

Justin B, Meghann R, Bradley R, Jeremy P, Steven F, Kenneth N (2009) Characterization of electrochemically active bacteria utilizing a high-throughput voltage-based screening assay. Biotechnol Bioeng 102:436–444. https://doi.org/10.1002/bit.22072

Kaewkannetra P, Imai T, Garcia-Garcia FJ, Chiu TY (2009) Cyanide removal from cassava mill wastewater using Azotobactor vinelandii TISTR 1094 with mixed microorganisms in activated sludge treatment system. J Hazard Mater 172:224–228. https://doi.org/10.1016/j.jhazmat.2009.06.162

Kalia VC, Purohit HJ (2011) Quenching the quorum sensing system: potential antibacterial drug targets. Crit Rev Microbiol 37:121–140

Kalia VC, Prakash J, Koul S (2016) Biorefinery for glycerol rich biodiesel industry waste. Indian J Microbiol 56:113–125

Kalia VC, Prakash J, Koul S, Ray S (2017) Simple and rapid method for detecting biofilm forming bacteria. Indian J Microbiol 57:109–111. https://doi.org/10.1007/s12088-016-0616-2

Kassongo J, Togo CA (2011) Performance improvement of whey-driven microbial fuel cells by acclimation of indigenous anodophilic microbes. Afr J Biotechnol 10(40):7846–7852. https://doi.org/10.5897/AJB11.206

Katuri KP, Enright AM, O'Flaherty V, Leech D (2012) Microbial analysis of anodic biofilm in a microbial fuel cell using slaughter house wastewater. Bioelectrochemistry 87:164–171. https://doi.org/10.1016/j.bioelechem.2011.12.002

Kolter R, Greenberg EP (2006) Microbial sciences: the superficial life of microbes. Nature 441(7091):300–302. https://doi.org/10.1038/441300a

Kotloski NJ, Gralnick JA (2013) Flavin electron shuttles dominate extracellular electron transfer by *Shewanella oneidensis*. MBio 4:e00553–e00512. https://doi.org/10.1128/mBio.00553-12

Kristen S, Brastad ZH (2013) Water softening using microbial desalination cell technology. Desalination 309:32–37. https://doi.org/10.1016/j.desal.2012.09.015

Lade H, Paul D, Kweon JH (2014) Quorum quenching mediated approaches for control of membrane biofouling. Int J Biol Sci 10:550. https://doi.org/10.7150/ijbs.9028

Logan BE, Hamelers B, Rozendal R, Schroder U, Keller J, Freguia S, Aelterman P, Verstraete W, Rabaey K (2006) Microbial fuel cells: methodology and technology. Environ Sci Technol 40:5181–5192 PMID: 16999087

Lojou E, Durand MC, Dolla A (2002) Hydrogenase activity control at *Desulfovibrio vulgaris* cell-coated carbon electrodes: biochemical and chemical factors influencing the mediated bioelectrocatalysis. Electroanalysis 14:913–922

Lovley DR (2006) Bug juice: harvesting electricity with microorganisms. Nat Rev Microbiol 4:497–508. https://doi.org/10.1038/nrmicro1442

Lovley DR (2008) The microbe electric: conversion of organic matter to electricity. Curr Opin Biotechnol 19:1–8. https://doi.org/10.1016/j.copbio.2008.10.005

Malvankar NS, Lovley DR (2012) Microbial nanowires:a new paradigm for biological electron transfer and bioelectronics. ChemSusChem 5:1039–1046. https://doi.org/10.1002/cssc.201100733

Mathis BJ, Marshall CW, Milliken CE, Makkar RS, Creager SE, May HD (2008) Electricity generation by thermophilic microorganisms from marine sediment. Appl Microbiol Biotechnol 78:147–155. https://doi.org/10.1007/s00253-007-1266-4

Orellana R, Leavitt JJ, Comolli LR, Csencsits R, Janot N, Flanagan KA, Gray AS, Leang C, Izallalen M, Mester T (2013) U(VI) reduction by a diversity of outer surface c-type cytochromes of *Geobacter sulfurreducens*. Appl Environ Microbiol 79:6369–6374. https://doi.org/10.1128/AEM.02551-13

Osman MH, Shah AA, Walsh FC (2010) Recent progress and continuing challenges in bio-fuel cells part II: microbial. Biosens Bioelectron 26:953–963

Park HS, Kim SK, Shin IH, Jeong YJ (2001) A novel electrochemically active and Fe (III)-reducing bacterium phylogenetically related to *Clostridium butyricum* isolated from a microbial fuel cell. Anaerobe 7:297–300

Parsek MR, Greenberg EP (2000) Acylhomoserine lactone quorum sensing in gram-negative bacteria: a signaling mechanism involved in associations with higher organisms. Proc Natl Acad Sci U S A 97:8789–8793. https://doi.org/10.1073/pnas.97.16.8789

Patel SKS, Kumar P, Mehariya S, Purohit HJ, Lee JK, Kalia VC (2014) Enhancement in hydrogen production by co-cultures of Bacillus and Enterobacter. Int J Hydrogen Energy 39(27):14663–14668. https://doi.org/10.1016/j.ijhydene.2014.07.084

Patel SKS, Kumar P, Singh M, Lee JK, Kalia VC (2015) Integrative approach to produce hydrogen and polyhydroxybutyrate from biowaste using defined bacterial cultures. Bioresour Technol 176:136–141. https://doi.org/10.1016/j.biortech.2014.11.029

Patrick DK, Roland C, Douglas FC, Priscilla AS, John MR, Bruce EL (2011) Anode microbial communities produced by changing from microbial fuel cell to microbial electrolysis cell operation using two different wastewaters. Bioresour Technol 102:388–394. https://doi.org/10.1016/j.biortech.2010.05.019

Pham TH, Boon N, Aelterman P, Clauwaert P, Schamphelaire LD, Vanhaecke L, Maeyer KD, Hofte M, Verstraete W, Rabaey K (2008) Metabolites produced by *Pseudomonas* sp. enables a gram positive bacterium to achieve extracellular electron transfer. Appl Microbiol Biotechnol 77:1119–1129. https://doi.org/10.1007/s00253-007-1248-6

Phuc TH, Beomseok T, Chang IS (2008) Performance and bacterial consortium of microbial fuel cell fed with formate. Energy Fuel 22:164–168. https://doi.org/10.1021/ef700294x

Prasertsung N, Reungsang A, Ratanatamskul C (2012) Alkalinity of cassava wastewater fed in anodic enhance electricity generation by a single chamber microbial fuel cells. Eng J 16:17–28

Qiao Y, Qiao YJ, Zou L, Ma CX, Liu JH (2015) Real-time monitoring of phenazines excretion in *Pseudomonas aeruginosa* microbial fuel cell anode using cavity microelectrodes. Bioresour Technol 198:1–6. https://doi.org/10.1016/j.biortech.2015.09.002

Rabaey K, Lissens G, Siciliano SD, Verstraete W (2003) A microbial fuel cell capable of converting glucose to electricity at high rate and efficiency. Biotechnol Lett 25:1531–1535 PMID: 14571978

Rabaey K, Boon N, Siciliano SD, Verhaege M, Verstraete W (2004) Biofuel cells select for microbial consortia that self-mediate electron transfer. Appl Environ Microbiol 70:5373–5382. https://doi.org/10.1128/AEM.70.9.5373-5382.2004

Rabaey K, Read ST, Clauwaert P, Freguia S, Bond PL, Blackall LL, Keller J (2008) Cathodic oxygen reduction catalyzed by bacteria in microbial fuel cells. ISME J 2:519–527

Recinos DA, Sekedat MD, Hernandez A, Cohen TS, Sakhtah H, Prince AS, Price-Whelan A, Dietrich LEP (2012) Redundant phenazine operons in *Pseudomonas aeruginosa* exhibit environment-dependent expression and differential roles in pathogenicity. Proc Natl Acad Sci U S A 109:19420–19425. https://doi.org/10.1073/pnas.1213901109

Rezaei F, Xing D, Wagner R, Regan JM, Richard TL, Logan BE (2009) Simultaneous cellulose degradation and electricity production by *Enterobacter cloacae* in a microbial fuel cell. Appl Environ Microbiol 75:3673–3678. https://doi.org/10.1128/AEM.02600-08

Rismani YH, Christy AD, Dehority BA, Morrison M, Yu Z, Tuovinen OH (2007) Electricity generation from cellulose by rumen microorganisms in microbial fuel cell. Biotechnol Bioeng 97:1398–1407. https://doi.org/10.1002/bit.21366

Rozendal RA, Jeremiasse AW, Hamelers HVM, Buisman CJN (2008) Hydrogen production with a microbial biocathode. Environ Sci Technol 42:629–634. https://doi.org/10.1021/es071720+

Schaefer AL, Greenberg EP, Oliver CM, Oda Y, Huang JJ, Banin G, Peres CM, Schimdt S, Juhaszova K, Sufrin JR, Harwood CS (2008) A new class of homoserine lactone quorum-sensing signals. Nature 454:595–599. https://doi.org/10.1038/nature07088

Sedky H, Hassan A, Kim YS, Oh S (2012) Power generation from cellulose using mixed and pure cultures of cellulose-degrading bacteria in a microbial fuel cell. Enzym Microbiol Technol 51:269–273. https://doi.org/10.1016/j.enzmictec.2012.07.008

Shrout JD, Nerenberg R (2012) Monitoring bacterial twitter: does quorum sensing determine the behavior of water and wastewater treatment biofilms? Environ Sci Technol 46:1995–2005. https://doi.org/10.1021/es203933h

Singh PK, Schaefer AL, Parsek MR, Moninger TO, Welsh MJ, Greenberg EP (2000) Quorum-sensing signals indicate that cystic fibrosis lungs are infected with bacterial biofilms. Nature 407:762–764. https://doi.org/10.1038/35037627

Tender LM, Reimers CE, Stecher HA, Holmes DE, Bond DR, Lowy DA et al (2002) Harnessing microbially generated power on the seafloor. Nat Biotechnol 20:821–825. https://doi.org/10.1038/nbt716

Vega CA, Fernandez I (1987) Mediating effect of ferric chelate compounds in microbial fuel cell with *Lactobacillus planetarium*, *Streptococcus lactis* and *Erwinia dissolvens*. Bioelectrochem Bioenerg 17:217–222

Venkata MS, Veer RS, Srikanth S, Sarma PN (2007) Bioelectricity production by mediatorless microbial fuel cell under acidophilic condition using wastewater as substrate: influence of substrate loading rate. Curr Sci 92(12):1720–1726

Venkata MS, Mohanakrishna G, Purushotham RB, Saravanan R, Sarma PN (2008) Bioelectricity generation from chemical wastewater treatment in mediatorless (anode) microbial fuel cell (MFC) using selectively enriched hydrogen producing mixed culture under acidophilic microenvironment. Biochem Eng J 39:121–130. https://doi.org/10.1016/j.bej.2007.08.023

Xi MY, Sun YP (2008) Preliminary study on *E. coli* microbial fuel cell and on-electrode taming of the biocatalyst. Chin J Process Eng 8(6):1179–1184

Yi Z, Defeng X, John MR, Bruce EL (2008) Isolation of exoeletrogenic bacterium *Ochrobactrum anthropi* YZ-1 by using a U-tube microbial fuel cell. Appl Environ Microbiol 74:3130–3137. https://doi.org/10.1128/AEM.02732-07

Yong YC, Wu XY, Sun JZ, Cao YX, Song H (2015) Engineering quorum sensing signaling of *Pseudomonas* for enhanced wastewater treatment and electricity harvest: a review. Chemosphere 140:18–25. https://doi.org/10.1016/j.chemosphere.2014.10.020

Yujie F, Xin W, Bruce EL, He L (2008) Brewery wastewater treatment using air-cathode microbial fuel cells. Appl Microbiol Biotechnol 78:873–880

Zhuang LS, Zhou YY, Liu T, Wu Z (2011) Development of *Enterobacter aerogenes* fuel cells: from in situ biohydrogen oxidization to direct electroactive biofilm. Bioresour Technol 102:284–289

Part III

Agriculture

Pseudomonas: A Quorum Sensing System for Improved Crop Production

12

Bhagwan Rekadwad and Pallab Kumar Ghosh

Abstract

The soil is a unique and ultimate home of the variety of beneficial and biotechnologically important inhabitant microorganisms. These microorganisms comply such a best services especially beneficial agriculture in turn farmers benefitted. Effective management of agriculture ecosystem and proper use of microorganisms such as *Pseudomonas* and *Pseudomonas*-like can improve crop health, increase crop yield and productivity, maintain the health of soil over a period of long time span. Bacterium *Pseudomonas* associated species in the six groups in the Pseudomonadaceae family capable of sensing and generating biomolecules having short and long chains. These include quorum sensing (QS) molecules, hydrolytic enzymes, proteins, siderophores, antibiotics and much many antibacterial and antifungal compounds under various environmental situations such as high or low temperature, high or low salt concentrations, in the presence and absence of contaminants (chemicals, bio-chemicals and hydrocarbons), in response to specific ions and in response to specific signalling molecules. All these characteristics possessed and activities carried out by *Pseudomonas* have biotechnological applications especially in agriculture. Pseudomonas have major application in crop production by acting as biocontrol agent due to its infection ability, recognising and sending QS molecules, as an antagonistic, as phytopathogens, as plant growth promoting (PGP) agent in the form of individual cell (solid, liquid, spray), in mixed culture and co-culture, as individual or mixed culture inoculums etc. All these characteristics of genus *Pseudomonas* make a suitable and biotechnologically important cellular model for the variety of application in agriculture and horticulture.

B. Rekadwad (✉)
National Centre for Microbial Resource, National Centre for Cell Science, Pune, India

P. K. Ghosh
Microbiology Laboratory, Department of Marine Science, Ballygunge Science College, Calcutta University, Kolkata, West Bengal, India

V. C. Kalia (ed.), *Quorum Sensing and its Biotechnological Applications*,
https://doi.org/10.1007/978-981-13-0848-2_12

Keywords
PGPR · Plant growth promotion activity · AHLs · Cell-cell communication · Crop yield

Abbreviations

VOCs Volatile organic compounds

12.1 Introduction

The genus *Pseudomonas* exhibits a broad spectrum of traits and it shows a remarkable adaptability in nature under environmental extremes. The family Pseuodomonadaceae currently has six major groups and more than 200 identified species. Quorum sensing (QS) is operated and regulated by collective behaviour through communication by molecules called as auto-inducers based on release, signal and recognition (Ng and Bassler 2009; Rutherford and Bassler 2012; O'Loughlin et al. 2013; Kalia 2014). Bacteria use quorum sensing to coordinate certain type of behaviours as biofilm formation, virulence activity and resistance capacity based on the density of bacterial population (Kalia et al. 2014; Kalia 2015; Kalia and Kumar 2015; Kaur et al. 2015; Pooja et al. 2015). QS can occur within same bacterial species or between different species. These bacteria have an ability to produce natural products with exceptional biological activity and curing properties (Krishna et al. 2015; Kumar et al. 2015). This process can regulate different processes of the host within the environment and different variety of molecules can be used as signalling. Many researchers worldwide working in this area to dig out answer that helps us to get better crop yield (Gewin 2010; Manogari and Daniel 2015; Rani et al. 2017). QS molecule's carry out signal transduction and trigger during plant-microbe interaction in the rhizospheric region which help to check plant-microbe i.e. cell to cell communication (Koul et al. 2016, Koul and Kalia 2017; Kalia et al. 2017). In this context, this review describes the roles of QS and bio-molecules produced by *Pseudomonas* having utility in crop production and in green biotechnology with selective emphasis on it as the model organism.

12.2 Application in Agriculture for Improved Crop Production

Pseudomonas has biotechnological application in improved production of crop plants and thereby increases a crop yield through the production of agriculturally important bio-molecules helpful. The roles of *Pseudomonas* spp. and molecules produced by them are illustrated.

12.2.1 Biocontrol Agent

Pseudomonas is well known to bio-control agent because of its well-regulated QS system. Variety of naturally occurring compounds are related to QS system of *Pseudomonas* viz. antibiotics and effective antifungal chemicals including hydrogen cyanide (HCN), 2, 4-di-acetyl-phloro-glucinol, pyrrolnitrin, pyoluteorin and phenazines (Mezaache-Aichour et al. 2016) and the *pcoIR* system in *Pseudomonas fluorescens*(Wu et al. 2010; Bauer et al. 2016; Pandin et al. 2017). In a similar fashion, the biocontrol capacity by *Pseudomonasaureofaciens*30–84 strain positively regulates the *phzFABCD* operon for synthesizing phenazine. *Serratia plymuthica* is universal inhabitant in the rhizospheric soils and a proven antagonist of soil-borne plant pathogens. The QS molecules – acylhomoserine lactones (AHLs) molecules produced by *Pseudomonas* act as triggering molecules, which is very crucial in biocontrol activity/response of strainHRO-C48.AHLs communicate information and allows *Pseudomonas* to exude hydrolytic enzymes-mainly chitin and protein, pyrrolnitrin and VOCs, which play a major role in anti-fungal activity (Kilani and Fillinger 2016; Shehata et al. 2016; Gupta et al. 2017; Yang et al. 2017). Several species of *Pseudomonas* spp. have been studied by different research groups worldwide (Table 12.1).

12.2.2 Plant Growth Promoting Agent (PGPR)

Pseudomonas acts as a good bio-control agent (Borges et al. 2016). Many strains of *Pseudomonas* were more effective and useful in co-culture or in mixed culture as the bio-control agent. For instance, bio-control agent *P. fluorescens* strain PF1 along with biofertilizer *Azospirillum brasilense*strain TNAU was developed as talc based bioformulation. This individual, as well as mixed bioformulation, can be used for soil application, seed application/treatment, as the spray for seedlings and foliar parts in groundnut. It was observed that *P. fluorescens* strain PF1 have the ability to enhance lateral root growth. The combination of both bacteria complies development of shoot. Collectively, *Pseudomonas* bacterium is useful both individual and in combination as for plant growth promotion and enhanced crop yield (Andhare and Babu 2017).

12.2.3 Beneficial Infections and Quorum Sensing in Gram Negative Bacteria for Survival in Cold Environments

Many useful plants found in the cold environments are capable of resistant to cold but not for productivity. Productivity in the sense growth of vegetative parts, flowering, fruit formation and production of biotechnologically important bio-molecules. They undergo in the condition somehow we called it as dormancy. Gram-negative psychrophilic bacteria, especially from glacier environment, have application in growth and development of plants due to their inherent ability to survive under extremely low temperatures.

Table 12.1 *Pseudomonas spp.* bio-controlling agents regulated by quorum sensing

Pseudomonas strains	Biocontrol molecules	Application	References
P. chlororaphis 30–84, *P. chlororaphis* GP72	Phenazine	Wheat disease	Liu et al. (2016)
*P.fluòrescens*D7	Cell free filtrate	Field crops	Banowetz et al. (2008), Caldwell et al. (2012), and Harding and Raizada (2015)
*P. fluorescens*BRG100	Granules	Plant protection	
*P.fluorescens*WH6	Cell free filtrate	Plant protection	
Pseudomonas species (*P. aeruginosa, P. fluorescens, P. fragi, P. syringae* and *P.putida*	Cell free extract	Biosurfactant	Awada et al. (2005)
Pseudomonas spp.	Phenasin 2,4- di-acetyl-phloro-glucinol, flagellin, Fe^{+3} chelating siderophores	Against *Fusarium* and *Fusarium*-like plant pathogens, insects	Wachowska et al. (2013), Lakshmi et al. (2015), and Kergunteuil et al. (2016)
P.fluorescens, *P. aureofaciens* 2–79, 30–84) *P.fluorescens*Pf-5, Q2–87, F113	Plt, Prn, DAPG, HCN	As antifungal agent against *Rhizoctonia solani, Pythium ultimum*	Weller (2007)
P.aureofaciens TX-1	Spot less, bio-jet	As antifungal agent against *R. solani, Pythium*	Junaid et al. (2013)
*P.fluorescens*A506	Frostban	Against Bunch rot, fire blight	
P.aureofaciens 30–84	Phenazine	Regulation of *phzFABCD* operon for synthesizing phenazine	Sun et al. (2016)
P. putida	Acylhomoserine lactones (AHLs)	Confer systemic resistance against *Alternaria alternate* in tomato plant	Chaturvedi and Kumar 2014 Shang et al. (2014), Vaikuntapu et al. (2014), Jayapradha and Yesu Raja (2016), and Meena et al. (2016)

Fluorescent *Pseudomonas*	Pyoluteorin/phenazine	Antagonist of a broad spectrum of phytopathogenic microorganisms isolated from the maize rhizosphere and soil borne phytopathogens such as *Fusarium*	Moubarak and Abdel-Monaim (2011), Mehrabi et al. (2016), Kunova et al. (2016), and Vacheron et al. (2016)
P. entomophila	Toxins	Entomopathogen	Ruiu (2015)
P. fluorescens EPS62e	Antagonistic	*Erwiniaamylovora,* causative of fire blight	Cabrefiga et al. (2007)
Pseudomonas spp.	Antagonistic	*Fusarium* causative of wilt of chickpea	Abed et al. (2016)
P. aeruginosa LV	Produce antibiotic like substance	Against plant pathogen*Xanthomonasar-boricola*	da Silva Vasconcellos et al. (2014)
P.chlororaphis PCL1391	Phenazine	Exhibit antagonistic of *F. oxysporum,* which causes tomato foot and root rot	Egamberdieva et al. (2017), and Jendoubi et al. (2017)
Pseudomonas spp.	Antagonistic	Against *Rhizoctonia solani* and *Sclerotium rolfsii*	Jani et al. (2015), Kotasthane et al. (2015), and van Lenteren et al. (2018)
P. putida, P.fluorescens 2–79, *P. protegens* Pf-5, and *P. brassicacearum*Q8r1–96	Bio-control activities	As an endophytes against plant pathogen	deBruijn and Raaijmakers (2009), Mavrodi et al. (2012), Khan et al. (2014), Oteino et al. (2015), and Kandel et al. (2017)
P. fluorescens 1–112, 2–28, 4–6	Volatiles such as phenazine-1-carboxylic acid and HCN	Application during storage of apple. Kills blue mold	Wallacea et al. (2017)

Psychrophilic bacteria such as Escherichia coli strain pJBA132, *Chromobacterium violaceum* strain CV026,*Pseudomonas putida* strain F117 and strain pKR-C12 could sense long and short AHLs and capable of producing diverse molecules under varied experimental conditions. These bacteria possess AHLs mediated QS is regulated by gene LuxR transcriptional regulator harboured in bacterial genome (De Maayer et al. 2014; Abraham and Thomas 2015; Dharmaprakash et al. 2016).

12.2.4 *Pseudomonas* as Plant Growth Promoting (PGP) Microbial Inoculums

Indian conventional agricultural method was very good and beneficial agricultural practice. The use of manure and original varieties which are highly resistant to pathogenic attack occurred during the cropping season and changes in weather. The current scenario is more painful and reflects some farmer are fell as prey due to uncontrolled use of chemical fertilizer. Global warming incurs the increase in environmental temperature. These artificial and natural calamity cause damage to the crop plants and hamper their yield. The use of manure and bio-inoculum are the best option to solve existing problem. The use of microbial inoculum such *Methylobacterium organophilum* capable of growth at elevated temperature(Rekadwad 2014), use of biopolymers and spores (Rekadwad et al. 2016), bacterial biofilms (Rekadwad and Khobragade 2017b), nutrient recycling bacteria (Rekadwad and Khobragade 2017b), salt and chemical tolerating bacteria (Rekadwad and Khobragade 2015, 2017c) etc.

The eco-friendly approaches such as the use of the whole bacterium, bacterial powdered form include talc and embedded bioformulation, solid and liquid bioinoculants etc., are the solutions to solve above-explained problems. Microorganisms such as *Pseudomonas, Bacillus, Rhizobium, Microbacterium, Agrobacterium, Chryseobacterium, Ensifer* and *Rhodococcus* many other include fungi are helpful microorganism to do best agriculture and horticulture practice. These microorganisms help plant through the variety of ways such as producing and sensing QS microorganisms, by growing plants, growing rhizospheric sites and as endophytes in diverse environments (Abbamondi et al. 2016; Papenfort and Bassler 2016).

12.2.5 Increased Crop Productivity by *Pseudomonas* Siderophore

The iron specific chelating agents i.e. siderophores are produced by bacteria in soil, water-fresh and marine and by endophytes facilitating plant growth (Fig. 12.1). The majority of bacteria found in the marine environment are belonging to alpha-proteobacteria and gamma-proteobacteria (Rekadwad and Khobragade 2017a). Siderophores produced by *Pseudomonas* species complies biological control of plant pathogens such as fungi (phytopathogens). The application of siderophore-producing bacteria in soil, on plants or as endophytes. Either of these practices should help to control the adverse effects of phytopathogens on plants. The siderophore has biotechnological applications and promotes plant growth by increasing

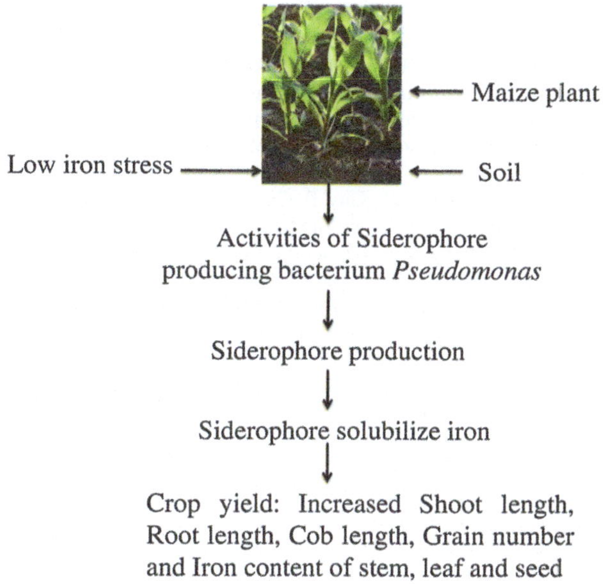

Fig. 12.1 *Pseudomonas* siderophores enhance crop yield

plant biomass (Sasirekha and Srividya 2016; Sah et al. 2017). Farmer will benefit due to increase in shoot length of the plant, the length of root help to absorb more nutrient and water, increase in cob length, increase in a number of healthy grains per plant and increase in iron content in the leaves, stem, roots and seeds.

Acknowledgement BR is thankful to University Grants Commission for funds (No.PDFSS-2013-14-ST-MAH-4350_website). PKG is grateful to DST-Science and Engineering Research Board (SERB), New Delhi (India) for financial aid in the form of National Postdoctoral Fellowship (N-PDF, Award letter No.: NPDF/2016/00323 dated. 05.07.2016).

References

Abbamondi GR, Tommonaro G, Weyens N, Thijs S, Sillen W, Gkorezis P, Iodice C, de Melo RW, Nicolaus B, Vangronsveld J (2016) Plant growth-promoting effects of rhizospheric and endophytic bacteria associated with different tomato cultivars and new tomato hybrids. Chem Biol Technol Agric 3:1. https://doi.org/10.1186/s40538-015-0051-3

Abed H, Rouag N, Mouatassem D, Rouabhi A (2016) Screening for *Pseudomonas* and *Bacillus* antagonistic rhizobacteria strains for the biocontrol of *Fusarium* wilt of chickpea. Eur J Soil Sci 5:182–191. https://doi.org/10.18393/ejss.2016.3.182-191

Abraham WP, Thomas S (2015) Draft genome sequence of *Pseudomonas psychrophila* MTCC 12324, isolated from the Arctic at 79°N. Genome Announc 3:e00578–e00515. https://doi.org/10.1128/genomeA.00578-15

Andhare AP, Babu S (2017) Compatibility of *Azospirillumbrasilense* and *Pseudomonas fluorescens* in growth promotion of groundnut (*Arachis hypogea* L.). Ann Braz Acad Sci 89:1027–1040. https://doi.org/10.1590/0001-3765201720160617

Awada S, Spendlove R, Awada M (2005) US patent: microbial biosurfactants as agents for controlling pests. Publication No. US 20050266036 A1, Application No. US 11/141,669, Date of filing 31 May 2005. http://www.google.co.in/patents/US20050266036

Banowetz GM, Azevedo MD, Armstrong DJ, Halgren AB, Mills DI (2008) Germination-arrest factor (GAF): biological properties of a novel, naturally-occurring herbicide produced by selected isolates of rhizosphere bacteria. Biol Control 46:380–390. https://doi.org/10.1016/j.biocontrol.2008.04.016

Bauer JS, Hauck N, Christof L, Mehnaz S, Gust B, Gross H (2016) The systematic investigation of the quorum sensing system of the biocontrol strain *Pseudomonas chlororaphis* subsp. *aurantiaca* PB-St2 Unveils aurI to be a biosynthetic origin for 3-Oxo-homoserine lactones. PLoS One 11:e0167002. https://doi.org/10.1371/journal.pone.0167002

Borges A, Abreu AC, Dias C, Saavedra MS, Borges F, Simões M (2016) New perspectives on the use of phytochemicals as an emergent strategy to control bacterial infections including biofilms. Molecules 21:877. https://doi.org/10.3390/molecules21070877

de Bruijn I, Raaijmakers JM (2009) Regulation of cyclic lipopeptide biosynthesis in *Pseudomonas fluorescens* by the ClpP Protease. J Bacteriol 191:1910–1923. https://doi.org/10.1128/JB.01558-08

Cabrefiga J, Bonaterra A, Montesinos E (2007) Mechanisms of antagonism of *Pseudomonas fluorescens* EPS62e against *Erwiniaamylovora*, the causal agent of fire blight. Int Microbiol 10:123–132. https://doi.org/10.2436/20.1501.01.18

Caldwell CJ, Hynes RK, Boyetchko SM, Korber DR (2012) Colonization and bioherbicidal activity on green foxtail by *Pseudomonas fluorescens* BRG100 in a pesta formulation. Can J Microbiol 58:1–9. https://doi.org/10.1139/w11-109

Chaturvedi V, Kumar A (2014) Metabolism dependent chemotaxis of *Pseudomonas aeruginosa* N1 towards anionic detergent sodium dodecyl sulfate. Indian J Microbiol 54:134–138. https://doi.org/10.1007/s12088-013-0426-8

De Maayer P, Anderson D, Cary C, Cowan DA (2014) Some like it cold: understanding the survival strategies of psychrophiles. EMBO Rep 15:508–517. https://doi.org/10.1002/embr.201338170

Dharmaprakash A, Reghunathan D, Sivakumar KC, Prasannakumar M, Thomas S (2016) Insights into the genome sequences of an *N*-acyl homoserine lactone molecule producing two *Pseudomonas* spp. isolated from the Arctic. Genome Announc 4:e00767–e00716. https://doi.org/10.1128/genomeA.00767-16

Egamberdieva D, Wirth S, Behrendt U, Ahmad P, Berg G (2017) Antimicrobial Activity of medicinal plants correlates with the proportion of antagonistic endophytes. Front Microbiol 8:199. https://doi.org/10.3389/fmicb.2017.00199

Gewin V (2010) Food: an underground revolution. Nature 466:552–553. https://doi.org/10.1038/466552a

Gupta V, Gupta N, Capalash N, Sharma P (2017) Bio-prospecting bacterial diversity of hot springs in Northern Himalayan region of India for laccases. Indian J Microbiol:1–7. https://doi.org/10.1007/s12088-017-0656-2

Harding DP, Raizada MN (2015) Controlling weeds with fungi, bacteria and viruses: a review. Front Plant Sci 6:659. https://doi.org/10.3389/fpls.2015.00659

Jani J, Parvez N, Mehta D (2015) Metabolites of *Pseudomonads*: a new avenue of plant health management. In: Chakravarthy AK (ed) New horizons in insect science: towards sustainable pest management. Springer, New Delhi, pp 61–69. https://doi.org/10.1007/978-81-322-2089-3_7

Jayapradha C, Yesu Raja I (2016) A Review of eco-friendly management of *Alternaria* species. Indian J Hill Farm 29:106–119

Jendoubi W, Bouhadida M, Boukteb A, Béji M, Kharrat M (2017) *Fusarium* wilt affecting chickpea crop. Agriculture 7:23. https://doi.org/10.3390/agriculture7030023

Junaid JM, Dar NA, Bhat TA, Bhat AH, Bhat MA (2013) Commercial biocontrol agents and their mechanism of action in the management of plant pathogens. Int J Mod Plant Ani Sci 1:39–57

Kalia VC, Kumar P, Pandian SK, Sharma P (2014) Biofouling control by quorum quenching. In: Kim SK (ed) Hb_25 Springer handbook of marine biotechnology 15. Springer, Berlin, pp 431–440

Kalia VC (2015) Microbes: the most friendly beings? In: Kalia VC (ed) Quorum sensing vs quorum quenching: a battle with no end in sight. Springer, India, pp 1–5. https://doi.org/10.1007/978-81-322-1982-8_1

Kalia VC, Kumar P (2015) Genome wide search for biomarkers to diagnose *Yersinia* infections. Indian J Microbiol 55:366–374. https://doi.org/10.1007/s12088-015-0552-6

Kalia VC, Kumar P, Pandian SK, Sharma P (2014) Biofouling control by quorum quenching. In: Kim SK (ed) Hb_25 Springer handbook of marine biotechnology 15. Springer, Berlin, pp 431–440

Kalia VC, Prakash J, Koul S, Ray S (2017) Simple and rapid method for detecting biofilm forming bacteria. Indian J Microbiol 57:109–111. https://doi.org/10.1007/s12088-016-0616-2

Kandel SL, Firrincieli A, Joubert PM, Okubara PA, Leston ND, McGeorge KM, Mugnozza GS, Harfouche A, Kim S-H, Doty SL (2017) An in vitro study of bio-control and plant growth promotion potential of Salicaceae endophytes. Front Microbiol 8:386. https://doi.org/10.3389/fmicb.2017.00386

Kaur G, Rajesh S, Princy SA (2015) Plausible drug targets in the *Streptococcus mutans* quorum sensing pathways to combat dental biofilms and associated risks. Indian J Microbiol 55:349. https://doi.org/10.1007/s12088-015-0534-8

Kergunteuil A, Bakhtiari M, Formenti L, Xiao Z, Defossez E, Rasmann S (2016) Biological control beneath the feet: a review of crop protection against insect root herbivores. Insects 7:70. https://doi.org/10.3390/insects7040070

Khan Z, Roman D, Kintz T, delas Alas M, Yap R, Doty S (2014) Degradation phytoprotection and phytoremediation of phenanthrene by endophyte *Pseudomonas putida* PD1. Environ Sci Technol 48:12221–12228. https://doi.org/10.1021/es503880t

Kilani J, Fillinger S (2016) Phenylpyrroles: 30 years, two molecules and (nearly) no resistance. Front Microbiol 7:2014. https://doi.org/10.3389/fmicb.2016.020142016

Kotasthane AS, Agrawal T, Zaidi NW, Singh US (2015) Identification of siderophore producing and cynogenic fluorescent *Pseudomonas* and a simple confrontation assay to identify potential bio-control agent for collar rot of chickpea. 3Biotech 7:137. https://doi.org/10.1007/s13205-017-0761-2

Koul S, Kalia VC (2017) Multiplicity of quorum quenching enzymes: a potential mechanism to limit quorum sensing bacterial population. Indian J Microbiol 57:100–108. https://doi.org/10.1007/s12088-016-0633-1

Koul S, Prakash J, Mishra A, Kalia VC (2016) Potential emergence of multi-quorum sensing inhibitor resistant (MQSIR) bacteria. Indian J Microbiol 56:1–18. https://doi.org/10.1007/s12088-015-0558-0

Krishna PS, Kumar BS, Raju P, Murty MSR, Rao TP, Charya MSA, Prakasham RS (2015) Fermentative production of pyranone derivate I from marine *Vibrio* sp. SKMARSP9: isolation, characterization and bioactivity evaluation. Indian J Microbiol 55:292–301. https://doi.org/10.1007/s12088-015-0521-0

Kumar P, Mehariya S, Ray S, Mishra A, Kalia VC (2015) Biodiesel industry waste: a potential source of bioenergy and biopolymers. Indian J Microbiol 55:1–7. https://doi.org/10.1007/s12088-014-0509-1

Kunova A, Bonaldi M, Saracchi M, Pizzatti C, Chen X, Cortesi P (2016) Selection of *Streptomyces* against soil borne fungal pathogens by a standardized dual culture assay and evaluation of their effects on seed germination and plant growth. BMC Microbiol 16:272. https://doi.org/10.1186/s12866-016-0886-1

Lakshmi V, Kumari S, Singh A, Prabha C (2015) Isolation and characterization of deleterious *Pseudomonas aeruginosa* KC1 from rhizospheric soils and its interaction with weed seedlings. J King Saud Uni Sci 27:113–119. https://doi.org/10.1016/j.jksus.2014.04.007

Liu K, Hu H, Wang W, Zhang X (2016) Genetic engineering of *Pseudomonas chlororaphis* GP72 for the enhanced production of 2-hydroxyphenazine. Microb Cell Factories 15:131. https://doi.org/10.1186/s12934-016-0529-0

Manogari R, Daniel DK (2015) Isolation, characterization and assessment of *Pseudomonas* sp. VITDM1 for electricity generation in a microbial fuel cell. Indian J Microbiol 55:8–12. https://doi.org/10.1007/s12088-014-0491-7

Mavrodi OV, Walter N, Elateek S, Taylor CG, Okubara PA (2012) Suppression of *Rhizoctonia* and *Pythium* root rot of wheat by new strains of *Pseudomonas*. BiolCont 62:93–102. https://doi.org/10.1016/j.biocontrol.2012.03.013

Meena M, Zehra A, Dubey MK, Aamir M, Gupta VK, Upadhyay RS (2016) Comparative evaluation of biochemical changes in tomato (*Lycopersicon esculentum* Mill.) infected by *Alternaria* alternate and its toxic metabolites (TeA, AOH, and AME). Front Pl Sci 7:1408. https://doi.org/10.3389/fpls.2016.01408

Mehrabi Z, McMillan VE, Clark IM, Canning G, Hammond-Kosack KE, Preston G, Hirsch PR, Mauchline TM (2016) *Pseudomonas* spp. diversity is negatively associated with suppression of the wheat take-all pathogen. Sci Rep 6:29905. https://doi.org/10.1038/srep29905

Mezaache-Aichour S, Haichour N, Guechi A, Nicklin J, Zerroug MM (2016) Bacteriocins contributing in rhizospheric competition among fluorescent *Pseudomonads*. Annual Res Rev Biol 11:1–9. https://doi.org/10.9734/ARRB/2016/30682

Moubarak MY, Abdel-Monaim MF (2011) Effect of bio-control agents on yield, yield components and root rot control in two wheat cultivars at New Valley region. Egypt J Cereals Oilseeds 2:77–87. https://doi.org/10.5897/JCO11.029

Ng WL, Bassler BL (2009) Bacterial quorum-sensing network architectures. Annu Rev Genet 43:197–222. https://doi.org/10.1146/annurev-genet-102108-134304

O'Loughlin CT, Miller LC, Siryaporn A, Drescher K, Semmelhack MF, Bassler BL (2013) A quorum-sensing inhibitor blocks *Pseudomonas aeruginosa* virulence and biofilm formation. Proc Natl AcadSci 110:17981–17986. https://doi.org/10.1073/pnas.1316981110

Oteino N, Lally RD, Kiwanuka S, Lloyd A, Ryan D, Germaine KJ, Dowling DN (2015) Plant growth promotion induced by phosphate solubilizing endophytic *Pseudomonas* isolates. Front Microbiol 6:1–9. https://doi.org/10.3389/fmicb.2015.00745

Pandin C, Coq DL, Canette A, Aymerich S, Briandet R (2017) Should the biofilm mode of life be taken into consideration for microbial biocontrol agents? Microb Biotechnol 10:719–734. https://doi.org/10.1111/1751-7915.12693

Papenfort K, Bassler B (2016) Quorum-sensing signal-response systems in gram-negative bacteria. Nat Rev Microbiol 11:576–588. https://doi.org/10.1038/nrmicro.2016.89

Pooja S, Pushpananthan M, Jayashree S, Gunasekaran P, Rajendhran J (2015) Identification of periplasmic a-amlyase from cow dung metagenome by product induced gene expression profiling (Pigex). Indian J Microbiol 55:57–65. https://doi.org/10.1007/s12088-014-0487-3

Rani P, Mahato NK, Sharma A, Narsimha Rao DL, Kamra K, Lal R (2017) Genome mining and predictive functional profiling of acidophilic rhizobacterium *Pseudomonas fluorescens* Pt14. Indian J Microbiol 57:155–161. https://doi.org/10.1007/s12088-017-0648-2

Rekadwad BN (2014) Growth promotion of crop plants by *Methylobacterium organophilum*: efficient bio-inoculant and bio-fertilizer isolated from mud. Res Biotechnol 5:1–6

Rekadwad BN, Khobragade CN (2015) A case study on effects of oil spills and tar-ball pollution on beaches of Goa (India). Mar Pollut Bull 100:567–570. https://doi.org/10.1016/j.marpolbul.2015.08.019

Rekadwad BN, Khobragade CN (2017a) Microbial biofilm: role in crop productivity. In: Kalia VC (ed) Microbial applications volume 2 – biomedicine, agriculture and industry. Springer, Cham, pp 107–118. https://doi.org/10.1007/978-3-319-52669-0_5

Rekadwad BN, Khobragade CN (2017b) Bacterial quorum sensing (QS) in rhizosphere (paddy soil): understanding soil signaling and N- recycling for increased crop production. In: Kalia VC (ed) Microbialapplications volume 2 – biomedicine, agriculture and industry. Springer, Cham, pp 119–131. https://doi.org/10.1007/978-3-319-52669-0_6

Rekadwad BN, Khobragade CN (2017c) Marine polyextremophiles and their biotechnological applications. In: Kalia VC, Kumar P (eds) Microbial applications volume 1 – bioremediation and bioenergy. Springer, Cham, pp 319–331. https://doi.org/10.1007/978-3-319-52666-9_15

Rekadwad BN, Khobragade CN, Jadhav VG, Kadam SU (2016) Enhancing growth of *Vigna radiata* in the presence of *Pseudomonas aeruginosa* bio-polymer and *Metarhizium anisopliae* spores. AdvAgri 4314958:7. https://doi.org/10.1155/2016/4314958

Ruiu L (2015) Insect pathogenic bacteria in integrated pest management. Insects 6:352–367. https://doi.org/10.3390/insects6020352

Rutherford ST, Bassler BL (2012) Bacterial quorum sensing: its role in virulence and possibilities for its control. Cold Spring Harb Perspect Med 2:a012427. https://doi.org/10.1101/cshperspect.a012427

Sah S, Singh N, Singh R (2017) Iron acquisition in maize (*Zea mays* L.) using *Pseudomonas* siderophore. 3 Biotech 7:121. https://doi.org/10.1007/s13205-017-0772-z

Sasirekha B, Srividya (2016) Siderophore production by Pseudomonas aeruginosa FP6, a biocontrol strain for *Rhizoctonia solani* and *Colletotrichum gloeosporioides* causing diseases in chilli. Agri Nat Resour 50:250e256. https://doi.org/10.1016/j.anres.2016.02.003

Shang Z, Wang H, Zhou S, Chu W (2014) Characterization of N-acyl-homoserine lactones (AHLs)-deficient clinical isolates of *Pseudomonas aeruginosa*. Indian J Microbiol 54:158–162. https://doi.org/10.1007/s12088-014-0449-9

Shehata HR, Ettinger CL, Eisen JA, Raizada MN (2016) Genes required for the anti-fungal activity of a bacterial endophyte isolated from a Corn Landrace grown continuously by subsistence farmers since 1000 BC. Front Microbiol 7:1548. https://doi.org/10.3389/fmicb.2016.01548

da Silva Vasconcellos FC, de Oliveira AG, Lopes-Santos L, de Oliveira Beranger JP, Cely MVT, Simionato AS, Pistori JF, Spago FR, de Mello JCP, San Martin JAB, de Jesus Andrade CGT, Andrade G (2014) Evaluation of antibiotic activity produced by *Pseudomonas aeruginosa* LV strain against *Xanthomonas arboricola* pv. pruni. Agric Sci 5:71–76. https://doi.org/10.4236/as.2014.51008

Sun S, Zhou L, Jin K, Jiang H, He YW (2016) Quorum sensing systems differentially regulate the production of phenazine-1-carboxylic acid in the rhizobacterium *Pseudomonas aeruginosa* PA1201. Sci Rep 6:30352. https://doi.org/10.1038/srep30352

Vacheron J, Moënne-Loccoz Y, Dubost A, Gonçalves-Martins M, Muller D, Prigent-Combaret C (2016) Fluorescent *Pseudomonas* strains with only few plant-beneficial properties are favored in the maize rhizosphere. Front Plant Sci 7:1212. https://doi.org/10.3389/fpls.2016.01212

Vaikuntapu PR, Dutta S, Samudrala RB, Rao VRVN, Kalam S, Podile AR (2014) Preferential promotion of *Lycopersicon esculentum* (Tomato) growth by plant growth promoting bacteria associated with tomato. Indian J Microbiol 54:403–412. https://doi.org/10.1007/s12088-014-0470-z

van Lenteren JC, Bolckmans K, Kohl J, Ravensberg WJ, Urbaneja A (2018) Biological control using invertebrates and microorganisms: plenty of new opportunities. Biol Control 63:1–21. https://doi.org/10.1007/s10526-017-9801-4

Wachowska U, Kucharska K, Jędryczka M, Łobik N (2013) Microorganisms as biological control agents against *Fusarium* pathogens in winter wheat. Pol J Environ Stud 22:591–597

Wallacea R, Hirkalab DL, Nelson LM (2017) Postharvest biological control of blue mold of apple by *Pseudomonas fluorescens* during commercial storage and potential modes of action. Postharvest Biol Technol 133:1–11. https://doi.org/10.1016/j.postharvbio.2017.07.003

Weller DM (2007) *Pseudomonas* biocontrol agents of soil borne pathogens: looking back over 30 years. Phytopathology 97:250–256. https://doi.org/10.1094/PHYTO-97-2-0250

Wu X, Duan H, Tian T, Yao N, Zhou H, Zhang L (2010) Effect of the *hfq*gene on 2,4-diacetylphloroglucinol production and the PcoI/PcoR quorum-sensing system in *Pseudomonas fluorescens*2P24. FEMS Microbiol Lett 309:16–24. https://doi.org/10.1111/j.1574-6968.2010.02009.x

Yang JH, Zhang WW, Zhuang YQ, Xiao T (2017) Biocontrol activities of bacteria from cowdung against the rice sheath blight pathogen. J Plant Disease Prot 124:131. https://doi.org/10.1007/s41348-017-0080-1

13 Significance and Application of Quorum Sensing in Food Microbiology

Venkadesaperumal Gopu, Sivasankar Chandran, and Prathapkumar Halady Shetty

Abstract

Food spoilage and food borne diseases are still a significant problem even with the application of contemporary food preservation techniques. Microbial spoilage of food has led to serious public health consequences and led to the emergence of multi drug resistant strains. Quorum signaling mechanism in many of these spoilage and pathogenic bacteria for the management of phenotypic characteristics, including virulence determination has been well documented. Appreciating the knowledge on microbial ecosystem of food and their QS mediated signaling mechanism may aid in combating the microbial infections in food and food processing industries. Owing to abundant literatures on basics of quorum sensing, in this chapter we have narrowed down to microbial ecosystem of food, food borne pathogens and their QS regulatory mechanism. In addition, we have also attempted to throw up light on (i) quorum signals in food spoilage, (ii) QS mediated biofilms in food and food processing industries, (iii) detection of food spoilage using QS-biosensors and, (iv) several food based QS inhibitors (QSIs) as food preservatives and eventual approach which has to be given interest to further reveal the underlying mechanism of QS signaling in food system and potential food based QS inhibitors for the management of microbial infections as a novel food intervention strategy to ensure food safety and quality.

First and second authors contributed equally to the chapter.

V. Gopu
Department of Food Science and Technology, Pondicherry University, Puducherry, India

University of Illinois at Chicago, Chicago, IL, USA

S. Chandran · P. H. Shetty (✉)
Department of Food Science and Technology, Pondicherry University, Puducherry, India
e-mail: shetty.fst@pondiuni.edu.in

V. C. Kalia (ed.), *Quorum Sensing and its Biotechnological Applications*,
https://doi.org/10.1007/978-981-13-0848-2_13

Keywords
Quorum sensing · Biofilm · Food spoilage · Preservative · Microbial spoilage

13.1 Introduction

In recent years, rapid strides have been made on the microbiology of food spoilage. Food spoilage is of major concern in the food industry owing to the severe public health consequences in the population and considerable economic losses. Food safety is one of the major challenges of food industry in the rapidly melding market with ecumenical shipping and consumption of fresh, raw and miserly processed food. Presently, foodborne diseases are a major worry for the public health in both developed and developing nations (Shi and Zhu 2009). The antibiotics that were discovered for managing these foodborne pathogens have brought relief from a huge number of life-threatening communicable diseases. The unwarranted and haphazard usage of these antibiotics has led to the surfacing of bacterial strains which were resistant to conventional antibiotics (Davies and Davies 2010). Therefore, there is a need for a substitute which possibly renders the control of bacterial infections without stressing the microbial cells and thus reducing the emergence of multiple drug resistant strains.

Various bacterial species known to manage their physiological expressions by communicating among each other in a population-reliant approach called Quorum sensing (QS). A diverse cluster of food associated bacteria use to stimulate and response correlated to population density by discharging small, dispersible signaling particles known as autoinducers (Waters and Bassler 2005). In Gram-negative bacteria, N-acyl homoserine lactones (AHLs) act as autoinducing elements produced by the components of LuxI homologues (autoinducer synthases). These signal molecules binds to the receptor protein (LuxR homologues) to activate the target gene expressions.

Biofilm formation by infectious bacteria is the prime concern that endures throughout the world. Most of the spoilage bacteria were found to form biofilms on the food matrix, or the processing surfaces leading to public health problems and economic losses. Many foodborne pathogens like *Yersinia* sp., *Pseudomonas* sp., *Klebsiella* sp., *Enterobacter* sp., *Erwinia* sp., and *Ralstonia* sp. produces metabolic end products like proteolytic, lipolytic, and chitinolytic enzymes which directly associated with the spoilage process of food products (Bai and Rai 2011). Biofilms are complex aggregates of microorganism sheath in a self-produced exopolysaccharide (EPS) matrix to mature and endure in structured communities on solid surfaces. These biofilms act as proficient blockades against antimicrobials and host immune system resulting in relentless colonization and infection at the site of formation. Also, it renders protection from external stress, biocides and dehydration (Annous et al. 2009).

The identification of quorum signals in food produce has grabbed the attention of food technologists to study the possible role of quorum signaling in the process of food spoilage. It has been reported that several bacterial phenotypes like EPS production, biofilm formation, proteolytic, and pectinolytic activities related with the food spoilage are governed by quorum sensing (Skandamis and Nychas 2012). Many physical and chemical techniques including low-pressure oxygen plasma and water soluble polymers to deal with the bacterial biofilms on processing equipment and other solid surfaces were established to be effectual. On the other hand, considering the surfacing of multi-drug resistant bacterial strains (Koul et al. 2016), unsettling the bacterial signaling apparatus might play a mean part in regulating microbial gene expression associated with food spoilage and foodborne infection.

The involvement of AHLs based quorum signaling in regulating various QS-dependent phenotypes in foodborne pathogens has led to the search of active components that can interrupt the AHL signaling in particular. It is envisaged that the emergence of multi-drug resistant strains may be decreased by such compounds which can chunk the expression of virulence genes without inhibiting the bacterial growth. Halogenated furanones from *Delisea pulchra* was found to exhibit pronounced QS inhibitory activity by competitively inhibiting the binding of AHL signals to the receptor proteins (Manefield et al. 2002). Rasmussen et al. (2000) evidenced that furanones have inhibited biofilm formation and the AHL-based regulation of virulence factors in *Pseudomonas aeruginosa*. However, furanones were found to be unstable in nature and due to its toxicity it cannot be used in mammalian cells has provoked the hunt for stable, nontoxic compounds from natural sources. The identification of plant-based quorum sensing inhibitors has elevated the prospects of finding a promising source of QS inhibitory compounds from the plenty of natural resources, and the examination of their toxicological nature may smooth the progress of using QS inhibitors as food preservatives.

Although quorum sensing bacteria and their communicating molecules have been identified in food systems, the precise function played by them in the process of food spoilage is not apparent. Several synthetic and natural compounds have been demonstrated for their QS inhibitory activity against a broad range of pathogens, but the mechanistic activity lies behind their inhibitory nature have not been well documented so far. The overall objective of this chapter is to discuss about the bacterial ecosystem and their communication mechanism in food system. In addition to disclose reported food based QS inhibitors and their potential application as food preservatives as a novel food intervention strategy.

13.2 Bacterial Communication

One of the remarkable recent advances in the study of microbial gene expression is the verity that many bacterial species correspond among their population through a dedicated intracellular signaling mechanism. This communication system enables bacteria to control the broad spectrum of activities by sensing and integrating the information from its surroundings and thereby activating or repressing the particular gene expression. This population dependent signaling mechanism of bacteria is known as quorum sensing (QS) (Li and Tian 2012).

Quorum sensing bacteria diverges in signaling mechanism, transduction of signaling molecule and regulating genes (Galloway et al. 2011). Many Gram-negative bacteria uses the signaling mechanism mediated by fatty acid derivative (N-acyl-homoserine lactone) detected by promoters and transcriptional factors. On the other hand cyclic peptides and a membrane-correlated factor retort regulatory structure drives the signaling mechanism in Gram-positive bacteria. Many known bacterial species including *Pseudomonas*, *Enterobacter*, *Yersinia*, *Klebsiella*, *Vibrio* and *Agrobacterium* utilizes QS circuit for the regulation of virulence factor synthesis. Quorum sensing mechanism found to regulate biofilm formation, exotoxins and anti-microbial peptides in bacterial genera like *Streptomyces*, *Enterococcus*, *Bacillus* and *Staphylococcus* (Rutherford and Bassler 2012). The nitrogen fixation in the *Rhizobium* genus is regulated through QS mechanism. Conversely, *Vibrio harveyi* shares the characteristics of both Gram-positive and Gram-negative QS systems. Like Gram-negative bacteria, it synthesize and sense acylated homoserine lactones whereas, the signal transduction of acylated homoserine lactones happens by membrane-bound histidine kinases like Gram-positive bacteria (Waters and Bassler 2005).

Food safety has become one of the top priorities in the era of modern globalization with global shipping and consumption of raw, unsullied and modestly processed foods. It has been estimated that in developing counties like India 30% of people suffered from outbreaks of food borne pathogens (Scallan et al. 2011). Fresh products such as fruits, vegetables, meat and sea foods were found to be contaminated most likely by food-borne pathogens such as *Campylobacter* spp., *Escherichia coli*, *Listeria monocytogenes*, *Salmonella* spp., *Staphylococcus aureus*, and others (Dhama et al. 2015). Microbial activity is measured to be of immense import for the materialization of food spoilage as they produce proteolytic, lipolytic and pectinolytic enzymes whose end metabolites are allied with food deterioration (Ragaert et al. 2007). The discovery of quorum signals in food products has grabbed the attention of food technologists to study the possible role of quorum signaling in the process of food spoilage. It has been reported that several bacterial phenotypes such as EPS production, biofilm formation and other phenotypic expressions related with the food spoilage are governed by quorum sensing. Recently, several research groups has started to focus on quorum sensing based strategies to combat microbial pathogenesis in the context of food spoilage. Owing to the preponderance of current literature on the other aspects of quorum sensing, in this chapter we have taken efforts to afford insights into the part of bacterial signaling in food putrefaction, signaling mechanisms in food system, food based QS inhibitors and its potential role as food preservatives, bio-sensors based detection of food spoilage and the future research perspectives to investigate the "gray" or "black" side of bacterial communication in food system to exploit its benefits to ensure food preservation and food safety.

13.3 Microbial Ecosystem of Food

Microbes are ubiquitous in nature, as food substances comprise lots of nutrients, it is acting as a milieu for a diverge variety of organisms. Food is an ideal niche for a dynamic microbial ecosystem with high degree of complexity. It comprises of a broad range of bacteria, yeasts and filamentous fungi, in which they live, interact and communicate each other. The microbial interaction among the consortia is supposed to be fundamental to arrive at desired characteristics of a food product. As customary, the microbial inhabitants of food are categorized into beneficial and pathogenic organisms that cause betterment and spoilage of food respectively.

Fermentation is one of the best technologies to develop nutritious food products using beneficial microbes from time immemorial. Lactic acid bacteria (LAB) have become indispensable in food industries to produce fermented foods. They rapidly acidify the food materials by producing organic acid especially lactic acid. They can also produce acetic acid, ethanol, antimicrobials, aromatic compounds, exopolysaccharides and industrially important enzymes (Pessione 2012). As they are much beneficial they have also been used as probiotics in food and pharmaceutical products. Moreover, food fermentation is mostly carried out by complex microbial consortia, which operate the process altogether and resulting in food product. As consortia the microbes can able to execute intricate activities and possess more versatility and robustness than pure cultures (Smid and Lacroix 2013).

Ercolini et al. (2012) have reported the microbiota involved in the manufacture of water buffalo mozzarella cheese. They observed that some thermophilic LAB could carry on fermentation, while the mesophilic LAB such as *Lactobacillus lactis* are relatively fewer during the production. Similarly in a parallel study, the ascendance of common cheese bacteria was demonstrated. In furtherance, the difference in microbiota of pasteurized and raw milk cheeses and association between *Lactobacillus* populations and cheese maturation was also exemplified (Quigley et al. 2012). A culture-independent study delineated the effect of NaOH treatment on the microbiota of fermentation of table olives. As reported in the study, usual fermentation of the olives was initiated by halophilic bacteria and subsequently replaced by *Lactobacillus* at the later phase of fermentation. Whereas in the NaOH treated olives, they were dominated by enterobacteria. Moreover, significant disruption was found in the diversity of *Lactobacillus plantarum* biotypes (Cocolin et al. 2013). In another report, ecological succession of microbiota of rye and wheat sourdough preparation was elaborated. Throughout the fermentation, variation of ratio between dominant and subdominant populations of *L. sakei*, *Leuconostoc* spp., *Weissella* spp., and *Lactococcus lactis* was described (Ercolini et al. 2013). In general, the application of high throughput sequencing is highly constructive to comprehend the microbial functions and behavior corresponding to various process conditions. Further, metagenomic approaches will help to scrutinize the molecular evolution of various strains and to ascertain the quality of product, process efficiency and food safety (De Filippis et al. 2017).

13.4 Food Spoilage and Pathogenic Bacteria

Food spoilage is a multifarious process done by microbes causing lose of 25% world's food supply and a vast degree of illness. The economical burden and food borne diseases by spoilage of food is continuing in the contemporary technical advancement. Any undesirable physical or chemical change in the texture, flavor or odor by microbial growth or its metabolism is defined as microbial food spoilage. In moist atmospheric conditions, *Pseudomonas* has been found to adhere swiftly to meat surfaces (Sohaib et al. 2016). In the meat stored at aerobic refrigeration conditions, the genera *Moraxella*, *Psychrobacter* and *Acinetobacter* causes spoilage. Although Gram-negative, motile and non-motile aerobic rods are the dominant members of meat spoilage microbiota, varying degree of Gram-positive bacteria majorly *Micrococci*, lactic acid bacteria and *Bronchothrix thermosphacta* are also present (Skandamis and Nychas 2012).

Raw milk spoilage by lactic acid bacteria has turned out to be unfamiliar since from the arrival of refrigeration. However, a wide range of psychrotrophic bacteria have been identified from milk. Gram negative bacteria such as *Pseudomonas*, *Acinetobacter*, *Enterobacter*, *Klebsiella*, *Achromobacter*, *Flavobacterium*, *Aeromonas*, *Alcaligenes* and Gram-positive such as *Bacillus*, *Corynebacterium*, *Micrococcus*, *Microbacterium* and *Clostridium* were identified as psychrotrophic milk spoilers. Similarly, psychrotrophic Gram-negative rods such as *Alcaligenes*, *Achromobacter*, *Flavobacterium* and *Pseudomonas* can cause intolerable odors and flavors in soft cheeses by producing proteolytic and lipolytic enzymes (Oliveira et al. 2015). Cereals are fouled with several yeasts such as *Candida* spp., *Debaryomyces* spp., *Hansenula* spp., *Pichia* spp., *Saccharomyces* spp., *Saccharomycopsis* spp. and filamentous fungi such as *Stemphylium* spp., *Ulocaldium* spp., *Penicillium* spp., *Aspergillus* spp., *Eurotium* spp. throughout harvest and post-harvest processing.

Food borne pathogens are not necessarily to cause food spoilage but, cause food borne illness to the consumers. Food borne diseases are caused predominantly by bacteria especially *Listeria monocytogenes*, *Campylobacter jejuni*, *Salmonella* spp., *Yersinia enterocolitica*, *Clostridium spp. S. aureus*, *Bacillus cereus* and *E. coli*. As reported by Wirtanen and Salo (2003) the foresaid pathogens are potential biofilm formers and not responding to disinfectants used to clean the surfaces in the food industry.

13.4.1 *Campylobacter* spp.

Campylobacter jejuni and *C. coli* are the leading etiological agents of Campylobacter enteritis in humans (Serichantalergs et al. 2017). *C. jejuni* is accountable for 80–90% of campylobacteriosis. In Europe, as an average 51.6 per 100,000 populations were affected with campylobacteriosis. Overall, it has been estimated that nearly 2.5 million cases per year are recorded with Campylobacter infection in the US, and 80% of them have arrived from food-borne transmission (Kaakoush et al. 2015).

13.4.2 *Listeria monocytogenes*

Fecal material of healthy birds is comprised of *L. monocytogenes* and that may contaminate the poultry meat while processing (Ishola et al. 2016). *L. monocytogenes* has been reported to be found in a broad range of food commodities such as raw and fermented meat products, raw and cooked poultry, raw and smoked fish, raw vegetables, raw and improperly pasteurised milk, ice creams and cheeses (Jemmi and Stephan 2006). Even though, *L. monocytogenes* is a non spore-forming bacterium it has the potential to persist in detrimental conditions such as freezing, drying and heating. It can survive and grow even up to 3 °C, pH 4.4 and at salinity of 14%. *L. monocytogenes* is chiefly causing nosocomial and food-borne illness and it also transmitted through animal contact. Nevertheless, food-borne *Listeria* infections are less frequent when compared with other food borne pathogens, but it can cause serious infections such as Listeriosis. It is a life threatening issue for elderly people with an exhausted immune system and it can even cause miscarriage pregnant women by infecting the fetus. *L. monocytogenes* is the one having the highest mortality rate of about 28% among all the food borne pathogens (Wesley 2009).

13.4.3 *Salmonella* spp.

Salmonella is the most common food borne pathogen of animal origin. It is ubiquitous in nature, overall nearly 2400 serovars are identified and all are found to be human or animal pathogens. The *S. enteritidis* and *S. typhimurium* are the prevalent serotypes mainly in poultry and meats. A vast range of foods have been found with various serotypes of Salmonella such as poultry, eggs, meats, fish, shrimp, milk and dairy products, cream-filled desserts, cake mixes, cocoa, chocolate, peanut butter, yeast, coconut, sauces, dried gelatine and in dried chillies. Generally *Salmonella* serotypes were identified from egg shells but *S. enteritidis* is used to dwell inside the egg especially in yolk (Kim et al. 2014). The most widespread food borne illness caused by this genus is Salmonellosis and developing as one of the foremost public health problems. Every year millions of patients are hospitalized and resulting deaths in thousands. In Europe, the spread of *S. enteritidis* and *S. typhimurium* makes increased disquiet. In furtherance, the development of multi drug resistance in Salmonella is emerging as a huge obstacle for clinicians to treat salmonella associated infections worldwide (Jackson et al. 2013; Doumith et al. 2016).

13.4.4 *Yersinia enterocolitica*

Yersiniosis caused by *Y. enterocolitica* is one the important food borne illnesses next to campylobacteriosis and salmonellosis. Developing countries in temperate zone is struggling with high prevalence of yersiniosis even leads fatality (Kanan and Abdulla 2009; Okwori et al. 2009). *Y. enterocolitica* has been isolated from various food sources such as meat products, mussels, egg products, shrimp, fish, poultry and vegetables such as carrots, mushrooms, celery and cabbage (Rahman et al. 2011). *Y. enterocolitica* infections are highly linked with ingestion of raw or improperly cooked meat especially pork and has also been isolated from pasteurized milk (Fredriksson-Ahomaa et al. 2006). Several biotypes of *Y. enterocolitica* were reported with drug resistance against commonly prescribed antibiotics such as carbenicillin, ticarcillin, cephalothin, amoxicillin/clavulanic acid and cefoxitin (Fredriksson-Ahomaa et al. 2001).

13.4.5 *Campylobacter*

Campylobacter is known as the prime etiological agent of foodborne gastroenteritis in the US and also one of the recurrent pathogens causing acute bacterial enteritis worldwide. In fact, it is too hard to discriminate the *Campylobacter* gastroenteritis from diarrheal disease caused by *Salmonella* and *Shigella*. Authoritative diagnosis needs isolation and identification of bacteria from stool sample from afflicted patients. Nearly 95% of campylobacter enteritis is caused by *C. coli* and *C. jejuni*, these both are clinically indistinguishable (Kaakoush et al. 2015). In Canada, the tetracycline resistance in *C. jejuni* has increased from 19% to 55% since 1985–1995. In US at 1999, 54% of *Campylobacter* isolates were found resistant to at least one antimicrobial agent and 20% of them were resistant to two or more antimicrobials. According to the report by national antimicrobial resistance monitoring system 49% of isolates were found to be resistant to tetracyclines, 22% to quinolones and 2% of the isolates were found to be resistant to macrolides (Iovine 2013).

13.4.6 *Escherichia coli*

In general, *E. coli* is a commensal organism that restrains the colonization of pathogenic bacteria and synthesizing considerable amount of vitamins. Nevertheless, many strains of *E. coli* are acting as a pathogen by causing variety of diseases. So far, pathogenic *E. coli* are sorted into six groups with distinct virulence characteristics. *E. coli* O157:H7 is the most frequently identified pathogenic *E. coli* strain causing significant mortality and morbidity in humans (Nguyen and Sperandio 2012). Quite a lot of outbreaks of pathogenic *E. coli* have been linked with consumption of meat or meat products. Other than beef, meat from goats, sheep, pigs, wild deer and seagulls, feral pigeons are also found with pathogenic *E. coli*. Cheeses have also been identified as an important source of outbreaks of pathogenic *E. coli* (Ferens and Hovde 2011).

13.5 Quorum Sensing Signals in Food Spoilage

Recently, the function of QS in food spoilage has been recognized by detecting the QS signaling molecule in spoiled food. Expression of several degradative enzymes in food spoilage such as protease, lipase, chitinase and pectinase has been identified to be controlled by quorum sensing. Lu et al. (2004) have identified high level of AI-2 activity in frozen fish samples and also in tomato, cantaloupe, carrots, tofu and milk by using a reporter strain *Vibrio harveyi* BB170. Major quorum sensing signaling molecules such as AI-1 and AI-2 have been identified in many essential food substances such as meat, milk and vegetables (Pinto et al. 2007). Christensen et al. (2003) identified that *sprI* gene mutant of *S. proteamaculans* is unable to spoil the pasteurized milk, whereas the same can lead the milk spoilage by addition of 3-oxo-C6-HSL. Thus the role of QS in milk spoilage was demonstrated. The role of QS in raw and pasteurized milk spoilage have been determined by identifying AHLs produced by psychrotrophic *Pseudomonas* spp., *Serratia* spp., *Enterobacter* spp., and *Hafnia alvei* (Pinto et al. 2007). A variety of AHL molecules have been detected and their role in spoilage of cold stored meat and poultry has been established (Jay et al. 2003). *H. alvei* and *Serratia* spp. are found to be the dominant organisms among AHL producing Enterobacteriaceae in the spoilage of vacuum-packed meat products (Ravn et al. 2003). Interestingly, Ammor et al. (2008) have reported that aerobically refrigerated spoiled minced pork have been detected with higher level of QS signaling molecules than meat stored at 20 °C. Gram et al. (1999) have demonstrated the ability of food spoilage bacteria to produce AHLs at very low cell density. Recently, Zhu et al. (2015) have reported that the two cyclic dipeptides have acted as the major QS signaling molecule of *Shewanella baltica* in spoilage of yellow croaker fish. Consequently, interrupting the quorum-sensing system can be an ideal way to control the microbial gene expression associated with food spoilage and infection. Further, in depth research on the function of QS signaling molecule in food deterioration will be helpful in the improvement of food preservation strategies using effective quorum-sensing inhibitors.

13.6 Biofilms in Food and Food Processing Industries

Bacterial biofilm is a ubiquitous, sedentary population of bacterial cells embedded in hydrated polymeric matrices to have the benefit of increased resistant to environmental invectives like antimicrobial agents and antibiotics (Christensen et al. 2012). In socio-microbiology it is revealed that, these bacterial communities express QS signal molecules for both inter-species and intra-species communication within the matrices. Biofilm community in a food processing settings may be dwelled with copious diverse species in slam vicinity (Habimana et al. 2010). Cell-to-cell synergisms have been exhibited to play a key part in formation of biofilm structure and resisting the communal bacterial species against antimicrobial proxies. Mono-species biofilms were found to be less stable than diverse species biofilms (Van der Veen and Abee 2011). Bacterial biofilm construction has major impact on several human deeds and they can form on any surfaces like stainless steel, wood, rubber, animal tissues, teeth, medical devices and so on (Bai and Rai 2011). In food industry biofilms formed by spoilage and pathogenic bacteria remains as continual source of serious health hazards, food contamination and severe economic losses (Skandamis and Nychas 2012).

The capability of bacteria to append to the inanimate planes and form biofilms is a cause of distress for many food industries, including aquaculture and meat processing industries. Several studies has evidenced the ability of food borne pathogens to attach to various food contact surfaces and from robust biofilms, including *Y. enterocolitica* (Venkadesaperumal et al. 2016a), *L. monocytogenes* (Renier et al. 2011; Skandamis and Nychas 2012), *S. enterica* (Giaouris et al. 2012), *E. coli* (Dourou et al. 2011; Simpson Beauchamp et al. 2012), *C. jejuni* (Hanning et al. 2008). In addition to food borne pathogens, the biofilm forming ability of several other bacterial genera which are concerned in the spoilage of fresh, treated and minimally processed foods includes, *P. aeruginosa* (Venkadesaperumal et al. 2016a), *K. pneumoniae* (Venkadesaperumal et al. 2016b), *Lactobacillus* (Bove et al. 2012), *Leuconostoc* (Leathers and Bischoff 2011) have been studied in great detail.

In meat and meat processing industries, the attachment and subsequent biofilm formation remains as a serious threat since it leads to cross-contamination and severe economic losses due to spoilage of the product. Chagnot et al. (2012) studied the adherence of *E. coli* to beef cadaver. Morild et al. (2011) studied the biofilm formation by *Y. enterocolitica* and *L. monocytogenes* on red meat coat. *Pseudomonas* was found to co-exist with *Listeria* and forms biofilms on stainless steel surface by producing profuse quantity of exopolysaccharides. In the cultivation of aquatic organisms like shrimp and molluscs aquatic pathogens like *V. harveyi*, *Aeromonas hydrophila* and *Aeromonas salmonichida* persists as infecting agents by forming biofilms in the tanks and pipes. Biofilms on food processing surfaces are different than those on fresh produce. Fett (2000) reported the biofilms on sprouts like broccoli and alfalfa. The biofilm formations on leafy produce were found to be predominant on veins, stomata and cell wall junctions. Biofilm formation on fresh produce like cabbage (Patel and Sharma 2010), spinach (Niemira and Cook 2010) and lettuce (Olmez and Temur 2010) were reported on many studies.

Various studies has supported that the bacterial infections predominates on the injured parts of fresh produce. It has been demonstrated by Khalil and Frank (2010) that the injured parts of leafy fresh produces hold up the growth of bacterial pathogens and initiate the discrete infection site in numerous fresh produce. It was also shown that the injured skin parts of pepper disks were infected by *Salmonella* spp. more than the uninjured skin Liao and Cooke (2001). Conversely, another group (Seo and Frank 1999) has studied the entrapment of *E. coli* O157:H7 in injured corners over stomata under confocal laser electron microscope. It has to be noted that the fresh produce has been contaminated by the bacterial pathogens at any time point from the farm to plate, and the infection may be severed by the existence of biofilms on either inured or non-injured surface of the fresh produce.

13.7 Quorum Sensing in Biofilm Formation

Quorum sensing has been shown to govern the biofilm formation and maturation in many bacterial species (Kalia et al. 2017; Brackman et al. 2011). The correspondence amid quorum sensing and biofilm formation in many food borne bacteria was reported by Yoon and Sofos (2010). The possible role of QS in inhibition of biofilm structure of *S. enteritidis* by the food spoilage bacterium *H. alvei* was evidenced by Chorianopoulos et al. (2010). Quorum sensing found to play a major role in regulating the nutritional demand in biofilms. Barrios et al. (2006) studied a novel QS regulator molecule MqsR, B3022 is responsible for biofilm formation in *E. coli*. It has been evidenced that non-QS mutants have been revealed to form slimmer and more nebulous biofilms against the wild types (Priha et al. 2011). The QS mechanism of *Serratia marcescens* manages its pathogenicity by fabrication of virulence factors and its biofilm formation (Bakkiyaraj et al. 2012). *H. alvei,* 071 *halI* mutant strains showed the loss of biofilm forming ability due to the loss of hall showing that the QS molecules are essential for the formation of biofilms (Tan et al. 2014).

Quorum signals were found to be implicated in the different stages of biofilm configuration and maturation. The metabolic programs of the bacterial species in the biofilm architecture are different from their free living counter parts. *S. pneumoniae* strain in the sessile state form biofilm establishes localized infections better than their free living forms (Oggioni et al. 2006). The QS expression in many food borne pathogens were found to be induced by various extraneous and innate food factors (Bai and Rai 2011). Quorum sensing signaling compound regulates the target gene expression for virulence factor such as biofilm formation, motility, bacteriocins production or sporulation when the signaling molecule reaches its threshold level. It is shown that the protease, pectinase and siderophore production in *Enterobacteriaceae* and *Pseudomonas* spp. on beans sprouts has been regulated by the 3- oxo-C6-HSL (Rasch et al. 2005). Numerous studies have evidenced the quorum sensing regulated biofilm development by food associated bacteria. It was presumed that the quorum signaling has involved in biofilm formation of *H. alvei* 071 as AHL molecules were detected in their biofilms. The mutant strain of *H. alvei* 071 halI was found to be non-biofilm producer. The differentiation of *H. alvei* 071 cells into complex, multicellular biofilm structure was found to be mediated by QS signaling mechanism (Viana et al. 2009). The biofilm formation in dairy isolate *H. alvei* is found to be synchronized by AHL mediated quorum sensing mechanism (Vivas et al. 2010). On the other hand few studies have also reported that there is no correlation between the quorum sensing mechanism and biofilm formation by food related bacteria. However, the quorum sensing signals were found to stimulate the increased resistance of biofilms against the antimicrobial agents.

Biofilm formation by bacterial pathogens in food processing environments is a persistent issue. Even though, quorum signals have been identified in various biofilms, defined part of these signals in the biofilm formation is not clear. The role of quorum signals in various stages of biofilm formation and maturation has to be elucidated by advanced studies which may contribute for the precise regulation of biofilm formation and thus to minimize the food spoilage (Annous et al. 2009). As discussed above the role of quorum sensing and its signaling molecules in the biofilm formation by food borne bacteria is evidenced strongly but the molecular mechanism and their exact role has to be studied in detail. It is clear that the many food spoilage compounds or enzymes like protease, lipase, chitinase, cellulase, pectin lyase and pectate lyase were regulated by quorum sensing mechanism. Hence, control over the food spoilage can be achieved by impeding the bacterial cross-talk and thereby ensuring the improved shelf-life and safety of the food products.

13.8 Food Based Quorum Sensing Inhibitors as Preservatives

The mechanism of food spoilage which renders the undesirable product for consumption, often associated with the microbial association which produces extracellular enzymes like cellulase, protease, chitinase, lipase and others. The phenomenon of these microbial associations is found to be regulated by QS (Liu et al. 2007; Van Houdt et al. 2007). Considering the vital role of QS in food spoilage and in regulating virulence factors of food borne pathogens, in the past few years immense importance has been given to the research of potential compounds or mechanism which can hinder the signaling mechanism. Quorum quenching (QQ) can be regarded as every feasible means of interrupting bacterial signaling which may be accomplished either by mimicking or degrading the signaling molecule, inhibiting the synthesis of signaling molecules like acylated homoserine lactones (AHLs) or by the analogues tumbling the activity of AHL cognate receptor proteins (Dembitsky et al. 2011).

The very first technique for disrupting the QS mechanism was aimed at inhibiting the biosynthesis of signaling molecule. The effective quorum sensing inhibitors should be stable and highly specific to the quorum sensing regulator. Numerous potent QS inhibitors (QSIs) have been known to date, many of which have been isolated from natural and ensuing organic chemistry. One such group of promising QSI is halogenated furanones isolated from *D. pulchara* represses the AHL-regulated phenotypes in *Streptococcus liquefaciens* and *P. aeruginosa* by interfering with the receptor proteins. It was evidenced that various food materials contains furanones structurally similar to the one isolated from *D. pulchara* (Hentzer et al. 2003). The identification of plant-based quorum sensing inhibitors has elevated the prospects of finding a promising source of QS inhibitory compounds from the plenty of natural resources, and the examination of their toxicological nature may smooth the progress of using QS inhibitors as food preservatives. Hence, identifying a novel food based quorum sensing inhibitor would be of immense potential in this regard.

Recently, many food based substances have been assessed for their ability to impede QS. Jakobsen et al. (2012) examined the extracts of several food products and edible plants for the QSI activity against the common opportunistic pathogen and demonstrated Iberin, an isothiocyanate from horseradish. Food phytochemicals like reservatorol, cinnamaldehyde, ellagic acid and rutin were demonstrated to reduce the AHLs concentration in *Y. enterocolitica* and *E. carotovora* (Truchado et al. 2012). Extract of fruits, vegetables, spices and herbs have been demonstrated

Table 13.1 lists the several fruits extracts which have been demonstrated as potential QS inhibitors against food borne pathogens and spoilage bacteria

Source	Targeted pathogen	Biological effect	References
Curcumin from Curcuma family	*Escherichia coli*, *Pseudomonas aeruginosa* PAO1, *Proteus mirabilis* and *Serratia marcescens*	Inhibition of Biofilm, exopolysaccharide production, alginate production, swimming and swarming motility	Kalia (2013) and Packiavathy et al. (2014)
Limonene, α-pinene, terpinene-4-ol and linalool	*Escherichia coli*, *Bacilius cereus* and *Pseudomonas putida*	Inhibition of Biofilm	Kerekes et al. (2013)
Albiza schimperiana root methanol extract	AHL-QQ activity in *E. coli* based reporter strain AI1-QQ.1	AHL induced cell lethality	Bacha et al. (2016)
Justica schimperiana seed petroleum ether extract			
Polyphenolic extract from *Rosa rugosa* tea	*Escherichia coli* and *Pseudomonas aeruginosa*	Inhibition of Biofilm	Zhang et al. (2014)
Essential oils of cumin, fennel and pepper	*Salmonella typhimurium*, *Escherichia coli* and *Klebsiella pneumoniae*	Inhibition of Biofilm	Venkadesaperumal et al. (2016c)
Quercetin	*Pseudomonas aeruginosa*, *Yersinia enterocolitica* and *Klebsiella pneumoniae*	Inhibition of biofilm formation, exopolysaccharide (EPS) production, motility and alginate production	Venkadesaperumal et al. (2015a)
Thyme essential oil, carvacrol, and thymol	*Pseudomonas fluorescens*	Inhibition of biofilm formation, and motility	Myszka et al. (2016)
Salvadora persica L. methanolic extracts	*Pseudomonas aeruginosa* PAO1 and Staphylococcus strains	Inhibition of biofilm formation, and motility	Noumi et al. (2017)
Phytol	*P. aeruginosa*	twitching motility, flagella motility and pyocyanin production	Pejin et al. (2015)
α-Terpineol and cis-3-nonen-1-ol	*P. aeruginosa*	pyocyanin inhibition	Ahmad et al. (2015)
Resveratrol	*Campylobacter jejuni*, *C. coli* and *A. butzleri*	Inhibition of biofilm	Duarte et al. (2015)

(continued)

Table 13.1 (continued)

Source	Targeted pathogen	Biological effect	References
Agaricus blazei hot water extract	*P. aeruginosa*	Inhibition of biofilm, pyocyanin production, twitching and swimming motility	Sokovic et al. (2014)
Lavandula angustifolia (lavender oil)	*E. coli* [pSB401] and *E. coli* [pSB1075]	Light production	Yap et al. (2014)
Centella asiatica L.	*P. aeruginosa* PAO1	Inhibition of pyocyanin, elastolytic and proteolytic enzyme production, swarming motility, and biofilm formation	Vasavi et al. (2016)
Quercetin and Quercetin-3-O-arabinoside from *Psidium guajava* L	*P. aeruginosa* PAO1	Inhibition of pyocyanin, elastolytic and proteolytic enzyme production, swarming motility, and biofilm formation	Vasavi et al. (2014)
Amomum tsaoko	*S. aureus*, *S. Typhimurium* and *P. aeruginosa*	Inhibition of swarming motility, and biofilm formation	Rahman et al. (2017a)
Zingerone	*P. aeruginosa* PAO1	Inhibition of swarming motility, and biofilm formation rhamnolipid, elastase, protease, pyocyanin, cell free and cell bound hemolysin	Kumar et al. (2015)
Linalool	*Acinetobacter baumannii*	Inhibition of biofilm formation	Alves et al. (2016)
Murraya koenigii	*Pseudomonas* species	Inhibition of EPS production, and preventing biofilm maturation	Bai and Vittal (2014)
Colostrum hexasaccharide	*Staphylococcus aureus*	Inhibition of biofilm formation	Srivastava et al. (2015)
Cinnamon bark oil	*P. aeruginosa, E. coli* O157:H7 (EHEC)	Inhibition of biofilm formation	Kim et al. (2015)
Carvacrol	*P. aeruginosa*	Inhibition of biofilm formation	Tapia-Rodriguez et al. (2017)
Phytol from *Piper betle*	*S. marcescens*	Prodigiosin, protease, biofilm and hydrophobicity	Srinivasan et al. (2016)
Punicalagin	*Salmonella*	Motility	Li et al. (2014)
[6]-gingerol, [6]-shogaol and zingerone	*P. aeruginosa*	Inhibition of biofilm formation	Kumar et al. (2014)

(continued)

Table 13.1 (continued)

Source	Targeted pathogen	Biological effect	References
Honey with curcumin	*P. aeruginosa* PAO1	Inhibition of pyocyanin, pyoverdin, pyochelin, LasA protease, LasB elastase, and hemolysin	Jadaun et al. (2015)
Pelargonidin, cyanidin and delphinidin	*P. aeruginosa* PAO1	Inhibition of biofilm formation and pyocyanin	Pejin et al. (2017)
Alpha-bisabolol *Padina gymnospora*	*S. marcescens*	Inhibition of biofilm, prodigiosin, protease and swarming	Sethupathy et al. (2016)

as potential quorum quenchers (Kalia 2013). The evidence of antimicrobial activity of several food extract are abundant but several QS inhibitors have been identified to date (Venkadesaperumal et al. 2016c). Table 13.1 lists the several fruits extracts which have been demonstrated as potential QS inhibitors against food borne pathogens and spoilage bacteria. The QS inhibitory activity of these fruits extracts were primarily attributed to their phytochemicals such as flavanoids, anthocyanins, tannins and lignans (Jakobsen et al. 2012). Recently, Venkadesaperumal et al. (2015a, b) evidenced the ability of black jamun (*Syzygium cumini*) rich in anthocyanins and flavanoids, inhibits the exopolysaccharide production and biofilm formation in *Y. enterocolitica* and *P. aeruginosa* by binding with AHL receptor proteins. It was also demonstrated that the QS inhibitory activity of black jamun attributes to the anthocyanin, quercetin which efficiently binds with LasR receptor protein than the natural ligand, evidenced through molecular docking and simulation studies (Venkadesaperumal et al. 2016d). The phenolic extract of *Eugenia brasiliensis,* inhibited the violacein production in *C. violaceum* and swarming motility in *Serratia marcescens* (Rodrigues et al. 2016)

Koh and Tham (2011) demonstrated that the fruits of *Moringa oleifera* Lam inhibited the violacein production in *C. violaceum*. Flavanones rich extract of orange has substantially curtailed the concenration of 3-Oxo-C6-hexanoyl homoserine lactones (HHL) and C6-HHL secreted by *Y. enterocolitica* (Truchado et al. 2012). Numerous research groups have not only examined the QS inhibitory and anti-biofilm activity of fruit extracts but also attempted to discover and separate the bioactive compounds accountable for the inhibitory action. Though several fruit extracts and their phytochemical compounds exert the QS inhibitory activity many of the fruits fails to demonstrate the positive result in the screening.

Extract of Amomum tsaoko was reported to inhibit QS mediated swarming, motility and biofilm formation by *Staphylococcus, Salmonella* and *Pseudomonas* (Rahman et al. 2017a). Recently, Rahman et al. (2017b) proved that the extract of star anise (*Illicium verum*) inhibited exopolysaccharide production, biofilm formation and swarming motility in *S. aureus* in a dose-dependent manner. Broccoli extracts not only lowered AI-2 production, but also swimming and swarming motility in *E. coli* O157:H7 in a dose-dependent behavior (Lee et al. 2011). The ability of a methanolic extract of caper (*Capparis spinosa*) as quorum quencher has been evidenced using the biosensor *C. violaceum* (Abraham et al. 2011). It was also proved to regulate several virulence elements such as swimming and swarming motility, biofilm formation, and exopolysaccharide production in *P. aeruginosa* PAO1, *E. coli*, *Proteus mirabilis* and *S. marcescens* in a dose-dependent way, with no change in the pattern of bacterial growth. Extracellular virulence components such as pyoverdin, alginate, phospolipase and hemolysin production in *P. aeruginosa* has been studied to be inhibited by the garlic extract (Harjai et al. 2010). Extract of clove has been demonstrated for its efficiency to mitigate the lecA gene expression, swarming and pyocyanin production in *P. aeruginosa* PAO1 (Krishnan et al. 2012).

On the investigation of herbal products as QS inhibitors, the essential oils (EOs) of several plant materials were found to be effective agents. Essential oils of juniper, vanilla, tea, rosemary, ginger and several others have illustrated restrained or severe QS inhibitory activity (Kerekes et al. 2013; Alvarez et al. 2012). It has been also evidenced that numerous EOs were potent against biofilms produced by pathogens like *Salmonella, Listeria, Pseudomonas* and *Staphylococcus* (Desai et al. 2012; Valeriano et al. 2012). Recently, Venkadesaperumal et al. (2016c) has demonstrated that enhanced QS inhibitory activity of essential oils in their nanoemulsion form. The nanoemulsions of pepper, cumin and fennel seeds were demonstrated for their efficient inhibition against the biofilm formation of *E. coli*, *S. typhimurium* and *K. pneumoniae* than their essential oils as such.

Bacterial enzyme which has the quorum quenching potential has been classified based on their mechanistic action to degrade the QS signals. Enzymes which cleave the acyl side chains were classified as AHL-acylase and deaminase whereas, enzymes which hydrolyse the lactone ring were classified as AHL-lactonase and decarboxylases. Most of these enzymes harbor pronounced QS inhibitory activity. AHL-lactonases has been reported from various *Bacillus* spp. (Huma et al. 2011). AHL lactonases of *B. thuringiensis*, *B. subtilis* and *B. cereus* has shared 90% homology of their amino acid sequences (Dong and Zhang 2005). *Pseudomonas* and *Ralstonia* sp. produces AHL acylases which were quite diverse in their amino acid sequences. Tait et al. (2009) has reported that, molecule specific activity of the lactonases shares 32–36% resemblance. These enzymes were also found to be species specific. Sio et al. (2006) has demonstrated the long chain AHL-acylase activity of *P. aeruginosa.* It was shown that the enzyme degraded the C12-HSL but it was found to be inactive against C4-HSL. Similarly, the AHL-acylase produced by *Streptomyces sp.* was found to be active against the six or more carbon AHLs (Park et al. 2005).

Shepherd and Lindow (2009) have reported Hac A and C, two potent acylases of *Pseudomonas syringae* B728a which degrades AHLs. Cirou et al. (2009) reported the AHL-oxidase of *Bacillus megaterium* which degrades 4-C and 12-C homoserine lactones. Few other species of Bacillus like, *B. megaterium* was known to produce cytochrome P450, a wide range AHL oxidase to degrade AHLs (Chowdhary et al. 2007). It is reported that the novel quorum quenching bacterium, *Bacillus macrorestinctum* which was noted to reduce *Pectobaceirum caroovorum* induced soft rotting of potato tubers. *Tenacibaculum maritimum*, a fish borne pathogen demonstrated for its biofilm formation through short chain HSLs were found to degrade long chain HSLs by means of AHL-acylase (Romero et al. 2012). In addition to bacterial species which show the presence of either AHL-acylases or lactonases, few organism were know to produce several degrading enzymes and on the other hand few organism were known to be both quorum sensors and quorum quenchers (Kalia 2014). Bentley et al. (2002) reported the non-quorum sensor *B. thuringiensis* shows AHL-lactonase activity. Perpetually, Kalia and Purohit (2011) have reported that the comparative genomic analysis of bacterial strains such as *Photorhabdus luminescens*, *Deinococcus radiodurans* and *Hyphomonas neptunium*. It was shown that these bacterial strains posses the genes which codes for both AHL-acylases and lactonases. Uroz and Heinonsalo (2008) have reported that *Rhodococcus erythropolis* which produces different AHL degrading enzymes.

Also, Zhang and Dong (2004) showed that the strain of *A. tumefaciens* produces mid range AHLs and also the AHL-lactonase which degrades the produced signal molecule under starvation. Similar conditions were also noted in *P. aeruginosa*, *Acinetobacter* and *Burkholderia* spp., in which the QS-dependent pathways get silenced during the AHL degradation activity (Chan et al. 2011). Venkadesaperumal et al. (2016a) has studied the AHL-lactonase activity of *E. ludwigii* isolated from beef against zoonotic pathogen *Y. enterocolitica.* The presence of aiiA gene which codes for lactonase of Metallo-β-lactamase super family was demonstrated. Also, the affinity of the enzymes towards the AHL-lactonase complex was found to be less than that of natural ligand was evidenced through the tertiary structure prediction and molecular docking analysis.

13.9 Detection of Food Spoilage Using QS Bio-sensors

As QS plays a key part in food spoilage, detecting the QS signaling molecule in food or microbes isolated from food can be a paragon method to improve the quality control strategies. It can be identified from extract of food or cell free culture supernatants (CFCS) of food bacteria (Venkadesaperumal and Shetty 2016). The biosensor is easier, economic and faster than analytical techniques to detect QS signals. In general, biosensor strains possess a functional protein of LuxR family cloned with a relevant target promoter. As they do not synthesis AHL, the reporter gene will get expressed only by exogenous AHL and thus resulting in phenotypic response. As reported by Steindler and Venturi (2007), cross streaking of test organism near to biosensor strain have responded well to detect the AHL released by the test strain streaked proximally to biosensor. The AHLs can be extracted from CFCS of late log phase culture by organic solvents and then the extracts can be separated by thin layer chromatography (TLC). The QS signals separated in TLC plate can be identified by formation of spot of phenotypic expression such as colour development as overlaying TLC plate with the biosensor strain (Turovskiy et al. 2007). Burmolle et al. (2003) have developed an *E. coli* biosensor strain with a high copy plasmid containing lux operon from *V. fischeri* combined with green fluorescent protein

(GFP). The expression of GFP by sensing the external AHL can be analysed by flow cytometry. Mostly, the biosensor strains exhibit a high degree of specificity to recognize AHL and are the detection of related AHL is bleak. Hence, many reporter strains are required to detect a range of AHLs. Reporter strains with less stringency to detect AHLs would be helpful to detect a broad range of signals. *A. tumefaciens* NT1 strain is one such reporter strain that possesses less specificity to C6 – C8 AHL with or without 3-oxo group. It is harboring plasmids pDCI41E33 and pDSK519 with traG::lacZ fusion and traR respectively.

As AI-2 is comparatively instable among QS signals detection of AI-2 by chemical method is quite difficult. Hence, at present the detection of AI-2 is relying on biosensors. *V. harveyi* BB170 reporter strain with luxN::Tn5 can produce all three important auto inducers but detects merely AI-2 (Henke and Bassler 2004). *E. coli* O157:H7 TEVS232 strain is used as a biosensor to detect AI-3 signals through lacZ reporter gene fused with LEE1 regulatory region (Sperandio et al. 2002). Detection of auto inducer peptides is mostly based on nisin inducible bioluminescence or fluorescence by bacterial luciferase genes (lux) or GFP gene respectively induced by nisin inducible nisA promoter (Immonen and Karp 2007). The 2-Alkyl-4-quinolones (AHQs) and 2-heptyl-4-quinolone synthesized by *Pseudomonas* spp. can be traced by using the lux based *P. aeruginosa* AHQ reporter strain. The signaling molecule can be detected by either TLC or microtiter assay (Fletcher et al. 2007).

13.10 Future Perspectives

Quorum sensing plays a prime function in regulating food spoilage and food borne diseases. Thus proper and in depth understanding of QS system in those microbes is very much essential. As discussed in the present chapter, many biosensor strains are available to detect the QS signaling molecule from food products as well as from the food borne bacteria. Nevertheless, a broad spectrum reporter strain possesses the ability to detect a variety of signals ranging from AHLs, AIs and AIP is highly needed. So far, the reporter strains are qualitative in terms of detecting QS signals. However, quantity of signaling molecule in a food sample is also a determining factor for food spoilage. Hence, a quantitative biosensor would be more efficient to identify food spoilage. Meanwhile, several food based anti QS agents have been reported, consequently synergistic interaction between those bioactive principles can be studied and novel natural preservative formulation can be developed to antagonize food spoilage and pathogenic bacteria. With increasing information becoming available, QS based strategies could be used in increasing the shelf life of perishable food stocks as wells as innovative strategies in managing food borne pathogens. QS circuit can also be integrated to develop rapid detection kits for food spoilage detection and screening. Leads in anti QS system in biofilm disruption and increasing susceptibility of pathogens to antibiotics could further be explored to develop combination medicines for better management of food borne pathogens.

References

Abraham S, Palani A, Ramaswamy BR, Shunmugiah KP, Arumugam VR (2011) Antiquorum sensing and antibiofilm potential of *Capparis spinosa*. Arch Med Res 42:6568. https://doi.org/10.1016/j.arcmed.2011.12.002

Ahmad A, Viljoen AM, Chenia HY (2015) The impact of plant volatiles on bacterial quorum sensing. Lett Appl Microbiol 60:8–19. https://doi.org/10.1111/lam.12343

Alvarez MV, Moreira MR, Ponce A (2012) Anti-quorum sensing and antimicrobial activity of natural agents with potential use in food. J Food Saf 32:379–387. https://doi.org/10.1111/j.1745-4565.2012.00390.x

Alves S, Duarte A, Sousa S, Domingues FC (2016) Study of the major essential oil compounds of *Coriandrum sativum* against *Acinetobacter baumannii* and the effect of linalool on adhesion, biofilms and quorum sensing. Biofouling 32:155–165. https://doi.org/10.1080/08927014.2015.1133810

Ammor MS, Flórez AB, Alvarez-Martín P, Margolles A, Mayo B (2008) Analysis of tetracycline resistance tet (W) genes and their flanking sequences in intestinal *Bifidobacterium* species. J Antimicrob Chemother 62:688–693. https://doi.org/10.1093/jac/dkn280

Annous BA, Fratamico PM, Smith JL (2009) Quorum sensing in biofilms: why bacteria behave the way they do. J Food Sci 74:24–37. https://doi.org/10.1111/j.1750-3841.2008.01022.x

Bacha K, Tariku Y, Gebreyesus F, Zerihun S, Mohammed A, Weiland-Bräuer N, Schmitz RA, Mulat M (2016) Antimicrobial and anti-quorum sensing activities of selected medicinal plants of Ethiopia: implication for development of potent antimicrobial agents. BMC Microbiol 16:139. https://doi.org/10.1186/s12866-016-0765-9

Bai AJ, Rai VR (2011) Bacterial quorum sensing and food industry. Compr Rev Food Sci 10:183–193. https://doi.org/10.1111/j.1541-4337.2011.00150.x

Bai AJ, Vittal RR (2014) Quorum sensing inhibitory and anti-biofilm activity of essential oils and their in vivo efficacy in food systems. Food Biotechnol 28:269–292. https://doi.org/10.1080/08905436.2014.932287

Bakkiyaraj D, Sivasankar C, Pandian SK (2012) Inhibition of quorum sensing regulated biofilm formation in *Serratia marcescens* causing nosocomial infections. Bioorg Med Chem Lett 22:3089–3094. https://doi.org/10.1016/j.bmcl.2012.03.063

Barrios AFG, Zuo R, Hashimoto Y, Yang L, Bentley WE, Wood TK (2006) Autoinducer 2 controls biofilm formation in *Escherichia coli* through novel motility quorum sensing regulator (MqsR, B3022). J Bacteriol 188:305–316. https://doi.org/10.1128/JB.188.1.305-316.2006

Bentley SD, Chater KF, Cerdeño-Tárraga AM, Challis GL, Thomson NR, James KD, Harris DE, Quail MA, Kieser H, Harper D, Bateman A, Brown S, Chandra G, Chen CW, Collins M, Cronin A, Fraser A, Goble A, Hidalgo J, Hornsby T, Howarth S, Huang CH, Kieser T, Larke L, Murphy L, Oliver K, O'Neil S, Rabbinowitsch E, Rajandream MA, Rutherford K, Rutter S, Seeger K, Saunders D, Sharp S, Squares R, Squares S, Taylor K, Warren T, Wietzorrek A, Woodward J, Barrell BG, Parkhill J, Hopwood DA (2002) Complete genome sequence of the model actinomycete *Streptomyces coelicolor* A3(2). Nature 417:141–147. https://doi.org/10.1038/417141a

Bove P, Capozzi V, Garofalo C, Rieu A, Spano G, Fiocco D (2012) Inactivation of the ftsH gene of *Lactobacillus plantarum* WCFS1: effects on growth, stress tolerance, cell surface properties and biofilm formation. Microbiol Res 167:187–193. https://doi.org/10.1016/j.micres.2011.07.001

Brackman G, Cos P, Maes L, Nelis HJ, Coenye T (2011) Quorum sensing inhibitors increase the susceptibility of bacterial biofilms to antibiotics in vitro and in vivo. Antimicrob Agents Chemother 55:2655–2661. https://doi.org/10.1128/AAC.00045-11

Burmolle M, Hansen HL, Oregaard G, Sorensen SJ (2003) Presence of N-acyl homoserine lactones in soil detected by a whole cell biosensor and flow cytometry. Microb Ecol 45:226–236. https://doi.org/10.1007/s00248-002-2028-6

Chagnot C, Listrat A, Astruc T, Desvaux M (2012) Bacterial adhesion to animal tissues: protein determinants for recognition of extracellular matrix components. Cell Microbiol 14:1687–1696. https://doi.org/10.1111/cmi.12002

Chan KG, Atkinson S, Mathee K, Sam CK, Chhabra SR, Cámara M, Koh CL, Williams P (2011) Characterization of N-acylhomoserine lactone-degrading bacteria associated with *Zingiber officinale* (ginger) rhizosphere: co-existence of quorum quenching and quorum sensing in Acinetobacter and Burkholderia. BMC Microbiol 11:51. https://doi.org/10.1186/1471-2180-11-51

Chorianopoulos NG, Giaouris ED, Kourkoutas Y, Nychas GJ (2010) Inhibition of the early stage of *Salmonella enterica* serovar Enteritidis biofilm development on stainless steel by cell-free supernatant of a *Hafnia alvei* culture. App Environ Microbiol 76:2018–2022. https://doi.org/10.1128/AEM.02093-09

Chowdhary PK, Keshvan N, Nguyen HQ, Peterson JA, González JE, Haines DC (2007) *Bacillus megaterium* CYP102A1 oxidation of acyl homoserine lactones and acyl homoserines. Biochemistry 46:14429–14437. https://doi.org/10.1021/bi701945j

Christensen AB, Riedel K, Eberl L, Flodgaard LR, Molin S, Gram L, Givskov M (2003) Quorum-sensing-directed protein expression in *Serratia proteamaculans* B5a. Microbiology 149:471–483. https://doi.org/10.1099/mic.0.25575-0

Christensen LD, van Gennip M, Jakobsen TH, Alhede M, Hougen HP, Høiby N, Bjarnsholt T, Givskov M (2012) Synergistic antibacterial efficacy of early combination treatment with tobramycin and quorum-sensing inhibitors against *Pseudomonas aeruginosa* in an intraperitoneal foreign-body infection mouse model. J Antimicrob Chemother 67:1198–1206. https://doi.org/10.1093/jac/dks002

Cirou A, Uroz S, Chapelle E, Latour X, Orange N, Faure D, Dessaux Y (2009) Quorum sensing as a target for novel biocontrol strategies directed at *Pectobacterium*. In: Gisi U, Chet I, Gullino ML (eds) Recent developments in management of plant diseases. Springer, London, pp 121–131. https://doi.org/10.1007/978-1-4020-8804-9 ISBN 978-1-4020-8804-9

Cocolin L, Alessandria V, Botta C, Gorra R, De Filippis F, Ercolini D, Rantsiou K (2013) NaOH-debittering induces changes in bacterial ecology during table olives fermentation. PLoS One 8:e69074. https://doi.org/10.1371/journal.pone.0069074

Davies J, Davies D (2010) Origins and evolution of antibiotic resistance. Microbiol Mol Biol Rev 74:417–433. https://doi.org/10.1128/MMBR.00016-10

De Filippis L, Zalfa C, Ferrari D (2017) Neural stem cells and human induced pluripotent stem cells to model rare CNS diseases. CNS Neurol Disord Drug Targets. https://doi.org/10.2174/1871527316666170615121753

Dembitsky VA, Al Quntar AAA, Srebnik M (2011) Natural and synthetic small boron containing molecules as potential inhibitors of bacterial and fungal quorum sensing. Chem Rev 111:209–237. https://doi.org/10.1021/cr100093b

Desai MA, Soni KA, Nannapaneni R, Schilling MW, Silva JL (2012) Reduction of *Listeria monocytogenes* in raw catfish fillets by essential oils and phenolic constituent carvacrol. J Food Sci 77:516–522. https://doi.org/10.1111/j.1750-3841.2012.02859.x

Dhama K, Malik YS, Karthik K, Tiwari R, Chakraborthy S (2015) Emerging therapies and their benefits to animal and humans. Indian Farm 64:40–42 doi: Not available

Dong Y-H, Zhang L-H (2005) Quorum sensing and quorum-quenching enzymes. J Microbiol 43:101–109

Doumith M, Godbole G, Ashton P, Larkin L, Dallman T, Day M, Day M, Muller-Pebody B, Ellington MJ, de Pinna E, Johnson AP, Hopkins KL, Woodford N (2016) Detection of the plasmid-mediated mcr-1 gene conferring colistin resistance in human and food isolates of *Salmonella enterica* and *Escherichia coli* in England and Wales. J Antimicrob Chemother 71:2300–2305. https://doi.org/10.1093/jac/dkw093

Dourou D, Beauchamp CS, Yoon Y, Geornaras I, Belk KE, Smith GC, Nychas GJ, Sofos JN (2011) Attachment and biofilm formation by *Escherichia coli* O157: H7 at different temperatures, on various food-contact surfaces encountered in beef processing. Int J Food Microbiol 149:262–268. https://doi.org/10.1016/j.ijfoodmicro.2011.07.004

Duarte A, Alves AC, Ferreira S, Silva F, Domingues FC (2015) Resveratrol inclusion complexes: antibacterial and anti-biofilm activity against *Campylobacter* spp. and *Arcobacter butzleri*. Food Res Int 77:244–250. https://doi.org/10.1016/j.foodres.2015.05.047

Ercolini D, De Filippis F, La Storia A, Iacono M (2012) "Remake" by high-throughput sequencing of the microbiota involved in the production of water buffalo Mozzarella cheese. Appl Environ Microbiol 78:8142–8145. https://doi.org/10.1128/AEM.02218-12

Ercolini D, Pontoniob D, Filippisa FD, Minervinib F, Storiaa AL, Gobbettib M, Cagnob RD (2013) Microbial ecology dynamics during rye and wheat sourdough preparation. Appl Environ Microbiol 24:7827–7836. https://doi.org/10.1128/AEM.02955-13

Ferens WA, Hovde CJ (2011) *Escherichia coli* O157:H7: animal reservoir and sources of human infection. Foodborne Pathog Dis 8:465–487. https://doi.org/10.1089/fpd.2010.0673

Fett WF (2000) Naturally occurring biofilms on alfalfa and other types of sprouts. J Food Prot 63:625–632. https://doi.org/10.4315/0362-028X-63.5.625

Fletcher MP, Diggle SP, Camara M, Williams P (2007) Biosensor-based assays for PQS, HHQ and related 2-alkyl-4-quinolone quorum sensing signal molecules. Nat Protoc 2:1254–1262. https://doi.org/10.1038/nprot.2007.158

Fredriksson-Ahomaa M, Hallanvuo S, Korte T, Siitonen A, Korkeala H (2001) Correspondence of genotypes of sporadic *Yersinia enterocolitica* 4/O:3 strains from human and porcine origin. Epidemiol Infect 127:37–47. https://doi.org/10.1017/S0950268801005611

Fredriksson-Ahomaa M, Stolle A, Korkeala H (2006) Molecular epidemiology of *Yersinia enterocolitica* infections. FEMS Pathog Dis 47:315–329. https://doi.org/10.1111/j.1574-695X.2006.00095.x

Galloway WR, Hodgkinson JT, Bowden SD, Welch M, Spring DR (2011) Quorum sensing in Gram-negative bacteria: small-molecule modulation of AHL and Al-2 quorum sensing pathways. Chem Rev 111:28–67. https://doi.org/10.1021/cr100109t

Giaouris E, Chorianopoulos N, Skandamis P, Nychas GJ (2012) Attachment and biofilm formation by *Salmonella* in food processing environments. In: Salmonella: a dangerous foodborne pathogen. Intech. Mahmoud BSM (ed) pp 157–180. doi: Not available

Gram L, Christensen AB, Ravn L, Molin S, Givskov M (1999) Production of acylated homoserine lactones by psy-chrotrophic *Enterobacteriaceae* isolated from foods. Appl Environ Microbiol 65:3458–3463

Habimana O, Møretrø T, Langsrud S, Vestby LK, Nesse LL, Heir E (2010) Micro ecosystems from feed industry surfaces: a survival and biofilm study of *Salmonella* versus host resident flora strains. BMC Vet Res 6:48. https://doi.org/10.1186/1746-6148-6-48

Hanning I, Jarquin R, Slavik M (2008) *Campylobacter jejuni* as a secondary colonizer of poultry biofilms. J Appl Microbiol 105:1199–1208. https://doi.org/10.1111/j.1365-2672.2008.03853.x

Harjai K, Kumar R, Singh S (2010) Garlic blocks quorum sensing and attenuates the virulence of *Pseudomonas aeruginosa*. FEMS Immunol Med Microbiol 58:161–168. https://doi.org/10.1111/j.1574-695X.2009.00614.x

Henke JM, Bassler BL (2004) Three parallel quorum-sensing systems regulate gene expression in *Vibrio harveyi*. J Bacteriol 186:6902–6914. https://doi.org/10.1128/JB.186.20.6902-6914.2004

Hentzer M, Wu H, Andersen JB, Riedel K, Rasmussen TB, Bagge N, Kumar N, Schembri MA, Song Z, Kristoffersen P, Manefield M, Costerton JW, Molin S, Eberl L, Steinberg P, Kjelleberg S, Høiby N, Givskov M (2003) Attenuation of *Pseudomonas aeruginosa* virulence by quorum sensing inhibitors. EMBO J 22:3803–3815. https://doi.org/10.1093/emboj/cdg366

Huma N, Shankar P, Kushwah J, Bhushan A, Joshi J, Mukherjee T, Raju S, Purohit HJ, Kalia VC (2011) Diversity and polymorphism in AHL-lactonase gene (aiiA) of *Bacillus*. J Microbiol Biotechnol 21:1001–1011. https://doi.org/10.4014/jmb.1105.05056

Immonen N, Karp M (2007) Bioluminescence-based bioassays for rapid detection of nisin in food. Biosens Bioelectron 22:1982–1987. https://doi.org/10.16/j.bios.2006.08.029

Iovine NM (2013) Resistance mechanisms in *Campylobacter jejuni*. Virulence 4:230–240. https://doi.org/10.4161/viru.23753

Ishola OO, Mosugu JI, Adesokan HK (2016) Prevalence and antibiotic susceptibility profiles of *Listeria monocytogenes* contamination of chicken flocks and meat in Oyo State, south-western Nigeria: public health implications. J Prev Med Hyg 57:E157–E163 doi: Not available

Jackson D, Cotter D, Newell J, McEvoy S, O'Donohoe P, Kane F, McDermott T, Kelly S, Drumm A (2013) Impact of *Lepeophtheirus salmonis* infestations on migrating Atlantic salmon, *Salmo salar* L., smolts at eight locations in Ireland with an analysis of lice-induced marine mortality. J Fish Dis 36:273–281. https://doi.org/10.1111/jfd.12054

Jadaun V, Singh BR, Paliya BS, Upreti DK, Rao CV, Rawat AKS, Singh BN (2015) Honey enhances the anti-quorum sensing activity and anti-biofilm potential of curcumin. RSC Adv 5:71060–71070. https://doi.org/10.1039/C5RA14427B

Jakobsen TH, Bragason SK, Phipps RK, Christensen LD, van Gennip M, Alhede M, Skindersoe M, Larsen TO, Høiby N, Bjarnsholt T, Givskov M (2012) Food as a source for quorum sensing inhibitors: iberin from horseradish revealed as a quorum sensing inhibitor of *Pseudomonas aeruginosa*. Appl Environ Microbiol 78:2410e2421. https://doi.org/10.1128/AEM.05992-11

Jay LS, Davos D, Dundas M, Frankish E, Lightfoot D (2003) Salmonella. In: Hocking AD (ed) Foodborne microorganisms of public health significance. Australian Institute of Food Science and Technology, Sydney, pp 207–266 doi: Not available

Jemmi T, Stephan R (2006) *Listeria monocytogenes*: food-borne pathogen and hygiene indicator. Rev Sci Tech 25:571–580 doi: Not available

Kaakoush NO, Castaño-Rodríguez N, Mitchell HM, Man SM (2015) Global epidemiology of *Campylobacter* infection. Clin Microbiol Rev 28:687–720. https://doi.org/10.1128/CMR.00006-15

Kalia VC (2013) Quorum sensing inhibitors: an overview. Biotechnol Adv 31:224e245. https://doi.org/10.1016/j.biotechadv.2012.10.004

Kalia VC (2014) In search of versatile organisms for quorum-sensing inhibitors: acyl homoserine lactones (AHL)-acylase and AHL-lactonase. FEMS Microbiol Lett 359:143. https://doi.org/10.1111/1574-6968.12585

Kalia VC, Purohit HJ (2011) Quenching the quorum sensing system: potential antibacterial drug targets. Crit Rev Microbiol 37:121–140. https://doi.org/10.3109/1040841X.2010.532479

Kalia VC, Prakash J, Koul S, Ray S (2017) Simple and rapid method for detecting biofilm forming bacteria. Indian J Microbiol 57:109–111. https://doi.org/10.1007/s12088-016-0616-2

Kanan TA, Abdulla ZA (2009) Isolation of *Yersinia* spp. from cases of diarrhoea in Iraqi infants and children. East Mediterr Health J 15:276–284 doi: Not available

Kerekes EB, Deák É, Takó M, Tserennadmid R, Petkovits T, Vágvölgyi C, Krisch J (2013) Anti-biofilm forming and anti-quorum sensing activity of selected essential oils and their main components on food-related micro-organisms. J Appl Microbiol 115:933–942. https://doi.org/10.1111/jam.12289

Khalil RK, Frank JF (2010) Behavior of *Escherichia coli* O157:H7 on damaged leaves of spinach, lettuce, cilantro, and parsley stored at abusive temperatures. J Food Protect 73:212–220. https://doi.org/10.4315/0362-028X-73.2.212

Kim S, Kim M, Ryu S (2014) Development of an engineered bioluminescent reporter phage for the sensitive detection of viable *Salmonella Typhimurium*. Anal Chem 86:5858–5864. https://doi.org/10.1021/ac500645c

Kim YG, Lee JH, Kim SI, Baek KH, Lee J (2015) Cinnamon bark oil and its components inhibit biofilm formation and toxin production. Int J Food Microbiol 195:30–39. https://doi.org/10.1016/j.ijfoodmicro.2014.11.028

Koh KH, Tham FY (2011) Screening of traditional Chinese medicinal plants for quorum sensing inhibitors activity. J Microbiol Immunol Infect 44:144–148. https://doi.org/10.1016/j.jmii.2009.10.001

Koul S, Prakash J, Mishra A, Kalia VC (2016) Potential emergence of multi-quorum sensing inhibitor resistant (MQSIR) bacteria. Indian J Microbiol 56:1–18. https://doi.org/10.1007/s12088-015-0558-0

Krishnan T, Yin WF, Chan KG (2012) Inhibition of quorum sensing-controlled virulence factor production in *Pseudomonas aeruginosa* PAO1 by ayurveda spice clove (*Syzygium Aromaticum*) bud extract. Sensors 12:4016–4030. https://doi.org/10.3390/s120404016

Kumar NV, Murthy PS, Manjunatha JR, Bettadaiah BK (2014) Synthesis and quorum sensing inhibitory activity of key phenolic compounds of ginger and their derivatives. Food Chem 159:451–457. https://doi.org/10.1016/j.foodchem.2014.03.039

Kumar L, Chhibber S, Kumar R, Kumar M, Harjai K (2015) Zingerone silences quorum sensing and attenuates virulence of *Pseudomonas aeruginosa*. Fitoterapia 102:84–95. https://doi.org/10.1016/j.fitote.2015.02.002

Leathers TD, Bischoff KM (2011) Biofilm formation by strains of *Leuconostoc citreum* and *L. mesenteroides*. Biotechnol Lett 33:517–523. https://doi.org/10.1007/s10529-010-0450-2

Lee KM, Lim J, Nam S, Yoon MY, Kwon YK, Jung BY, Park Y, Park S, Yoon SS (2011) Inhibitory effects of broccoli extract on *Escherichia coli* O157:H7 quorum sensing and in vivo virulence. FEMS Microbiol Lett 321:67–74. https://doi.org/10.1111/j.1574-6968.2011.02311.x

Li YH, Tian X (2012) Quorum sensing and bacterial social interactions in biofilms. Sensors 12:2519–2538. https://doi.org/10.3390/s120302519

Li G, Yan C, Xu Y, Feng Y, Wu Q, Lv X, Yang B, Wang X, Xia X (2014) Punicalagin inhibits *Salmonella* virulence factors and has anti-quorum-sensing potential. Appl Environ Microbiol 80:6204–6211. https://doi.org/10.1128/AEM.01458-14

Liao CH, Cooke PH (2001) Response to trisodium phosphate treatment of *Salmonella Chester* attached to fresh-cut green pepper slices. Can J Microbiol 47:25–32 doi: Not available

Liu M, Wang H, Griffiths MW (2007) Regulation of alkaline metalloprotease promoter by N-acyl homoserine lactone quorum sensing in *Pseudomonas fluorescens*. J Appl Microbiol 103:2174–2184. https://doi.org/10.1111/j.1365-2672.2007.03488.x

Lu L, Hume ME, Pillai SD (2004) Autoinducer-2-like activity associated with foods and its interaction with food additives. J Food Prot 67:1457–1462. https://doi.org/10.4315/0362-028X-67.7.1457

Manefield M, Rasmussen TB, Kumar N, Hentzer M, Andersen JB, Steinberg P, Kjelleberg S, Givskov M (2002) Halogenated furanones inhibit quorum sensing through accelerated LuxR turnover. Microbiology 148:1119–1127. https://doi.org/10.1099/00221287-148-4-1119

Morild RK, Olsen JE, Aabo S (2011) Change in attachment of *Salmonella Typhimurium*, *Yersinia enterocolitica*, and *Listeria monocytogenes* to pork skin and muscle after hot water and lactic acid decontamination. Int J Food Microbiol 145:353–358. https://doi.org/10.1016/j.ijfoodmicro.2010.12.018

Myszka K, Schmidt MT, Majcher M, Juzwa W, Olkowicz M, Czaczyk K (2016) Inhibition of quorum sensing-related biofilm of *Pseudomonas fluorescens* KM121 by *Thymus vulgare* essential oil and its major bioactive compounds. Int Biodeterior Biodegrad 114:252–259. https://doi.org/10.1016/j.ibiod.2016.07.006

Nguyen Y, Sperandio V (2012) Enterohemorrhagic *E. coli* (EHEC) pathogenesis. Front Cell Infect Microbiol 2:90. https://doi.org/10.3389/fcimb.2012.00090

Niemira BA, Cooke PH (2010) *Escherichia coli* O157:H7 biofilm formation on Romaine lettuce and spinach leaf surfaces reduces efficacy of irradiation and sodium hypochlorite washes. J Food Sci 75:M270–M277. https://doi.org/10.1111/j.1750-3841.2010.01650.x

Noumi E, Snoussi M, Merghni A, Nazzaro F, Quindós G, Akdamar G, Mastouri M, Al-Sieni A, Ceylan O (2017) Phytochemical composition, anti-biofilm and anti-quorum sensing potential of fruit, stem and leaves of *Salvadora persica* L. methanolic extracts. Microb Pathog 109:169–176. https://doi.org/10.1016/j.micpath.2017.05.036

Oggioni MR, Trappetti C, Kadioglu A, Cassone M, Iannelli F, Ricci S, Andrew PW, Pozzi G (2006) Switch from planktonic to sessile life: a major event in pneumococcal pathogenesis. Mol Microbiol 61:1196–1210. https://doi.org/10.1111/j.1365-2958.2006.05310.x

Okwori AE, Martínez PO, Fredriksson-Ahomaa M, Agina SE, Korkeala H (2009) Pathogenic *Yersinia enterocolitica* 2/O: 9 and *Yersinia pseudotuberculosis* 1/O: 1 strain isolated from human and non-human sources in the plateau state of Nigeria. Food Microbiol 26:872–875. https://doi.org/10.1016/j.fm.2009.06.001

Oliveira GB, Favarin L, Luchese RH, McIntosh D (2015) Psychrotrophic bacteria in milk: how much do we really know? Braz J Microbiol 46:313–321. https://doi.org/10.1590/S1517-838246220130963

Ölmez H, Temur SD (2010) Effects of different sanitizing treatments on biofilms and attachment of *Escherichia coli* and *Listeria monocytogenes* on green leaf lettuce. LWT Food Sci Technol 43:964–970. https://doi.org/10.1016/j.lwt.2010.02.005

Packiavathy IA, Priya S, Pandian SK, Ravi AV (2014) Inhibition of biofilm development of uropathogens by curcumin – an anti-quorum sensing agent from *Curcuma longa*. Food Chem 148:453–460. https://doi.org/10.1016/j.foodchem.2012.08.002

Park SY, Kang HO, Jang HS, Lee JK, Koo BT, Yum DY (2005) Identification of extracellular N-acylhomoserine lactone acylase from a *Streptomyces* sp. and its application to quorum quenching. Appl Environ Microbiol 71:2632–2641. https://doi.org/10.1128/AEM.71.5.2632-2641.2005

Patel J, Sharma M (2010) Differences in attachment of *Salmonella enterica serovars* to cabbage and lettuce leaves. Int J Food Microbiol 139:41–47. https://doi.org/10.1016/j.ijfoodmicro.2010.02.005

Pejin B, Ciric A, Glamoclija J, Nikolic M, Sokovic M (2015) In vitro anti-quorum sensing activity of phytol. Nat Prod Res 29:374–377. https://doi.org/10.1080/14786419.2014.945088

Pejin B, Ciric A, Dimitric Markovic J, Glamoclija J, Nikolic M, Sokovic M (2017) An insight into anti-biofilm and anti-quorum sensing activities of the selected anthocyanidins: the case study of *Pseudomonas aeruginosa* PAO1. Nat Prod Res 31:1177–1180. https://doi.org/10.1080/14786419.2016.1222386

Pessione E (2012) Lactic acid bacteria contribution to gut microbiota complexity: lights and shadows. Front Cell Infect Microbiol 2:86. https://doi.org/10.3389/fcimb.2012.00086

Pinto UM, de Souza VE, Martins ML, Vanetti MCD (2007) Detection of acylated homoserine lactones in gram-negative proteolytic psychrotrophic bacteria isolated from cooled raw milk. Food Control 18:1322–1327. https://doi.org/10.1016/j.foodcont.2006.09.005

Priha O, Juvonen R, Tapani K, Storgårds E, Brew JI (2011) Acyl homoserine lactone production of brewery process surface bacteria. J Inst Brew 117:182–187. https://doi.org/10.1002/j.2050-0416.2011.tb00458.x

Quigley L, O'Sullivan O, Beresford TP, Ross RP, Fitzgerald GF, Cotter PD (2012) High-throughput sequencing for detection of subpopulations of bacteria not previously associated with artisanal cheeses. Appl Environ Microbiol 78:5717–5723. https://doi.org/10.1128/AEM.00918-12

Ragaert P, Devlieghere F, Debevere J (2007) Role of microbiological and physiological spoilage mechanisms during storage of minimally processed vegetables. Postharvest Biol Technol 44:185–194. https://doi.org/10.1016/j.postharvbio.2007.01.001

Rahman A, Bonny TS, Stonsaovapak S, Ananchaipattana C (2011) *Yersinia enterocolitica*: epidemiological studies and outbreaks. J Pathog 2011:239391. https://doi.org/10.4061/2011/239391

Rahman MR, Lou Z, Yu F, Wang P, Wang H (2017a) Anti-quorum sensing and anti-biofilm activity of Amomum tsaoko (Amommum tsao-ko Crevost et Lemarie) on foodborne pathogens. Saudi J Biol Sci 24:324–330. https://doi.org/10.1016/j.sjbs.2015.09.034

Rahman MR, Lou Z, Zhang J, Yu F, Timilsena YP, Zhang C, Zhang Y, Bakry AM (2017b) Star anise (*Illicium verum Hook. f.*) as quorum sensing and biofilm formation inhibitor on foodborne bacteria: study in milk. J Food Prot 80:645–653. https://doi.org/10.4315/0362-028X.JFP-16-294

Rasch M, Andersen JB, Nielsen KF, Flodgaard LR, Christensen H, Givskov M, Gram L (2005) Involvement of bacterial quorum-sensing signals in spoilage of bean sprouts. Appl Environ Microb 71:3321–3330. https://doi.org/10.1128/AEM.71.6.3321-3330.2005

Rasmussen TB, Manefield M, Andersen JB, Eberl L, Anthoni U, Christophersen C, Steinberg P, Kjelleberg S, Givskov M (2000) How *Delisea pulchra* furanones affect quorum sensing and swarming motility in *Serratia liquefaciens* MG1. Microbiology 146:3237–3244. https://doi.org/10.1099/00221287-146-12-3237

Ravn DB, Ng Y, Hjelm M, Christiansen JN, Johansen C, Gram L (2003) The microbial ecology of processing equipment in different fish industries – analysis of the microflora during processing and following cleaning and disinfection. Int J Food Microbiol 87:239–250. https://doi.org/10.1016/S0168-1605(03)00067-9

Renier S, Hébraud M, Desvaux M (2011) Molecular biology of surface colonization by *Listeria monocytogenes*: an additional facet of an opportunistic gram-positive foodborne pathogen. Environ Microbiol 13:835–850. https://doi.org/10.1111/j.1462-2920.2010.02378.x

Rodrigues AC, Oliveira BD, Silva ERD, Sacramento NTB, Bertoldi MC, Pinto UM (2016) Anti-quorum sensing activity of phenolic extract from Eugenia brasiliensis (Brazilian cherry). Food Sci Technol 36:337–343. https://doi.org/10.1590/1678-457X.0089

Romero M, Acuña L, Otero A (2012) Patents on quorum quenching: interfering with bacterial communication as a strategy to fight infections. Recent Pat Biotechnol 6:2–12. https://doi.org/10.2174/187220812799789208

Rutherford ST, Bassler BL (2012) Bacterial quorum sensing: its role in virulence and possibilities for its control. Cold Spring Harb Perspect Med 2:a012427. https://doi.org/10.1101/cshperspect.a012427

Scallan E, Hoekstra RM, Angulo FJ, Tauxe RV, Widdowson M, Roy SL, Jones JL, Griffin PM (2011) Foodborne illness acquired in the United States-major pathogens. Emerg Infect Dis 17:7–15. https://doi.org/10.3201/eid1701.P11101

Seo KH, Frank JF (1999) Attachment of *Escherichia coli* O157:H7 to lettuce leaf surface and bacterial viability in response to chlorine treatment as demonstrated by using confocal scanning laser microscopy. J Food Protect 62:3–9. https://doi.org/10.4315/0362-028X-62.1.3

Serichantalergs O, Ruekit S, Pandey P, Anuras S, Mason C, Bodhidatta L, Swierczewski B (2017) Incidence of Campylobacter concisus and C. ureolyticus in traveler's diarrhea cases and asymptomatic controls in Nepal and Thailand. Gut Pathog 9:47. https://doi.org/10.1186/s13099-017-0197-6

Sethupathy S, Shanmuganathan B, Kasi PD, Pandian SK (2016) Alpha-bisabolol from brown macroalga *Padina gymnospora* mitigates biofilm formation and quorum sensing controlled virulence factor production in *Serratia marcescens*. J Appl Phycol 28:1987–1996. https://doi.org/10.1007/s10811-015-0717-z

Shepherd RW, Lindow SE (2009) Two dissimilar N-acyl-homoserine lactone acylases of *Pseudomonas syringae* influence colony and biofilm morphology. Appl Environ Microbiol 75:45–53. https://doi.org/10.1128/AEM.01723-08

Shi X, Zhu X (2009) Biofilm formation and food safety in food industries. Trends Food Sci Technol 20:407–413. https://doi.org/10.1016/j.tifs.2009.01.054

Simpson Beauchamp C, Dourou D, Geornaras I, Yoon Y, Scanga JA, Belk KE, Smith GC, Nychas GJ, Sofos JN (2012) Transfer, attachment, and formation of biofilms by *Escherichia coli* O157:H7 on meat-contact surface materials. J Food Sci 77:M343–M347. https://doi.org/10.1111/j.1750-3841.2012.02695.x

Sio CF, Otten LG, Cool RH, Diggle SP, Braun PG, Bos R, Daykin M, Cámara M, Williams P, Quax WJ (2006) Quorum quenching by an N-Acyl-homoserine lactone acylase from *Pseudomonas aeruginosa* PAO1. Infect Immun 74:1673–1682. https://doi.org/10.1128/IAI.74.3.1673-1682.2006

Skandamis PN, Nychas GJ (2012) Quorum sensing in the context of food microbiology. App Environ Microbiol 78:5473–5482. https://doi.org/10.1128/AEM.00468-12

Smid EJ, Lacroix C (2013) Microbe-microbe interactions in mixed culture food fermentations. Curr Opin Biotechnol 24:148–154. https://doi.org/10.1016/j.copbio.2012.11.007

Sohaib M, Anjum FM, Arshad MS, Rahman UU (2016) Postharvest intervention technologies for safety enhancement of meat and meat based products; a critical review. J Food Sci Technol 53:19–30. https://doi.org/10.1007/s13197-015-1985-y

Soković M, Ćirić A, Glamočlija J, Nikolić M, van Griensven LJ (2014) *Agaricus blazei* hot water extract shows anti quorum sensing activity in the nosocomial human pathogen *Pseudomonas aeruginosa*. Molecules 19:4189–4199. https://doi.org/10.3390/molecules19044189

Sperandio V, Li CC, Kaper JB (2002) Quorum-sensing *Escherichia coli* regulator A: a regulator of the LysR family involved in the regulation of the locus of enterocyte effacement pathogenicity island in enterohemorrhagic *E. coli*. Infect Immun 70:3085–3093. https://doi.org/10.1128/IAI.70.6.3085-3093.2002

Srinivasan R, Devi KR, Kannappan A, Pandian SK, Ravi AV (2016) Piper betle and its bioactive metabolite phytol mitigates quorum sensing mediated virulence factors and biofilm of nosocomial pathogen *Serratia marcescens* in vitro. J Ethnopharmacol 193:592–603. https://doi.org/10.1016/j.jep.2016.10.017

Srivastava A, Singh BN, Deepak D, Rawat AK, Singh BR (2015) Colostrum hexasaccharide, a novel *Staphylococcus aureus* quorum-sensing inhibitor. Antimicrob Agents Chemother 59:2169–2178. https://doi.org/10.1128/AAC.03722-14

Steindler L, Venturi V (2007) Detection of quorum-sensing N-acyl homoserine lactone signal molecules by bacterial biosensors. FEMS Microbiol Lett 266:1–9. https://doi.org/10.1111/j.1574-6968.2006.00501.x

Tait K, Williamson H, Atkinson S, Williams P, Cámara M, Joint I (2009) Turnover of quorum sensing signal molecules modulates cross-kingdom signalling. Environ Microbiol 11:1792–1802. https://doi.org/10.1111/j.1462-2920.2009.01904.x

Tan JY, Yin WF, Chan KG (2014) Quorum sensing activity of *Hafnia alvei* isolated form packed food. Sensors 14:6788–6796. https://doi.org/10.3390/s140406788

Tapia-Rodriguez MR, Hernandez-Mendoza A, Gonzalez-Aguilar GA, Martinez-Tellez MA, Martins CM, Ayala-Zavala JF (2017) Carvacrol as potential quorum sensing inhibitor of *Pseudomonas aeruginosa* and biofilm production on stainless steel surfaces. Food Control 75:255–261. https://doi.org/10.1016/j.foodcont.2016.12.014

Truchado P, Gil A, Tomas-Barberan FA, Larrosa M, Allende A (2012) Food phytochemicals act as quorum sensing inhibitors reducing production and/or degrading autoinducers of *Yersinia enterocolitica* and *Erwinia Carotovora*. Food Control 24:78–75. https://doi.org/10.1016/j.foodcont.2011.09.006

Turovskiy Y, Kashtanov D, Paskhover B, Chikindas ML (2007) Quorum sensing: fact, fiction, and everything in between. Adv Appl Microbiol 62:191–234. https://doi.org/10.1016/S0065-2164(07)62007-3

Uroz S, Heinonsalo J (2008) Degradation of N-acyl homoserine lactone quorum sensing signal molecules by forest root-associated fungi. FEMS Microbiol Ecol 65:271–278. https://doi.org/10.1111/j.1574-6941.2008.00477.x

Valeriano C, de Oliveria TLC, de Carvalho SM, Cardoso MG, Alves E, Piccoli RH (2012) The sanitizing action of essential oil-based solutions against *Salmonella enterica* serotype *Enteritidis* S64 biofilm formation on AISI 304 stainless steel. Food Control 25:673–677. https://doi.org/10.1016/j.foodcont.2011.12.015

van der Veen S, Abee T (2011) Mixed species biofilms of *Listeria monocytogenes* and *Lactobacillus plantarum* show enhanced resistance to benzalkonium chloride and peracetic acid. Int J Food Microbiol 144:421–431. https://doi.org/10.1016/j.ijfoodmicro.2010.10.029

Van Houdt R, Moons P, Aertsen A, Jansen A, Vanoirbeek K, Daykin M, Williams P, Michiels CW (2007) Characterization of a luxI/luxR-type quorum sensing system and N-acyl-homoserine lactone-dependent regulation of exo-enzyme and antibacterial component production in *Serratia plymuthica* RVH1. Res Microbiol 158:150–158. https://doi.org/10.1016/j.resmic.2006.11.008

Vasavi HS, Arun AB, Rekha PD (2014) Anti-quorum sensing activity of *Psidium guajava* L. flavonoids against *Chromobacterium violaceum* and *Pseudomonas aeruginosa* PAO1. Microbiol Immunol 58:286–293. https://doi.org/10.1111/1348-0421.12150

Vasavi HS, Arun AB, Rekha PD (2016) Anti-quorum sensing activity of flavonoid-rich fraction from *Centella asiatica* L. against *Pseudomonas aeruginosa* PAO1. J Microbiol Immunol Infect 49:8–15. https://doi.org/10.1016/j.jmii.2014.03.012

Venkadesaperumal G, Shetty PH (2016) Cyanidin inhibits quorum signaling pathway of foodborne opportunistic pathogen. J Food Sci Technol 53:968–976. https://doi.org/10.1007/s13197-015-2031-9

Venkadesaperumal G, Meena CK, Shetty PH (2015a) Quercetin influences quorum sensing in foodborne bacteria: in-vitro and in-silico evidence. PLoS One 10:e0134684. https://doi.org/10.1371/journal.pone.0134684

Venkadesaperumal G, Sundar K, Shetty PH (2015b) Quorum quenching activity of Syzygium *cumini* (L.) Skeels and its anthocyanin malvidin against *Klebsiella pneumoniae*. Microb Pathog 79:61–69. https://doi.org/10.1016/j.micpath.2015.01.010

Venkadesaperumal G, Meena CK, Shetty PH (2016a) Cell-free lysate from *E. ludwigii* regulates QS-dependent phenotypes of *Yersinia enterocolitica*. RSC Adv 6:21277–21284. https://doi.org/10.1039/C5RA25440J

Venkadesaperumal G, Meena CK, Murali A, Shetty PH (2016b) Petunidin as a competitive inhibitor of acylated homoserine lactones in *Klebsiella pneumoniae*. RSC Adv 6:2592–2601. https://doi.org/10.1039/C5RA20677D

Venkadesaperumal G, Rucha S, Sundar K, Shetty PH (2016c) Anti-quorum sensing activity of spice oil nanoemulsions against foodborne pathogens. LWT Food Sci Technol 66:225–231. https://doi.org/10.1016/j.lwt.2015.10.044

Venkadesaperumal G, Vishnuvardhan R, Shetty PH (2016d) Regulation of acylated homoserine lactones (AHLs) in beef by spice marination. J Food Sci Technol 53:2686–2694. https://doi.org/10.1007/s13197-016-2240-x

Viana ES, Campos MEM, Ponce AR, Mantovani HC, Vanetti MCD (2009) Biofilm formation and acyl homoserine lactone production in *Hafnia alvei* isolated from raw milk. Biol Res 42:427–436. https://doi.org/10.4067/S0716-97602009000400004

Vivas J, Padilla D, Real F, Bravo J, Grasso V, Acosta F (2010) Influence of environmental conditions on biofilm formation by *Halfnia alvei* strains. Vet Microbiol 129:150–155. https://doi.org/10.1016/j.vetmic.2007.11.007

Waters CM, Bassler BL (2005) Quorum sensing: cell-to-cell communication in bacteria. Ann Rev Cell Dev Biol 21:319–346. https://doi.org/10.1146/annurev.cellbio.21.012704.131001

Wesley I (2009) Public health impact of food-borne illness: impetus for the international food safety effort. In: Heredia N, Wesley I, Garcia S (eds) Microbiologically safe foods. Wiley, New Jersey, pp 3–13 doi:Not available

Wirtanen G, Salo S (2003) Disinfection in food processing – efficacy testing of disinfectants. Rev Environ Sci Biotechnol 2:293–306. https://doi.org/10.1023/B:RESB.0000040471.15700.03

Yap PS, Krishnan T, Yiap BC, Hu CP, Chan KG, Lim SH (2014) Membrane disruption and anti-quorum sensing effects of synergistic interaction between *Lavandula angustifolia* (lavender oil) in combination with antibiotic against plasmid-conferred multi-drug-resistant *Escherichia coli*. J Appl Microbiol 116:1119–1128. https://doi.org/10.1111/jam.12444

Yoon Y, Sofos JN (2010) Absence of association of autoinducer-2-based quorum sensing with heat and acid resistance of *Salmonella*. J Food Sci 75:M444–M448. https://doi.org/10.1111/j.1750-3841.2010.01744.x

Zhang LH, Dong YH (2004) Quorum sensing and signal interference: diverse implications. Mol Microbiol 53:1563–1571. https://doi.org/10.1111/j.1365-2958.2004.04234.x

Zhang J, Rui X, Wang L, Guan Y, Sun X, Dong M (2014) Polyphenolic extract from *Rosa rugosa* tea inhibits bacterial quorum sensing and biofilm formation. Food Control 42:125–131. https://doi.org/10.1016/j.foodcont.2014.02.001

Zhu J, Huang X, Zhang F, Feng L, Li J (2015) Inhibition of quorum sensing, biofilm, and spoilage potential in *Shewanella baltica* by green tea polyphenols. J Microbiol 53:829–836. https://doi.org/10.1007/s12275-015-5123-3

Part IV

Health

14 CRISPR-Cas Systems Regulate Quorum Sensing Genes and Alter Virulence in Bacteria

Qinqin Pu and Min Wu

Abstract

Although understanding how CRISPR and its associated systems are controlled is at the infant stages, recent studies present exciting discoveries in this fast-moving field. A number of studies find that CRISPR-Cas systems regulate quorum sensing (QS) genes by targeting and degrading *lasR* mRNA, while QS systems can also modulate the CRISPR-Cas as revealed in more recent reports. The importance of the QS system for bacterial pathogenicity is well recognized and the indispensable features of CRISPR-Cas in adaptive immunity and biotechnology application are gaining great attention. Analyzing interaction between QS and CRISPR-Cas systems represents an interesting field as CRISPR-Cas systems are not only the adaptive immunity of bacteria, but also the regulators of their own genes. Undoubtedly, the continued understanding of molecular basis of CRISPR-Cas action and regulation may indicate novel strategies for treatment of bacterial infections.

Keywords

CRISPR-Cas systems · Quorum sensing · Endogenous gene targeting

Q. Pu
Department of Biomedical Sciences, University of North Dakota, Grand Forks, ND, USA

State Key Laboratory of Biotherapy, West China Hospital, Sichuan University, Chengdu, China

M. Wu (✉)
Department of Biomedical Sciences, University of North Dakota, Grand Forks, ND, USA
e-mail: min.wu@med.und.edu

V. C. Kalia (ed.), *Quorum Sensing and its Biotechnological Applications*,
https://doi.org/10.1007/978-981-13-0848-2_14

14.1 Introduction

The quorum sensing (QS) system in many bacterial species plays a significant role in cell growth, development and differentiation as an intercellular signaling system that relies on cell density (Hurley and Bassler 2017). Recent studies found that QS regulates a spectrum of critical functions of bacteria, such as synthesis of extracellular enzymes and toxins from pathogens, formation of biofilm and production of drug resistance (Atkinson and Williams 2009; Das et al. 2015; Tan et al. 2015; Vuotto et al. 2017). Hence, research into QS signaling may indicate targets to design novel and effective anti-microbial therapeutics. Despite a great deal of efforts in understanding QS regulation, the detailed mechanism of QS gene expression and function has remained largely unknown.

Clustered regularly interspaced short palindromic repeats (CRISPR) and CRISPR-associated (Cas) systems are the essential adaptive immunity to microbes. It is commonly believed that more than 90% archaea and 40% eubacteria possess one or more types of CRISPR-Cas systems (Marraffini and Sontheimer 2010). CRISPR loci are a special DNA sequence located in the bacterial and archaea genomes, often consisting of a leader, a plurality of highly conserved repeats and multiple spacers that were derived from the predators, bacteriophages. Analyzing the flanking sequence of the CRISPR site revealed that there was a polymorphic family gene in the vicinity thereof. The family encodes a group of proteins that contain a functional domain that interacts with the nucleic acid and acts together with the CRISPR region and is named CRISPR associated gene, abbreviated as Cas (Song 2017). When the phage containing the DNA that matches with any of the spacer invades again, the CRISPR-Cas system will cleave the nucleic acids to protect the bacterium itself from killing (Puschnik et al. 2017). CRISPR-Cas systems are categorized to two main classes, 6 types and 19 subtypes, based on the signature Cas protein and other nomenclature features. Currently, there is outstanding interest in understanding the CRISPR-Cas9 biology and function of bacterial *S. pyogenes* due to the powerful gene-editing function and potential application in medicine and biotechnology; however, the original roles of CRISPR-Cas in various aspects of bacterial physiology and immunity are relatively understudied.

Although the best-known Type II CRISPR-Cas (Cas9) was reported to play roles in mammalian host defense (Heidrich and Vogel 2013), it remains unknown whether other CRISPR-Cas systems, such as Type I and Type III CRISPR-Cas, are also involved in host immunity by targeting bacterial endogenous genes. We have recently revealed a new function for Type I-F CRISPR-Cas system which consists of 6 important proteins (Csy1, Csy2, Csy3, Csy4 and Cas1/Cas3 survey complex) in the control of QS associated gene expression in *Pseudomonas aeruginosa*. This study shows that the Type I-F CRISPR immune system of PA14 (a strain of *P. aeruginosa*) modulates the master QS regulator LasR by degrading its mRNA, thus protecting the bacteria to evade recognition by host Toll-like receptor 4 (TLR4), and inhibiting host pro-inflammatory response. Hence, targeting the CRISPR-Cas system which regulates quorum-sensing and alters pathogenesis may open a new avenue to tackle the drug resistance of *P. aeruginosa* (Li et al. 2016). Here, we discuss the potential role of CRISPR-Cas systems in regulating endogenous genes to alter virulence.

14.2 CRISPR-Cas in Regulating Virulence and Altering Mammalian Defense

The first study reporting that Type II CRISPR-Cas, namely Cas9, is involved in regulating bacterial endogenous genes was observed in *Francisella novicida* (Sampson et al. 2013). This study demonstrated that Cas9 is necessary for the bacterium to evade detection by Toll-like receptor 2 (TLR2) and cause serious disease. In particular, the Cas9 functions were associated with tracrRNA and crRNA. The authors show that Cas9 alters the stability of endogenous transcript encoding bacterial lipoprotein (BLP) that is essential for its virulence and can be recognized by TLR2. CRISPR-Cas mediates repression of BLP expression by degrading the mRNA and decreasing transcript levels. Hence for the first time, scientists may recognize the CRISPR-Cas components play a key role in pathogenesis of bacteria in causing disease in mammalian systems through self-regulation of their gene expression (Heidrich and Vogel 2013).

Since then, several studies demonstrated that different types of CRISPR-Cas systems were involved in bacterial physiology by targeting the endogenous genes besides silencing of foreign nucleic acids (Yosef et al. 2012; Heussler et al. 2016; Fu et al. 2017). One of the most remarkable examples is the CRISPR-Cas system of *P. aeruginosa* was reported to regulate its biofilm formation (Cady and O'Toole 2011; Heussler et al. 2016; Li et al. 2016). The detailed mechanism is not yet understood but conventional wisdom is that the CRISPR-Cas systems interact with the target genes in the chromosomally integrated prophage to abolish the generation of biofilm. *P. aeruginosa* is lysogenized by bacteriophages and it is clear that the process requires the Cas proteins (Zegans et al. 2009). Biofilm and QS have inseparable relationship. They are two important information departments for bacterial communication. One is for surface-associated communities and the other is for intercellular signaling. How they communicate with each other is known thanks to many recent intense studies but there is much to be learned. In addition to *Vibrio harveyi* and yellow *Myxococcus xanthus*, the population effect of Gram-negative bacteria is generally regulated by LuxR/I-ty information systems (Schaefer et al. 2013). Studies reported that QS systems may play key roles in regulating the biofilm formation for many bacterial species (Shao et al. 2012; Yu et al. 2012). Recently, Lan and his collaborators found a cascade regulatory pathway to regulate the Rhl population induction system, Crc-Hfq / Lon / RhlI in *P. aeruginosa,* which added knowledge of QS in regulating biofilm (Cao et al. 2014). In recent years, the discoveries of microbial QS system and its relationship with some drug resistance via biofilm provide new perspective and means for the study of drug resistance mechanism. Altogether, these studies provide good theoretical basis for the hypothesis that CRISPR-Cas regulates QS and biofilm to alter pathogenesis. CRISPR-Cas systems may be a new target for bacterial resistant treatments.

Studies reveal that the link between the CRISPR-Cas and QS systems are increasing. A recent paper reported that the QS regulation leads to enhanced expression of the CRISPR-Cas systems in *Serratia* especially for high cell density situation (Patterson et al. 2016; Semenova and Severinov 2016). On the contrary, Zuberi's team found that CRISPR interference (CRISPRi) inhibited biofilm by repressing the luxS QS gene expression in *E. coli* (Zuberi et al. 2017). Consistently, Li et al., demonstrate that self-targeting CRISPR spacers bear sequences for degrading transcription factor mRNA of *lasR* and that CRISPR-Cas systems can control the QS response in some cases (Li et al. 2016). The LasR/LasI system consists of transcriptional activator LasR and acetyl homoserine lactone (AHL) synthase LasI protein in the Gram-negative bacteria QS system with AHL as a self-inducing agent. LasI guides the synthesis of 3-OXO-C-HSL and is secreted into the extracellular spaces by active transport, which binds to LasR at a certain threshold and activates gene transcription, including alkaline protease, exotoxin A, elastase, and other virulence factors. Hence, LasR plays a key role in increasing the expression of *P. aeruginosa* virulence related genes (Lee and Zhang 2015). Li et al. revealed mechanistically that CRISPR-mediated mRNA degradation needs the "5′-GGN-3′" (protospacer adjacent motif [PAM]) sequence and the HD and DExD/H domains of Cas3 protein for recognition in *lasR* mRNA (Li et al. 2016). As the consequence, LasR is decreased, the PA14 strain with CRISPR-Cas shows decreased bacterial phagocytosis by host alveolar macrophages and lower mouse survival than the CRISPR-Cas deleted one. This implicates that the CRISPR-Cas regulates innate immunity, which is exerted via TLR4-initiated signaling as upstream events. These studies have opened up new fields to elucidate the interaction between QS and CRISPR-Cas system and develop new drugs for treating infection (Fig. 14.1).

Once CRISPR-Cas components are activated, the crRNA12 structure and Cascade (Csy1–4 complex) will interact with *lasR* mRNA (or other potential genes) through a sequence matching with crRNA12. Then *lasR* mRNA will be cleaved by Cas3 protein. The degradation of *lasR* mRNA and the changes of downstream genes alter bacterial behaviors and subsequent host inflammatory responses.

A potential function of CRISPR-Cas systems in endogenous gene regulation as well as in pathogenesis may be to acquire self-targeting crRNAs with spacer sequences complementary to chromosomally encoded genes. By analyzing CRISPRs from 330 organisms, Stern et al., found that only approximately 0.4% spacers are potentially self-targeting and that frequent targeting non-moving genes occurs in 18% of all bacteria containing a type of CRISPR-Cas (Stern et al. 2010). The result of self-targeting is likely deleterious chromosomal cleavage and deletion, thus is considered to be detrimental effects to the body of bacteria, hence termed "autoimmunity". In terms of existing knowledge, the Cas proteins, such as Cas9, bind with target genes and inhibit transcription capacity (Qi et al. 2013), while other Cas proteins like Cas1 protein and Cas2 protein are reported to prevent the acquisition of new crRNAs to protect the loss of previously acquired crRNAs (Westra and Brouns 2012). This suggests that the inhibition and destruction of Cas proteins or the process of autoimmunity may be involved in endogenous gene regulation, which may also potentially impact virulence and pathogenesis in some aspects. crRNAs are originally known to target DNA, but a recent study reported that RNA also can be targeted in *F. novicida* (Sampson et al. 2013). The mechanisms and consequences of autoimmunity as well as their types and substrate specificities in bacteria are totally unknown. Much needs to be done to clearly elucidate how autoimmunity and

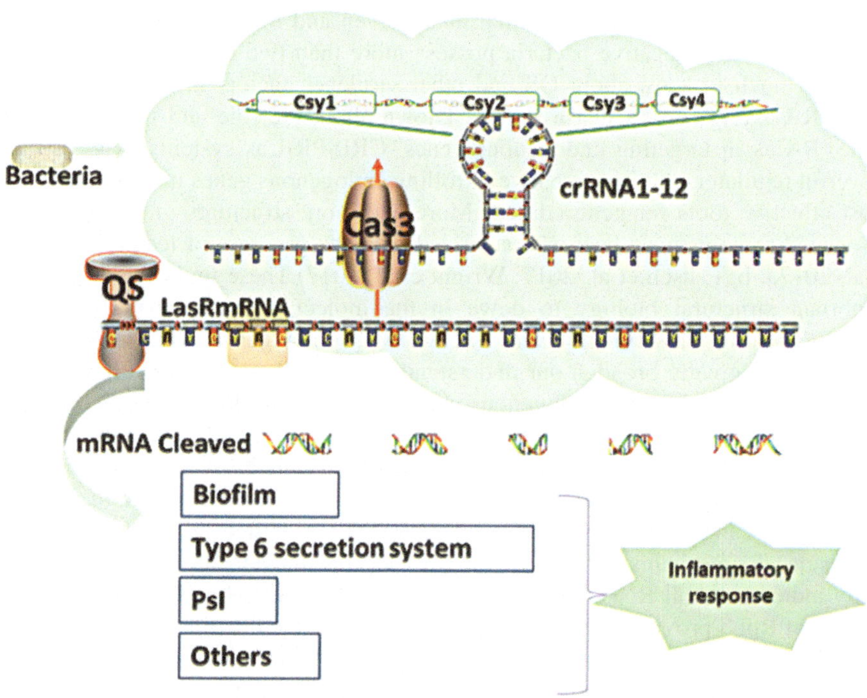

Fig. 14.1 Model of type I-F CRISPR-Cas modulates inflammatory response by temporal mRNA repression of QS genes (*lasR*)

CRISPR-Cas systems contribute to the regulation of bacterial genes (such as virulence, biofilm and proliferation associated genes) and pathogenesis. Accomplishing these pursues will be very helpful for designing new strategies for anti-infective therapy.

14.3 Discussion

CRISPR-Cas systems as the newly emerging genome editing tool to identify and edit target genes have built an extremely efficient platform for gene placement technology. It is well publicized in scientific research and application due to its high efficiency, simple and economic features, indicating enormous biotechnology and therapeutic values. However, as an adaptive immune system for prokaryotes and archaea, CRISPR-Cas' own superior functions in virulence regulation are at much slower paces and require strong efforts to further explore. The function of CRISPR-Cas systems in regulating endogenous genes and altering pathogenesis is undoubtedly the great entry point of research. The release of virulence factors is the main form of bacterial infection especially for Gram-negative bacteria that have up to six (type I, II, III, IV, V and VI) types of special secretion systems to transfer virulence factors to neighboring cells or animal host (Green and Mecsas 2016). Moreover, most of the Gram-negative bacteria possess more than two types of QS systems. There is much to learn about QS and other virulence factors in interaction with CRISPR-Cas systems as just a little is known about the role and mechanism of CRISPR-Cas in targeting endogenous genes. CRISPR-Cas systems may also be powerful regulators in virulence by controlling endogenous genes in addition to the most effective tools for gene editing. More and more structures of CRISPR-Cas associated proteins have been characterized with the progress of technology (Liu et al. 2017a, b; Pausch et al. 2017; Wright et al. 2017).These finding will provide important structural biology to delve in the molecular mechanism by which CRISPR-Cas systems function and interact with other genes. The discovery of structure will greatly broaden our understanding about how the CRISPR-Cas systems target to QS or other endogenous genes and improve the transformation and utilization value of CRISPR-Cas systems. For example, the revelation of the Cas13a structure (Liu et al. 2017a, b) may add value in developing RNA research tools and extending CRISPR's application in gene editing because Cas13a is one of the few proteins that can degrade RNA in Class II CRISPR-Cas system.

Besides the mechanism that CRISPR-Cas systems degrade QS mRNA. A recent study indicates small RNA ReaL regulates *P. aeruginosa* QS networks due to the activity of RpoS provides a new perspective to explore the relationship between QS and CRISPR-Cas systems that may be through small RNAs (Carloni et al. 2017). The role of small non-coding RNAs in mammals has been recognized while its function in prokaryote is almost unknown. The research prospect of small RNAs for bacterial physiology is immense and much less is known about their interactions with CRISPR-Cas systems. The roles of QS in diverse fields, especially human health and disease, are being dissected. Inhibiting the bacterial QS by deactivating their chemical signaling molecules or by producing competitors may help in designing approaches to treating infectious diseases. On contrary, bacteria constantly evolve new strategies to battle with drugs and antibiotics. Like the CRISPR-Cas systems, the QS is diverse with strains and species, therefore unravelling the molecular detail in the cross-roads may open new avenues for fighting bacteria.

References

Atkinson S, Williams P (2009) Quorum sensing and social networking in the microbial world. J R Soc Interface 6:959–978. https://doi.org/10.1098/rsif.2009.0203

Cady KC, O'Toole GA (2011) Non-identity-mediated CRISPR-bacteriophage interaction mediated via the Csy and Cas3 proteins. J Bacteriol 193:3433–3445. https://doi.org/10.1128/JB.01411-10

Cao Q, Wang Y, Chen F, Xia Y, Lou J, Zhang X, Yang N, Sun X, Zhang Q, Zhuo C, Huang X, Deng X, Yang CG, Ye Y, Zhao J, Wu M, Lan L (2014) A novel signal transduction pathway that modulates rhl quorum sensing and bacterial virulence in *Pseudomonas aeruginosa*. PLoS Pathog 10:e1004340. https://doi.org/10.1371/journal.ppat.1004340

Carloni S, Macchi R, Sattin S, Ferrara S, Bertoni G (2017) The small RNA ReaL: a novel regulatory element embedded in the *Pseudomonas aeruginosa* quorum sensing networks. Environ Microbiol. https://doi.org/10.1111/1462-2920.13886

Das T, Kutty SK, Tavallaie R, Ibugo AI, Panchompoo J, Sehar S, Aldous L, Yeung AW, Thomas SR, Kumar N, Gooding JJ, Manefield M (2015) Phenazine virulence factor binding to extracellular DNA is important for *Pseudomonas aeruginosa* biofilm formation. Sci Rep 5:8398. https://doi.org/10.1038/srep08398

Fu Q, Su Z, Cheng Y, Wang Z, Li S, Wang H, Sun J, Yan Y (2017) Clustered, regularly interspaced short palindromic repeat (CRISPR) diversity and virulence factor distribution in avian *Escherichia coli*. Res Microbiol 168:147–156. https://doi.org/10.1016/j.resmic.2016.10.002

Green ER, Mecsas J (2016) Bacterial secretion systems: an overview. Microbiol Spectr 4:1–19. https://doi.org/10.1128/microbiolspec.VMBF-0012-2015

Heidrich N, Vogel J (2013) CRISPRs extending their reach: prokaryotic RNAi protein Cas9 recruited for gene regulation. EMBO J 32:1802–1804. https://doi.org/10.1038/emboj.2013.141

Heussler GE, Miller JL, Price CE, Collins AJ, O'Toole GA (2016) Requirements for *Pseudomonas aeruginosa* type I-F CRISPR-Cas adaptation determined using a biofilm enrichment assay. J Bacteriol 198:3080–3090. https://doi.org/10.1128/JB.00458-16

Hurley A, Bassler BL (2017) Asymmetric regulation of quorum-sensing receptors drives autoinducer-specific gene expression programs in *Vibrio cholerae*. PLoS Genet 13:e1006826. https://doi.org/10.1371/journal.pgen.1006826

Lee J, Zhang L (2015) The hierarchy quorum sensing network in *Pseudomonas aeruginosa*. Protein Cell 6:26–41. https://doi.org/10.1007/s13238-014-0100-x

Li R, Fang L, Tan S, Yu M, Li X, He S, Wei Y, Li G, Jiang J, Wu M (2016) Type I CRISPR-Cas targets endogenous genes and regulates virulence to evade mammalian host immunity. Cell Res 26:1273–1287. https://doi.org/10.1038/cr.2016.135

Liu L, Li X, Ma J, Li Z, You L, Wang J, Wang M, Zhang X, Wang Y (2017a) The molecular architecture for RNA-guided RNA cleavage by Cas13a. Cell 170:714–726 e710. https://doi.org/10.1016/j.cell.2017.06.050

Liu L, Li X, Wang J, Wang M, Chen P, Yin M, Li J, Sheng G, Wang Y (2017b) Two distant catalytic sites are responsible for C2c2 RNase activities. Cell 168:121–134 e112. https://doi.org/10.1016/j.cell.2016.12.031

Marraffini LA, Sontheimer EJ (2010) CRISPR interference: RNA-directed adaptive immunity in bacteria and archaea. Nat Rev Genet 11:181–190. https://doi.org/10.1038/nrg2749

Patterson AG, Jackson SA, Evans TC, Salmond GB, Przybilski GP, Staals RH R, Fineran PC (2016) Quorum sensing controls adaptive immunity through the regulation of multiple CRISPR-Cas systems. Mol Cell 64:1102–1108. https://doi.org/10.1016/j.molcel.2016.11.012

Pausch P, Muller-Esparza H, Gleditzsch D, Altegoer F, Randau L, Bange G (2017) Structural variation of type I-F CRISPR RNA guided DNA surveillance. Mol Cell 67:622–632. https://doi.org/10.1016/j.molcel.2017.06.036

Puschnik AS, Majzoub K, Ooi YS, Carette JE (2017) A CRISPR toolbox to study virus-host interactions. Nat Rev Microbiol 15:351–364. https://doi.org/10.1038/nrmicro.2017.29

Qi LS, Larson MH, Gilber LA, Doudna JA, Weissman JS, Arkin AP, Lim WA (2013) Repurposing CRISPR as an RNA-guided platform for sequence-specific control of gene expression. Cell 152:1173–1183. https://doi.org/10.1016/j.cell.2013.02.022

Sampson TR, Saroj SD, Llewellyn AC, Tzeng YL, Weis DS (2013) A CRISPR/Cas system mediates bacterial innate immune evasion and virulence. Nature 497:254–257. https://doi.org/10.1038/nature12048

Schaefer AL, Lappala CR, Morlen RP, Pelletier DA, Lu T-YS, Lankford PK, Harwood CS, Peter Greenberg E (2013) LuxR- and LuxI-Type Quorum-Sensing Circuits Are Prevalent in Members of the Populus deltoides Microbiome. Appl Environ Microbiol 79(18):5745–5752

Semenova E, Severinov K (2016) Come together: CRISPR-Cas immunity senses the quorum. Mol Cell 64:1013–1015. https://doi.org/10.1016/j.molcel.2016.11.037

Shao C, Shang W, Yang Z, Sun Z, Li Y, Guo J, Wang X, Zou D, Wang S, Lei H, Cui Q, Yin Z, Li X, Wei X, Liu W, He X, Jiang Z, Du S, Liao X, Huang L, Wang Y, Yuan J (2012) LuxS-dependent AI-2 regulates versatile functions in *Enterococcus faecalis* V583. J Proteome Res 11:4465–4475. https://doi.org/10.1021/pr3002244

Song M (2017) The CRISPR/Cas9 system: their delivery, in vivo and ex vivo applications and clinical development by startups. Biotechnol Prog 33:1035–1043. https://doi.org/10.1002/btpr.2484

Stern A, Keren L, Wurtze O, Amitai G, Sorek R (2010) Self-targeting by CRISPR: gene regulation or autoimmunity? Trends Genet 26:335–340. https://doi.org/10.1016/j.tig.2010.05.008

Tan CH, Koh KS, Xie C, Zhang J, Tan XH, Lee GP, Zhou Y, Ng WJ, Rice SA, Kjelleberg S (2015) Community quorum sensing signalling and quenching: microbial granular biofilm assembly. NPJ Biofilms Microbiomes 1:15006. https://doi.org/10.1038/npjbiofilms.2015.6

Vuotto C, Longo F, Pascolini C, Donell G, Balice MP, Libori MF, Varaldo PE (2017) Biofilm formation and antibiotic resistance in *Klebsiella pneumoniae* urinary strains. J Appl Microbiol. 123:1003–1018. https://doi.org/10.1111/jam.13533

Westra ER, Brouns SJ (2012) The rise and fall of CRISPRs – dynamics of spacer acquisition and loss. Mol Microbiol 85:1021–1025. https://doi.org/10.1111/j.1365-2958.2012.08170.x

Wright AV, Liu JJ, Knott GJ, Doxzen KW, Nogales E, Doudna JA (2017) Structures of the CRISPR genome integration complex. Science 64:eaao0679. https://doi.org/10.1126/science.aao0679

Yosef I, Goren MG, Qimron U (2012) Proteins and DNA elements essential for the CRISPR adaptation process in *Escherichia coli*. Nucleic Acids Res 40:5569–5576. https://doi.org/10.1093/nar/gks216

Yu D, Zhao L, Xue T, Sun B (2012) *Staphylococcus aureus* autoinducer-2 quorum sensing decreases biofilm formation in an icaR-dependent manner. BMC Microbiol 12:288. https://doi.org/10.1186/1471-2180-12-288

Zegans ME, Wagner JC, Cady KC, Murphy DM, Hammond JH, O'Toole GA (2009) Interaction between bacteriophage DMS3 and host CRISPR region inhibits group behaviors of *Pseudomonas aeruginosa*. J Bacteriol 191:210–219. https://doi.org/10.1128/JB.00797-08

Zuberi A, Misb L, Khan AU (2017) CRISPR interference (CRISPRi) inhibition of luxS gene expression in *E. coli*: an approach to inhibit biofilm. Front Cell Infect Microbiol 7:214. https://doi.org/10.3389/fcimb.2017.00214

Understanding the Connect of Quorum Sensing and CRISPR-Cas System: Potential Role in Biotechnological Applications

15

Anoop Singh, Mohita Gaur, and Richa Misra

Abstract

Quorum sensing is employed by bacteria to control gene expression, by communicating through signals, necessary to confer advantageous traits in a community. We now understand that to counter susceptibility to phages in communities, bacteria evolved specialized adaptive immune system called CRISPR-Cas (clustered regularly interspaced short palindromic repeats and CRISPR-associated proteins) system that may use quorum sensing for its regulation. As a countermeasure, phages have also evolved diverse mechanisms to evade the defense strategies. The chapter discusses the dynamics of this co-evolutionary war, understanding of which will help pave way for many biotechnological applications. An important aspect includes refining tools such as quorum-sensing inhibition and phage therapy that are utilized to control many biofilm-forming bacterial infections.

Keywords

Quorum sensing · CRISPR-CAS system · Biotechnology

15.1 Introduction

Although bacteria are unicellular organisms, they are capable of organizing themselves in a community for mutual benefit, for which they need to work in a coordinated manner. They live in diverse niches, from sporadic population to dense

A. Singh · M. Gaur
Department of Zoology, University of Delhi, Delhi, India

R. Misra (✉)
Department of Zoology, University of Delhi, Delhi, India

Department of Zoology, Sri Venkateswara College, University of Delhi, Delhi, India

V. C. Kalia (ed.), *Quorum Sensing and its Biotechnological Applications*,
https://doi.org/10.1007/978-981-13-0848-2_15

communities, like biofilms, cell chains and microcolonies (Hall-Stoodley et al. 2004), which impact various physiological processes. Though bacterial cells collectively provide several advantages to its population, a major drawback could be the increased vulnerability to phage infections (Abedon 2012) and invading foreign mobile genetic elements (Pinedo and Smets 2005). Thus, the formation of bacterial colonies is of advantage only in case of low phage densities, or if the threat can be countered by heightened bacterial defense response (Abedon 2012).

In bacterial context, "Quorum" word is frequently used to indicate the minimal density of cells that leads to altered gene expression in a population of bacteria. In other words, it describes a phenomenon of bacterial cell-cell communication wherein collective gene regulation happens in a bacterial community in reaction to alteration in bacterial cell-population density (Platt and Fuqua 2010). Quorum sensing bacteria secrete signaling molecules known as 'autoinducers' that fluctuate in its level as a response to bacterial cell-population density. Once autoinducers attain a threshold level, at that phase the bacterial community is contemplated to be 'quorate', which leads to activation of downstream pathways that direct the bacteria jointly to modulate the gene expression and, therefore, group behavior adaptation (Kalia 2014a).

15.2 Difference in Quorum Sensing Systems

The process of quorum sensing regulates a wide range of physiological deeds such as motility, conjugation, sporulation, production of antibiotics, competence, virulence, and biofilm formation in both Gram-negative and Gram-positive bacteria. However, while Gram-positive bacteria utilizes oligo-peptides and two-component systems for quorum sensing, the Gram-negative bacteria use N-Acyl-L-homoserine lactones (AHLs) as autoinducers to communicate with each other (Miller and Bassler 2001; Huma et al. 2011; Rutherford and Bassler 2012; Kalia 2014b).

15.2.1 LuxI- and LuxR-Type Quorum Sensing: Example in Gram-Negative Bacteria

The simple signal-response system described in Gram-negative bacteria based on cell-density-dependent mechanism generally belongs to the LuxI/LuxR quorum type sensing system (Whitehead et al. 2001). This system consists of autoinducer synthases belonging to LuxI family and transcriptional regulators that falls in LuxR family. LuxI-like enzymes are responsible to produce the specific signaling molecule known as acylated homoserine lactone signaling molecule (acyl-HSL) which functions like an autoinducer. This AHL synthase enzyme, a product of *luxI* gene, catalyzes the acylation and lactonization reactions by coupling the acyl-side chain of a defined acyl-acyl carrier protein (acyl-ACP), an intermediary in fatty acid biosynthesis pathway, with the homocysteine moiety of *S*-adenosylmethionine (SAM).

This leads to formation of an amide bond between acyl-ACP and the amino group of the homocysteine molecule of SAM. The lactonisation of the ligated intermediate thus form the acyl-HSL with the release of the methylthioadenosine as a by-product of the reaction. Acyl-HSL produced in the reaction act as an autoinducer and gets freely diffused across the cell membrane, leading to its increase in concentration in the external environment in concomitance with increased cell population (Fuqua et al. 1994; Parsek and Greenberg 2000; Schauder and Bassler 2001).

The transcription of *luxI* is subjected to autoregulation, i.e. gene product increases as AHL accumulates in the cell. This is achieved through another protein called LuxR, a transcriptional activator as well as cytoplasmic receptor which bind cognate HSL autoinducer which has reached a critical threshold concentration (forming the desired cell density i.e. 'quorum' of bacteria). This autoinducer binding to LuxR converts the LuxR into an active form. In the absence of HSL, the LuxR protein is not stable and may degrade rapidly (Whitehead et al. 2001). The LuxR-AHL complex which is thus formed recognizes a consensus sequence (*lux* box), upstream of the *luxICDABE* operon and activates the expression of various genes (Stevens et al. 1994). It has been found that LuxR proteins appear to homodimerize in presence of a cognate signal during binding to the *lux* box and the amount of this dimerization is dependent on the concentration of the ligand (Li and Nair 2012). This process of activation of LuxI by LuxR has been shown to be tightly regulated by the rapid degradation of the LuxR receptor when unbound to autoinducer, allowing for a positive-feedback loop (Gray and Garey 2001).

Many of the Gram-negative bacteria are shown to possess *luxI* and *luxR* homologs and are known to use LuxI/LuxR-type quorum sensing, with some members having an additional complexity in these LuxI/LuxR systems to regulate a wide range of cellular processes. Unique AHLs with specific concentration are produced by different species; as a consequence, members belonging to the same species recognize and respond to its own signal molecule (Bassler 1999). This specificity is determined by the interaction of the correct acyl side chain moiety of the acyl-ACP with a specific LuxI-type protein. Thus, the LuxI/LuxR gene products form a functional pair, with LuxI as an autoinducer synthase and LuxR as the receptor (Rutherford and Bassler 2012; Schaefer et al. 2013). Quorum sensing was first shown to regulate the bioluminescence of marine bacterium *Vibrio fischeri*, a phenomenon where transcriptional regulation of luciferase genes responsible for luminescence was connected to the attainment of threshold cell density in the system (Nealson et al. 1970). With regards to control of pathogenicity via the quorum sensing systems, one of the most well-characterized examples is of *Pseudomonas aeruginosa*, an opportunistic pathogen, which primarily infects immunocompromised individuals. It produces an extensive array of virulence factors many of which are under control of a complex network of quorum sensing regulators. Currently, four quorum sensing pathways are known in *P. aeruginosa*: two LuxR and LuxI-type systems called LasR and LasI and RhlR and RhlI, and few alternative non-AHL signaling molecule generating PqsR-controlled quinolone system and the IQS system (Papenfort and Bassler 2016).

15.2.2 Oligo-Peptides: Example in Gram-Positive Bacteria

Unlike Gram-negative bacteria, Gram-positive bacteria mainly exploit oligopeptides that are modified structurally as autoinducers in quorum-sensing based communication for guarded gene expression system (Ji et al. 1995; Ng and Bassler 2009). As these peptides are impermeable, i.e. cannot freely move in and out from biological membranes, their secretion is usually mediated with the help of some specialized transporters. The other prime dissimilarities between LuxI/R-type and peptide-based quorum-sensing system is the localization of the cognate receptors; for example, in Gram-positive bacteria the sensors for the oligopeptide autoinducers are membrane-bound, whereas it is cytoplasmic for Gram-negative bacteria. These membrane-bound receptors are members of the two-component signaling system and involve series of phosphorylation events for signal transduction. The secretion of these oligopeptide autoinducer increases with the increase in the cell density. The autophosphorylation activity of the membrane-bound receptor is stimulated by these oligopeptides, resulting in ATP-driven phosphorylation of the conserved residue which thus transfers the phosphate group to the response regulator in order to activate it, allowing further activation of the target genes (Miller and Bassler 2001; Ng and Bassler 2009). The typical example of a two-component system is a histidine kinase receptor, which is membrane-bound along with a cognate cytoplasmic response regulator. In most cases, genes producing this oligopeptide precursor, the histidine kinase receptor, and the response regulator is a part of an operon and the expression of these genes are auto-induced by quorum sensing (Ji et al. 1997). The oligo-peptide autoinducers of Gram-positive bacteria don't show variations on a single core molecule, which usually is employed by Gram-negative bacterial AHLs. Rather, these are genetically encoded, and each species is thus capable of producing a unique peptide signal with a unique sequence (Novick and Geisinger 2008). Examples include *agr* system that regulates pathogenesis in *Staphylococcus aureus*; *com* system that modulates competence and development in *Streptococcus pneumonia*; ComP/ComA system that regulates competence and sporulation in *Bacillus subtilis* and *fsr* system of *Enterococcus faecalis* (Ng and Bassler 2009).

15.3 CRISPR-Cas Systems Utilize Quorum Sensing

One of the potential drawbacks to a community life is that as the bacterial population attains a high cell density, they become more prone to foreign invasion. To encounter susceptibility to the phage infections, bacteria have evolved various defense mechanisms to limit the foreign invaders such as prevention of phage attachment to the host cell wall, assembly interference, restriction-modification and CRISPR-Cas system (Seed 2015). The discovery of CRISPR-Cas system in the past decade has unraveled newer aspects of bacterial defense strategies and opened up our understanding of bacterial adaptive immunity, a mechanism earlier associated with only vertebrates. CRISPR-Cas system is known to impart adaptive immunity to many prokaryotes against viral infections or invading foreign genetic elements, but there are many shreds of evidence that show these systems also regulate other processes like gene regulation, group behavior, and virulence (Westra et al. 2014). CRISPR might also provide information regarding bacterial adaptation to a particular niche and evolution. A CRISPR locus contains an ordered series of CRISPR repeats, which is interrupted by the short variable DNA sequences known as 'spacers', and the presence of the nearby diverse set of *cas* genes. Repeats with no *cas* genes are known as orphan CRISPRs. The characteristic features of a CRISPR repeat are the presence of a leader sequence, spacers, dyad symmetry, characteristic secondary structure, containing a stretch of three to four identical bases, and related *cas* genes (Bolotin et al. 2005). CRISPR-Cas system involves three different mechanistic steps: adaptation, expression, and interference. Adaption requires the acquisition of foreign DNA fragment as a new spacer into a CRISPR array. The spacer sequences behave as a memory and give protection against future invasion from the same virus or plasmid. The expression stage involves transcription of CRISPR RNAs (crRNAs). Interference stage involves recognition and target cleavage by crRNAs aided by Cas proteins (Barrangou 2013; van der Oost et al. 2014).

CRISPR-Cas systems have been classified into two different classes on the bases of effector module organization. Effector module is mainly involved in Interference. Class I systems are composed of multi-subunit effector complex, whereas Class II systems utilize single subunit effector complex. Each class is divided into three types and each type is further subdivided into many subtypes (Makarova et al. 2015). It is now known that in many bacteria CRISPR-Cas systems and quorum sensing are the two major mechanisms, that work together at a community level to structure, protect, and provide various advantages to a bacterial population at a high cell density.

15.3.1 Quorum Sensing Controls CRISPR-Based Adaptive Immunity in *Serratia sp.*

Serratia sp. is a Gram-negative bacterium, possessing commonly found LuxI/LuxR-type quorum sensing circuit and three types of CRISPR-Cas systems i.e., type I-E, I-F, and III-A each with minimum one CRISPR locus (Thomson et al. 2000). In *Serratia, luxI-luxR* homologs, *smaI* and *smaR* are known to regulate secondary synthesis, biofilm formation, and motility (Fineran et al. 2005). SmaI produces an autoinducer, N-butanoyl-L-homoserine lactone and SmaR produces a transcriptional modulator acting as a DNA binding repressor. It has been earlier reported in *Serratia* that *smaI* transcription and AHL level increases with increased cell density (Patterson et al. 2016). High cell density causes accumulation of AHLs, which in turn bind to SmaR, thereby suppressing its ability to bind to DNA. Inhibition of SmaR results in enhanced gene expression via a de-repression process (Fineran et al. 2005). The effect of quorum sensing signaling on CRISPR-Cas system has also been examined in *Serratia sp.* ATCC39006 (Patterson et al. 2016). In this study, CRISPR and *cas* operons' expression was assessed in wild-type and *smaI* mutant strains. Significant reduction in expression of type I-E CRISPR1 and type I-F

CRISPR2, and their *cas* operons including Type III-A *cas* operon was observed in the absence of AHL signal production in *smaI* mutants. Further, deletion of *smaR* gene results in restored expression of CRISPR array and *cas* operons in *smaI* mutants throughout growth, indicating that SmaR functions as a suppressor of CRISPR-Cas systems in absence of AHL production (Slater et al. 2003). These results strengthened the notion that defense system at high cell density needs to be elevated against invaders like phages or foreign genetic elements (Abedon 2012). To ascertain whether the changes in quorum sensing signal correlated with the CRISPR-Cas modulation, Patterson *et al.* introduced *Serratia* cells propagating in a highly dense population to donor bacteria which transfers its plasmids via conjugation process. These plasmids possess sequences mimicking the previously exposed intruders. These sequences were complementary to the spacer one of CRISPR1 (type I-E), CRISPR2 (type I-F), and CRISPR3 (type III-A) systems. In the wild-type population, CRISPR-Cas system mediated interference of their respective targets was robust. However, *smaI* mutants displayed a remarkable decline in interference capabilities for all three types. Surprisingly, type I-E system showed the lowest correlation between interference and quorum sensing signal despite having highest effect on the *cas8e* promoter. It might be due to the requirement of other type I-E components to provide overall interference. The weakened interference in all types was restored on the addition of exogenous quorum sensing signals (Patterson et al. 2016). Together, these results demonstrated that the quorum sensing signals regulates interferences and elevate the defense system at high cell density. Since CRISPR-Cas interference is very costly to cell (Vale et al. 2015), its expression is needed to be reduced in low cell density as horizontal gene transfer or phage spread is less likely in such conditions, whereas in case of high cell density risk to the phage infections increase and it becomes necessary for cells to heighten the CRISPR-Cas mediated immunity. Therefore, cell-cell communication is important in simulating population level immunity (Fig. 15.1).

15.3.2 Quorum Sensing Controls CRISPR-Dependent Adaptive Immunity and CRISPR Controls Biofilm Regulation in *P. aeruginosa*

In another example of adaptive immunity regulation, it is seen that bacterium *P. aeruginosa* uses quorum sensing to activate CRISPR-Cas system to target foreign DNA at high cell densities (Hoyland-Kroghsbo et al. 2017). It has also been earlier demonstrated that Type I-F CRISPR-Cas system in *P. aeruginosa* might be involved in controlling virulence (Zegans et al. 2009). CRISPR regions have been known to provide bacterial resistance to many lytic bacteriophage infections (Barrangou et al. 2007). In *P. aeruginosa,* the targeting of phage DMS3 by CRISPR also displays alternate functions like inhibition of swarming motility and biofilm formation without affecting the normal planktonic growth of bacteria and the efficiency of plaque formation by the DMS3 phage (Cady and O'Toole 2011). It is seen that several spacers in CRISPR loci are complementary to the bacteriophage genome. While

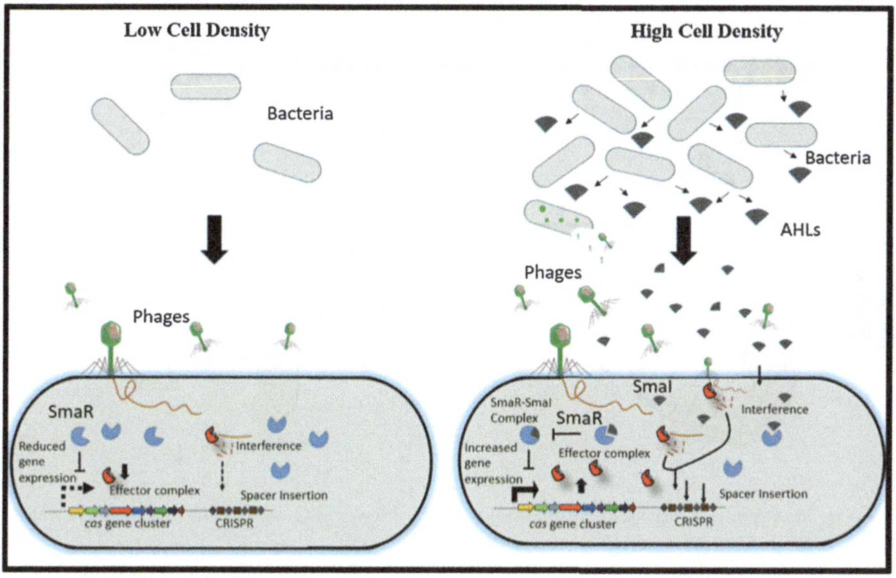

Fig. 15.1 Modulation of CRISPR-Cas system via quorum sensing in *Serratia sp*. Downregulation of *cas* genes at low cell density by SmaR (left). Inhibition of SmaR by AHLs i.e., SmaI at high cell density leads to upregulation of *cas* genes (right). Expression of the Cas protein complex helps the bacterium to cleave the incoming viral DNA (orange threads)

some of them are perfectly complementary to the target sequence known as protospacer, others have many mismatches (Zegans et al. 2009). Further investigation revealed that mismatches present on CR2_sp1 (CRISPR2_spacer1) to the integrated phage genome were involved in phage-dependent loss of swarming motility as well as biofilm formation. Earlier it was thought that this might be due to the crRNA-guided cleavage of RNA, but it was later proven that the phage-dependent inhibition of swarming and biofilm formation is a result of an imperfect base pairing between integrated prophage DNA and crRNA, which causes DNA damage and apoptosis during biofilm formation. Interestingly, the same crRNA with reduced mismatches provides partial resistance to DMS3 phage and the entire removal of mismatches results in complete phage resistance (Cady et al. 2012).

In *P. aeruginosa,* four Cas proteins (Csy 1-4) along with crRNA assembles to form an effector complex called Csy complex or CRISPR-associated complex for antiviral defense (Cascade) (Wiedenheft et al. 2011). Csy complex identifies and binds to foreign DNA in a sequential manner: at first it identifies a protospacer recognition motif (PAM) consisting of two consecutive GC base pairs adjacent to the protospacer followed by subsequent base pairing of the seed region (first 8 bases of crRNA) to the target DNA, which is crucial for high-affinity binding (Hoyland-Kroghsbo et al. 2017). Binding of Csy complex to the target with precise base pairing to the PAM, seed region, and rest of the protospacer recruits trans-acting Cas3 protein, which acts as a helicase and an HD family of nucleases required for the

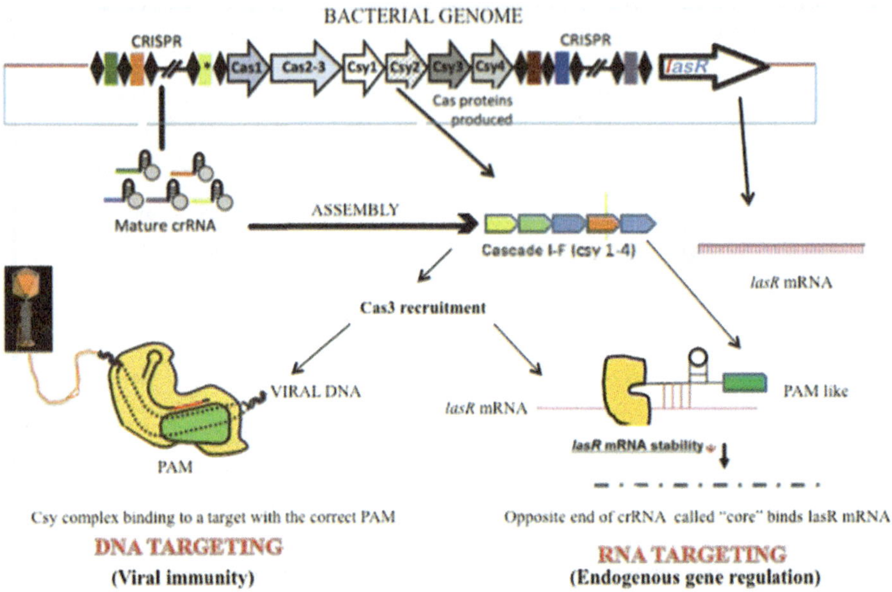

Fig. 15.2 Eradication of viruses by CRISPR/Cas system in *P. aeruginosa*. Targeting by CRISPR/Cas system is mainly done by two mechanisms. DNA targeting occurs through binding of DNA by CRISPR RNA-guided surveillance complex (Cascade I-F or Csy), which depends on recognition of a double-stranded PAM and base pairing at the seed. For RNA targeting such as repression of lasR mRNA (lasR gene associated with virulence) requires base pairing at the opposite end of the CRISPR RNA (crRNA), a region called the "core", and recognition of a single-stranded PAM-like sequence. In both cases, target binding recruits Cas3 for degradation

target cleavage (Wiedenheft et al. 2011). Nonetheless, the imperfect base pairing of crRNA to target although not lethal but is sufficient enough to recruit Cas3 and trigger DNA damage till a threshold enough to trigger RecA, a DNA-repair and SOS response protein (Heussler et al. 2015). Activated RecA induces activation of lysis genes such as *pyocin,* which is generally switched off in the bacterial cells growing on a surface but due to RecA activation, these bacterial cells are unable to repress these lysis genes, leading to apoptosis, which in turn inhibits swarming and biofilm formation (Fig. 15.2). In this case, we could say this kind of alternative CRISPR-Cas function is a type of side effect of the main immunity function and perhaps not a direct result of natural selection process. Since the biofilm formation and swarming in *P. aeruginosa* is monitored by several other pathways, the suggested mechanism of CRISPR-Cas modulation via the imperfect base pairing of spacer sequence to the target offers poor regulation of quorum sensing and appears to be dispensable for the bacteria (Karatan and Watnick 2009).

Though foreign DNA recognition and recruitment of Cas3 has been well-established, RNA targeting by this complex has not been discovered yet. A recent study in *Francisella novicida* has shown that this system is also efficient in crRNA guided mRNA targeting and monitors the expression of some virulence genes (Li et al. 2016). Similarly, in *P. aeruginosa* Li et al. tried to demonstrate the function of crRNA in modulating virulence gene expression. They demonstrated that deletion of certain components of CRISPR-Cas systems is associated with upregulation of certain virulence genes and enhanced pathogenicity. Particularly, they recognized a spacer with partial complementarity to the *lasR* gene involved in detecting quorum-sensing signals. Deletion of CRISPR or its associated genes results in upregulation of *lasR* and other factors important for virulence (Li et al. 2016). Zegans et al. suggested that perhaps a CRISPR-mediated alteration of group behavior in *P. aeruginosa* is a mechanism by which bacterium prevent its community from bacteriophage dissemination (Zegans et al. 2009).

15.3.3 Regulation of Fruiting Body Formation by CRISPR-Cas in *Myxococcus xanthus*

Myxococcus xanthus is a type of Gram-negative, predatory, and a soil-dwelling bacterium. It swarms over a solid surface to take up other bacteria and detritus from its surroundings as a source of nutrition. Though *M. xanthus* communicates using AHL-mediated quorum sensing signaling, it is unable to produce AHLs itself. It depends on exogenous AHLs produced during signaling among other bacteria, known as xenic quorum sensing signaling (Lloyd and Whitworth 2017). It uses AHLs as a marker of a nutrition rich environment and it is also known to activate collective germination of spores and release of proteases to digest its prey. In starvation conditions i.e., low exogenous AHLs, sporulation is triggered. Sporulation process involves among many intercellular signals, activation of an A-signal which induces the expression of *fruA*, a transcription factor involved in fruiting body generation (Lloyd and Whitworth 2017). A-signaling includes secretion of density-dependent diffusible signals consisting of mixtures of proteases, peptides, and amino acids.

Regulation of *fruA* by CRISPR-Cas systems has also been reported in *M. xanthus* (Boysen et al. 2002). The importance of type I-C CRISPR-Cas systems in sporulation was evident from a transposon mutagenesis study, in which it was shown that *cas7* and *cas5* gene disruption by the transposon lead to reduced spore formation (Thony-Meyer and Kaiser 1993). Further, *cas8c* mutants also showed slow aggregation and reduced sporulation due to the decrease in *fruA* expression (Boysen et al. 2002). Additionally, it has been seen that expression of *cas* gene is only restricted to fruiting body and absent in the cells present in peripheral layer (Julien et al. 2000). Expression of *cas* gene is probably activated through a positive feedback loop, which initiates on binding of FruA to one of the modulatory units linked with *cas* locus. Currently, the mechanism of action is uncertain but it is strongly evident that *cas* gene expression and fruiting body formation regulatory circuit is connected (Viswanathan et al. 2007). So, we can say that decrease in concentration of exogenous AHLs act as an indirect signal for the starvation condition, which may in turn result in activation of A-signaling and further downstream pathways but it is currently unclear how *M. xanthus* responds to AHLs. Currently,

Haliangium ochraceum is the only known myxobacterium that produces complete LuxR homolog with an N-terminal autoinducer binding site. The P2TF DNA-binding protein database (Ortet et al. 2012) enlists 12 proteins in *M. xanthus* that possess a LuxR DNA binding domain at its C-terminal. Six of these proteins have no identifiable sensory domain and five of these contain two-component system receiver domains at N-terminal, including FruA (Lloyd and Whitworth 2017). Hence, the presence of a receiver domain on FruA provides a strong evidence that there might be a strong connection between quorum sensing and CRISPR-Cas system in fruiting body generation in *M. xanthus*.

15.3.4 Quorum Sensing and CRISPR-Cas Regulation in Other Bacteria

Since quorum sensing and CRISPR-Cas system gene homologs are widely distributed throughout the bacterial world, their interdependence would also be prevalent. In favor of this statement, Patterson et al. used published microarray data of *Pectobacterium atrosepticum* (Bowden et al. 2013) for further analysis and observed a significant decrease in the expression of type I-F *cas* genes in an AHL synthase mutant (Patterson et al. 2016). Similarly, in a transcriptomic study of *Burkholderia glumae,* mutant strain of *luxI* displayed reduced expression of type I-F *cas* genes (Gao et al. 2015). In *Escherichia coli*, the decrease in AHL-dependent receptors limits the infections from λ and phages. As *E. coli* do not produce AHLs, it encodes a LuxR sensor, which might impart protection against phages preying on nearby bacteria (Hoyland-Kroghsbo et al. 2013).

On the other hand, the importance of quorum sensing and its role in physiological processes such as biofilm formation remains relatively less understood in mycobacteria, one of the clinically most significant genera, which include the highly infectious pathogen, *Mycobacterium tuberculosis*. An indication of quorum sensing based regulation is mainly indirect in *Mycobacterium.* Given the fact that quorum sensing is generally associated with biofilm formation and several species belonging to *Mycobacterium* genus such as *M. tuberculosis*, have been shown to produce drug-resistant biofilms *in vitro*, it is hypothesized that quorum sensing must be existing in both the non-pathogenic and pathogenic mycobacteria. This hypothesis gains strength from the fact that LuxR homologs are present in many mycobacterial species along with the presence of second messengers such as c-di-GMP, facilitating biofilm formation; however, direct experimental evidence of quorum sensing mechanism is still lacking (Sharma et al. 2014; Polkade et al. 2016). The phylogenetic discovery of CRISPR-Cas system in *Mycobacterium* in recent years (He et al. 2012; Freidlin et al. 2017), has refueled the interest to explore the functional significance of occurrence of these systems and questions their role in persistence and antibiotic resistance frequently encountered in pathogenic species. Table 15.1 summarizes the information on bacteria where association of CRISPR-Cas system and quorum sensing has been established.

Table 15.1 Associated functions of quorum sensing and CRISPR-Cas system

Species	Gram staining	Quorum sensing type	CRISPR-Cas type	AHLs Source	Function	Experimental evidence	Reference
Serratia sp.	Gram negative	LuxI-LuxR type	Type I-E, Type I-F, and Type III-A	Self producing-SmaI	Adaptive immunity regulation	Yes	Patterson et al. (2016)
Pseudomonas aeruginosa	Gram negative	LuxI-LuxR type	Type I-F	Self producing-LasI	Biofilm inhibition and appoptosis	Yes	Heussler et al. (2015)
Myxococcus xanthus	Gram negative	LuxI-LuxR and A-signaling	Type I-C	Exogenous AHLs	Fruiting body formation	No	Westra et al. (2014) and Lloyd and Whitworth (2017)
Pectobacterium atrosepticum	Gram negative	LuxI-LuxR type	Type I-F	Self producing-ExpI	Adaptive immunity regulation	No	Bowden et al. (2013) and Patterson et al. (2016)
Burkholderia glumae	Gram negative	LuxI-LuxR type	Type I-F	Self producing-TofI	Adaptive immunity regulation	No	Gao et al. (2015) and Patterson et al. (2016)
Escherichia coli	Gram negative	LuxI-LuxR type	Type I-E	Exogenous AHLs	Adaptive immunity regulation	Yes	Hoyland-Kroghsbo et al. (2013)

15.4 Smartness of the Phages

In a microbial community, there is a constant evolutionary pressure between phages and bacteria. As quorum sensing can regulate host defense, phages have also evolved diverse mechanisms to influence the bacterial defense system and use it to its own advantage (Hargreaves et al. 2014). A common way for phages to evade the CRISPR-Cas immunity is by random mutagenesis which can interfere crRNA interaction with the protospacer or the PAM recognition (Samson et al. 2013). Some distinct 'anti-CRISPR' genes have also been discovered in the genome of *P. aeruginosa* which interfere with the working of CRISPR-Cas complexes (Bondy-Denomy et al. 2013). Surprisingly, *Vibrio cholerae* serogroup O1phages express their own CRISPR-Cas machinery for targeting phage-inhibitory chromosomal islands present in its bacterial host. During infection, the phage releases crRNAs and Cas proteins that can inactivate the bacterial CRISPR–Cas systems and help in the infection process (Seed et al. 2013).

15.5 CRISPR-Cas and Quorum Sensing: Applications and Future Direction

As we have understood, quorum-sensing signals contributes to pathogenesis through synchronized production of toxins and other such virulence factors. Furthermore, it adds to the ability of bacteria to resist antimicrobial drugs by modulating community behaviour such as biofilm formation. Discovery of quorum sensing inhibitors (or quorum quenchers) paved way for its varied applications in biotechnology and has been extensively reviewed over time (Kalia and Purohit 2011; Kalia 2013, 2015a, b; LaSarre and Federle 2013; Kalia and Kumar, 2015a, b; Kalia et al. 2015; Kumar et al. 2015). Nonetheless, evidence suggests that quorum-sensing inhibition has variable efficacy against different pathogens. A better understanding of how quorum sensing is regulated in microbial populations has had a significant impact on utilizing and developing quorum-quenching molecules in a more efficacious manner. In the recent past, the area of CRISPR biology research has also evolved quickly and has shown its merit in diverse biotechnological applications. In a recent study, CRISPR-derived interference system was utilized to downregulate a quorum-sensing gene (*luxS*) in *E. coli* to hinder the biofilm formation. This can a major impact in tackling *E. coli* population during enteric infections, neonatal sepsis, urinary tract infections, and medical devices-associated nosocomial infections (Zuberi et al. 2017). This approach has a potential to tackle other such biofilm-forming bacteria in environmental or healthcare settings. Another way by which biofilms can be tackled is exemplified by the use of quorum-sensing inhibitors in tackling *P. aeruginosa*. The use of anti-quorum sensing compounds not only can repress virulence of *P. aeruginosa* making it more susceptible for clearance by the human immune system, but also dispose it to be killed by phage control by inhibition of the CRISPR-Cas immune system (Hoyland-Kroghsbo et al. 2017). Such synergistic efficacy can only be achieved if we have in-depth knowledge about the co-evolutionary strategy of both the systems in microbes. In such cases, a combination quorum-sensing–inhibition-phage therapy can be readily utilized for biotechnological applications.

Using phages as a biocontrol agent requires in-depth knowledge of bacterial antiphage strategies so that inadvertently selection of phage-resistant bacteria does not happen. These agents are especially useful in bacterial infections that do not respond well to antibiotics. The potential use of phages as biocontrol mediators has been tested in skin ulcers, purulent infections, burns, wounds, methicillin-resistant *S. aureus* (MRSA) mediated infections, respiratory tract infections, gastrointestinal ailments among many others (Abedon et al. 2011; Arora et al. 2017). The Food and Drug Administration, USA accepted phage-based control of *Listeria monocytogenes* by spraying meat and cheese with phages in 2006. Nonetheless, the use of phages in clinical practice is not very popular in all parts of the world, considering the fear of adverse reactions in absence of complete knowledge about the microbe-phage interaction. In this direction, knowledge about various counter strategies employed by both phages and bacteria is essential for the development of biotechnological tools for therapy.

Acknowledgments We sincerely acknowledge the support and guidance provided by Prof. Yogendra Singh, Department of Zoology, University of Delhi, for this work and Dr. Anupam Kumar Mondal, CSIR-IGIB, India for insights into the field of CRISPR biology.

References

Abedon ST (2012) Spatial vulnerability: bacterial arrangements, microcolonies, and biofilms as responses to low rather than high phage densities. Viruses 4:663–687. https://doi.org/10.3390/v4050663

Abedon ST, Kuhl SJ, Blasdel BG, Kutter EM (2011) Phage treatment of human infections. Bacteriophage 1:66–85. https://doi.org/10.4161/bact.1.2.15845

Arora G, Misra R, Sajid A (2017) Synthetic solutions to drug resistance. In: Arora G, Sajid A, Kalia VC (eds) Drug resistance in bacteria, fungi, malaria, and cancer. Springer International Publishing, Cham, pp 595–608. https://doi.org/10.1007/978-3-319-48683-3_26

Barrangou R (2013) CRISPR-Cas systems and RNA-guided interference. Wiley Interdiscip Rev RNA 4:267–278. https://doi.org/10.1002/wrna.1159

Barrangou R, Fremaux C, Deveau H, Richards M, Boyaval P, Moineau S, Romero DA, Horvath P (2007) CRISPR provides acquired resistance against viruses in prokaryotes. Science 315:1709–1712. https://doi.org/10.1126/science.1138140

Bassler BL (1999) How bacteria talk to each other: regulation of gene expression by quorum sensing. Curr Opin Microbiol 2:582–587. https://doi.org/10.1016/S1369-5274(99)00025-9

Bolotin A, Quinquis B, Sorokin A, Ehrlich SD (2005) Clustered regularly interspaced short palindrome repeats (CRISPRs) have spacers of extrachromosomal origin. Microbiology 151:2551–2561. https://doi.org/10.1099/mic.0.28048-0

Bondy-Denomy J, Pawluk A, Maxwell KL, Davidson AR (2013) Bacteriophage genes that inactivate the CRISPR/Cas bacterial immune system. Nature 493:429–432. https://doi.org/10.1038/nature11723

Bowden SD, Eyres A, Chung JC, Monson RE, Thompson A, Salmond GP, Spring DR, Welch M (2013) Virulence in *Pectobacterium atrosepticum* is regulated by a coincidence circuit involving quorum sensing and the stress alarmone, (p)ppGpp. Mol Microbiol 90:457–471. https://doi.org/10.1111/mmi.12369

Boysen A, Ellehauge E, Julien B, Sogaard-Andersen L (2002) The DevT protein stimulates synthesis of FruA, a signal transduction protein required for fruiting body morphogenesis in *Myxococcus xanthus*. J Bacteriol 184:1540–1546. https://doi.org/10.1128/JB.184.6.1540-1546.2002

Cady KC, O'Toole GA (2011) Non-identity-mediated CRISPR-bacteriophage interaction mediated via the Csy and Cas3 proteins. J Bacteriol 193:3433–3445. https://doi.org/10.1128/JB.01411-10

Cady KC, Bondy-Denomy J, Heussler GE, Davidson AR, O'Toole GA (2012) The CRISPR/Cas adaptive immune system of *Pseudomonas aeruginosa* mediates resistance to naturally occurring and engineered phages. J Bacteriol 194:5728–5738. https://doi.org/10.1128/JB.01184-12

Fineran PC, Slater H, Everson L, Hughes K, Salmond GP (2005) Biosynthesis of tripyrrole and beta-lactam secondary metabolites in *Serratia*: integration of quorum sensing with multiple new regulatory components in the control of prodigiosin and carbapenem antibiotic production. Mol Microbiol 56:1495–1517. https://doi.org/10.1111/j.1365-2958.2005.04660.x

Freidlin PJ, Nissan I, Luria A, Goldblatt D, Schaffer L, Kaidar-Shwartz H, Chemtob D, Dveyrin Z, Head SR, Rorman E (2017) Structure and variation of CRISPR and CRISPR-flanking regions in deleted-direct repeat region *Mycobacterium tuberculosis* complex strains. BMC Genomics 18:168. https://doi.org/10.1186/s12864-017-3560-6

Fuqua WC, Winans SC, Greenberg EP (1994) Quorum sensing in bacteria: the LuxR-LuxI family of cell density-responsive transcriptional regulators. J Bacteriol 176:269–275

Gao R, Krysciak D, Petersen K, Utpatel C, Knapp A, Schmeisser C, Daniel R, Voget S, Jaeger KE, Streit WR (2015) Genome-wide RNA sequencing analysis of quorum sensing-controlled regulons in the plant-associated *Burkholderia glumae* PG1 strain. Appl Environ Microbiol 81:7993–8007. https://doi.org/10.1128/AEM.01043-15

Gray KM, Garey JR (2001) The evolution of bacterial LuxI and LuxR quorum sensing regulators. Microbiology 147:2379–2387. https://doi.org/10.1099/00221287-147-8-2379

Hall-Stoodley L, Costerton JW, Stoodley P (2004) Bacterial biofilms: from the natural environment to infectious diseases. Nat Rev Microbiol 2:95–108. https://doi.org/10.1038/nrmicro821

Hargreaves KR, Kropinski AM, Clokie MR (2014) Bacteriophage behavioral ecology: how phages alter their bacterial host's habits. Bacteriophage 4:e29866. https://doi.org/10.4161/bact.29866

He L, Fan X, Xie J (2012) Comparative genomic structures of *Mycobacterium* CRISPR-Cas. J Cell Biochem 113:2464–2473. https://doi.org/10.1002/jcb.24121

Heussler GE, Cady KC, Koeppen K, Bhuju S, Stanton BA, O'Toole GA (2015) Clustered regularly interspaced short palindromic repeat-dependent, biofilm-specific death of *Pseudomonas aeruginosa* mediated by increased expression of phage-related genes. MBio 6:e00129–e00115. https://doi.org/10.1128/mBio.00129-15

Hoyland-Kroghsbo NM, Maerkedahl RB, Svenningsen SL (2013) A quorum-sensing-induced bacteriophage defense mechanism. MBio 4:e00362–e00312. https://doi.org/10.1128/mBio.00362-12

Hoyland-Kroghsbo NM, Paczkowski J, Mukherjee S, Broniewski J, Westra E, Bondy-Denomy J, Bassler BL (2017) Quorum sensing controls the *Pseudomonas aeruginosa* CRISPR-Cas adaptive immune system. Proc Natl Acad Sci U S A 114:131–135. https://doi.org/10.1073/pnas.1617415113

Huma N, Shankar P, Kushwah J, Bhushan A, Joshi J, Mukherjee T, Raju S, Purohit HJ, Kalia VC (2011) Diversity and polymorphism in AHL-lactonase gene (aiiA) of *Bacillus*. J Microbiol Biotechnol 21:1001–1011. https://doi.org/10.4014/jmb.1105.05056

Ji G, Beavis RC, Novick RP (1995) Cell density control of staphylococcal virulence mediated by an octapeptide pheromone. Proc Natl Acad Sci U S A 92:12055–12059

Ji G, Beavis R, Novick RP (1997) Bacterial interference caused by autoinducing peptide variants. Science 276:2027–2030. https://doi.org/10.1126/science.276.5321.2027

Julien B, Kaiser AD, Garza A (2000) Spatial control of cell differentiation in *Myxococcus xanthus*. Proc Natl Acad Sci U S A 97:9098–9103. https://doi.org/10.1073/pnas.97.16.9098

Kalia VC (2013) Quorum sensing inhibitors: an overview. Biotechnol Adv 31:224–245. https://doi.org/10.1016/j.biotechadv.2012.10.004

Kalia VC (2014a) Microbes, antimicrobials and resistance: the battle goes on. Indian J Microbiol 54:1–2. https://doi.org/10.1007/s12088-013-0443-7

Kalia VC (2014b) In search of versatile organisms for quorum-sensing inhibitors: acyl homoserine lactones (AHL)-acylase and AHL-lactonase. FEMS Microbiol Lett 359:143. https://doi.org/10.1111/1574-6968.12585

Kalia VC (2015a) Microbes: the most friendly beings? In: Kalia VC (ed) Quorum sensing vs quorum quenching: a battle with no end in sight. Springer, New Delhi, pp 1–5. https://doi.org/10.1007/978-81-322-1982-8_1

Kalia VC (2015b) Quorum sensing vs quorum quenching: a battle with no end in sight. Springer India, New Delhi. https://doi.org/10.1007/978-81-322-1982-8

Kalia VC, Kumar P (2015a) The battle: quorum-sensing inhibitors versus evolution of bacterial resistance. In: Kalia VC (ed) Quorum sensing vs quorum quenching: a battle with no end in sight. Springer, New Delhi, pp 385–391. https://doi.org/10.1007/978-81-322-1982-8_31

Kalia VC, Kumar P (2015b) Potential applications of quorum sensing inhibitors in diverse fields. In: Kalia VC (ed) Quorum sensing vs quorum quenching: a battle with no end in sight. Springer India, New Delhi, pp 359–370. https://doi.org/10.1007/978-81-322-1982-8_29

Kalia VC, Purohit HJ (2011) Quenching the quorum sensing system: potential antibacterial drug targets. Crit Rev Microbiol 37:121–140. https://doi.org/10.3109/1040841X.2010.532479

Kalia VC, Kumar P, Pandian STK, Sharma P (2015) Biofouling control by quorum quenching. In: Kim S-K (ed) Springer handbook of marine biotechnology. Springer, Berlin/Heidelberg, pp 431–440. https://doi.org/10.1007/978-3-642-53971-8_15

Karatan E, Watnick P (2009) Signals, regulatory networks, and materials that build and break bacterial biofilms. Microbiol Mol Biol Rev 73:310–347. https://doi.org/10.1128/MMBR.00041-08

Kumar P, Koul S, Patel SKS, Lee J-K, Kalia VC (2015) Heterologous expression of quorum sensing inhibitory genes in diverse organisms. In: Kalia VC (ed) Quorum sensing vs quorum quenching: a battle with no end in sight. Springer, New Delhi, pp 343–356. https://doi.org/10.1007/978-81-322-1982-8_28

LaSarre B, Federle MJ (2013) Exploiting quorum sensing to confuse bacterial pathogens. Microbiol Mol Biol Rev 77:73–111. https://doi.org/10.1128/MMBR.00046-12

Li Z, Nair SK (2012) Quorum sensing: how bacteria can coordinate activity and synchronize their response to external signals? Protein Sci 21:1403–1417. https://doi.org/10.1002/pro.2132

Li R, Fang L, Tan S, Yu M, Li X, He S, Wei Y, Li G, Jiang J, Wu M (2016) Type I CRISPR-Cas targets endogenous genes and regulates virulence to evade mammalian host immunity. Cell Res 26:1273–1287. https://doi.org/10.1038/cr.2016.135

Lloyd DG, Whitworth DE (2017) The myxobacterium *Myxococcus xanthus* can sense and respond to the quorum signals secreted by potential prey organisms. Front Microbiol 8:439. https://doi.org/10.3389/fmicb.2017.00439

Makarova KS, Wolf YI, Alkhnbashi OS, Costa F, Shah SA, Saunders SJ, Barrangou R, Brouns SJ, Charpentier E, Haft DH, Horvath P, Moineau S, Mojica FJ, Terns RM, Terns MP, White MF, Yakunin AF, Garrett RA, van der Oost J, Backofen R, Koonin EV (2015) An updated evolutionary classification of CRISPR-Cas systems. Nat Rev Microbiol 13:722–736. https://doi.org/10.1038/nrmicro3569

Miller MB, Bassler BL (2001) Quorum sensing in bacteria. Annu Rev Microbiol 55:165–199. https://doi.org/10.1146/annurev.micro.55.1.165

Nealson KH, Platt T, Hastings JW (1970) Cellular control of the synthesis and activity of the bacterial luminescent system. J Bacteriol 104:313–322

Ng WL, Bassler BL (2009) Bacterial quorum-sensing network architectures. Annu Rev Genet 43:197–222. https://doi.org/10.1146/annurev-genet-102108-134304

Novick RP, Geisinger E (2008) Quorum sensing in staphylococci. Annu Rev Genet 42:541–564. https://doi.org/10.1146/annurev.genet.42.110807.091640

Ortet P, De Luca G, Whitworth DE, Barakat M (2012) P2TF: a comprehensive resource for analysis of prokaryotic transcription factors. BMC Genomics 13:628. https://doi.org/10.1186/1471-2164-13-628

Papenfort K, Bassler BL (2016) Quorum sensing signal-response systems in gram-negative bacteria. Nat Rev Microbiol 14:576–588. https://doi.org/10.1038/nrmicro.2016.89

Parsek MR, Greenberg EP (2000) Acyl-homoserine lactone quorum sensing in gram-negative bacteria: a signaling mechanism involved in associations with higher organisms. Proc Natl Acad Sci U S A 97:8789–8793. https://doi.org/10.1073/pnas.97.16.8789

Patterson AG, Jackson SA, Taylor C, Evans GB, Salmond GPC, Przybilski R, Staals RHJ, Fineran PC (2016) Quorum sensing controls adaptive immunity through the regulation of multiple CRISPR-Cas systems. Mol Cell 64:1102–1108. https://doi.org/10.1016/j.molcel.2016.11.012

Pinedo CA, Smets BF (2005) Conjugal TOL transfer from *Pseudomonas putida* to *Pseudomonas aeruginosa*: effects of restriction proficiency, toxicant exposure, cell density ratios, and conjugation detection method on observed transfer efficiencies. Appl Environ Microbiol 71:51–57. https://doi.org/10.1128/AEM.71.1.51-57.2005

Platt TG, Fuqua C (2010) What's in a name? The semantics of quorum sensing. Trends Microbiol 18:383–387. https://doi.org/10.1016/j.tim.2010.05.003

Polkade AV, Mantri SS, Patwekar UJ, Jangid K (2016) Quorum sensing: an under-explored phenomenon in the phylum actinobacteria. Front Microbiol 7:131. https://doi.org/10.3389/fmicb.2016.00131

Rutherford ST, Bassler BL (2012) Bacterial quorum sensing: its role in virulence and possibilities for its control. Cold Spring Harb Perspect Med 2:a012427. https://doi.org/10.1101/cshperspect.a012427

Samson JE, Magadan AH, Sabri M, Moineau S (2013) Revenge of the phages: defeating bacterial defences. Nat Rev Microbiol 11:675–687. https://doi.org/10.1038/nrmicro3096

Schaefer AL, Lappala CR, Morlen RP, Pelletier DA, Lu TY, Lankford PK, Harwood CS, Greenberg EP (2013) LuxR- and LuxI-type quorum-sensing circuits are prevalent in members of the *Populus deltoides* microbiome. Appl Environ Microbiol 79:5745–5752. https://doi.org/10.1128/AEM.01417-13

Schauder S, Bassler BL (2001) The languages of bacteria. Genes Dev 15:1468–1480. https://doi.org/10.1101/gad.899601

Seed KD (2015) Battling phages: how bacteria defend against viral attack. PLoS Pathog 11:e1004847. https://doi.org/10.1371/journal.ppat.1004847

Seed KD, Lazinski DW, Calderwood SB, Camilli A (2013) A bacteriophage encodes its own CRISPR/Cas adaptive response to evade host innate immunity. Nature 494:489–491. https://doi.org/10.1038/nature11927

Sharma IM, Petchiappan A, Chatterji D (2014) Quorum sensing and biofilm formation in mycobacteria: role of c-di-GMP and methods to study this second messenger. IUBMB Life 66:823–834. https://doi.org/10.1002/iub.1339

Slater H, Crow M, Everson L, Salmond GP (2003) Phosphate availability regulates biosynthesis of two antibiotics, prodigiosin and carbapenem, in *Serratia* via both quorum-sensing-dependent and -independent pathways. Mol Microbiol 47:303–320. https://doi.org/10.1046/j.1365-2958.2003.03295.x

Stevens AM, Dolan KM, Greenberg EP (1994) Synergistic binding of the *Vibrio fischeri* LuxR transcriptional activator domain and RNA polymerase to the lux promoter region. Proc Natl Acad Sci U S A 91:12619–12623

Thomson NR, Crow MA, McGowan SJ, Cox A, Salmond GP (2000) Biosynthesis of carbapenem antibiotic and prodigiosin pigment in *Serratia* is under quorum sensing control. Mol Microbiol 36:539–556. https://doi.org/10.1016/j.molcel.2016.11.012

Thony-Meyer L, Kaiser D (1993) devRS, an autoregulated and essential genetic locus for fruiting body development in *Myxococcus xanthus*. J Bacteriol 175:7450–7462

Vale PF, Lafforgue G, Gatchitch F, Gardan R, Moineau S, Gandon S (2015) Costs of CRISPR-Cas-mediated resistance in *Streptococcus thermophilus*. Proc Biol Sci 282:20151270. https://doi.org/10.1098/rspb.2015.1270

van der Oost J, Westra ER, Jackson RN, Wiedenheft B (2014) Unravelling the structural and mechanistic basis of CRISPR-Cas systems. Nat Rev Microbiol 12:479–492. https://doi.org/10.1038/nrmicro3279

Viswanathan P, Murphy K, Julien B, Garza AG, Kroos L (2007) Regulation of *dev*, an operon that includes genes essential for *Myxococcus xanthus* development and CRISPR-associated genes and repeats. J Bacteriol 189:3738–3750. https://doi.org/10.1128/JB.00187-07

Westra ER, Buckling A, Fineran PC (2014) CRISPR-Cas systems: beyond adaptive immunity. Nat Rev Microbiol 12:317–326. https://doi.org/10.1038/nrmicro3241

Whitehead NA, Barnard AM, Slater H, Simpson NJ, Salmond GP (2001) Quorum-sensing in gram-negative bacteria. FEMS Microbiol Rev 25:365–404. https://doi.org/10.1111/j.1574-6976.2001.tb00583

Wiedenheft B, van Duijn E, Bultema JB, Waghmare SP, Zhou K, Barendregt A, Westphal W, Heck AJ, Boekema EJ, Dickman MJ, Doudna JA (2011) RNA-guided complex from a bacterial immune system enhances target recognition through seed sequence interactions. Proc Natl Acad Sci U S A 108:10092–10097. https://doi.org/10.1073/pnas.1102716108

Zegans ME, Wagner JC, Cady KC, Murphy DM, Hammond JH, O'Toole GA (2009) Interaction between bacteriophage DMS3 and host CRISPR region inhibits group behaviors of *Pseudomonas aeruginosa*. J Bacteriol 191:210–219. https://doi.org/10.1128/JB.00797-08

Zuberi A, Misba L, Khan AU (2017) CRISPR Interference (CRISPRi) Inhibition of luxS gene expression in *E. coli*: an approach to inhibit biofilm. Front Cell Infect Microbiol 7:214. https://doi.org/10.3389/fcimb.2017.00214

Quorum Sensing in Life Support Systems: The MELiSSA Loop

16

Sandra C. Catachura, Natalie Leys, and Felice Mastroleo

Abstract

Currently space exploration is possible thanks to the advanced technology that allow humans to survive on Space. However, for future long space mission it is necessary to investigate new technologies to ensure human life. Nowadays humans can survive at Space in the International Space Station (ISS) for a limited period of time i.e. almost 6 months at ISS whereas 40 days is foreseen for the Chinese Space Laboratory to be ready by 2020. Longer times of space exploration can be achieved if food oxygen and water (among other products) could be produced continuously without resupplying products from Earth. Several research groups have investigated about this possibility using Controlled Ecological Life-Support Systems (CELSS). Among those systems is the MELiSSA project that uses microorganism such as bacteria, cyanobacteria and higher plants to use human waste and convert it into water, oxygen and food.

The use of microorganism in these recycling systems needs special attention at different levels e.g. technical, environmental and biological parameters. In the frame work of the MELiSSA project some of the technical challenges include bioreactors design, the monitoring and control systems. Microorganisms behavior at space can be affected by environmental conditions such microgravity, space ionizing radiation as well as intrinsic biological behavior such genetic instability, metabolism and cell-to-cell communication also termed as quorum sensing. The aim of this chapter is to focus on that microbiological behavior and its possible effects on the MELiSSA loop.

Keywords

MELiSSA · Quorum sensing · Life support · Space · Microbes · Photosynthetic

S. C. Catachura · N. Leys · F. Mastroleo (✉)
Microbiology Unit, Interdisciplinary Biosciences (BIO), Institute for Environment, Health and Safety, Belgian Nuclear Research Centre (SCK•CEN), Mol, Belgium
e-mail: fmastrol@sckcen.be

V. C. Kalia (ed.), *Quorum Sensing and its Biotechnological Applications*,
https://doi.org/10.1007/978-981-13-0848-2_16

16.1 Introduction

Currently humans can live in space at the International Space Station (ISS) for a maximum period of 1 year. The daily demand of an astronaut is 1 k of oxygen, 2.8 k of drinking water, 2.7 kg of food (freeze-dried and canned food) and produces daily 1.2 kg of CO_2 (Farges et al. 2008). Astronauts needs such as food, fuel and other supplies are sent from Earth (and return to it) every 6 months approximately. However, for future long-term human manned exploration, it is needed to rely on life support systems that can continuously provide all the requirements to sustain the crew for years since resupplying from Earth will not be possible.

In this context, during the past years leading institutions have been studying the feasibility of live support systems. Several projects have been developed around the world e.g. The BIOS-3 experiments from Russia (Bartsev et al. 1996), The Advanced Life Support-ALS from United States (Erickson et al. 1996), The Micro-Ecological Life Support System Alternative-MELiSSA project from Europe (Gòdia et al. 2002), The Closed Equilibrated Biological Aquatic System-CEBAS from Germany (Bluem and Paris 2003), The Closed Ecological Experiment Facility-CEEF from Japan (Tako et al. 2010). Detailed information of each project can be found elsewhere (Nelson et al. 2009; Pycke 2009). With the exception of the MELiSSA program reports of active quorum sensing or the influence of it on each life support program is currently not available.

16.1.1 MELiSSA Project

MELiSSA stands for Micro-Ecological Life Support System Alternative, this system was inspired by a natural lake ecosystem (Mergeay et al. 1988). The concept of this loop is to rely on five interconnected compartments that will use organic and inorganic waste produced by the astronauts and transform it into fresh water, oxygen and food (Fig. 16.1). The main objective of the MELiSSA project was to acquire insights for developing regenerative life support systems for long term space missions (Lasseur et al. 2010).

Compartment I (CI) – The thermophilic compartment: the aim of this compartment is to degrade crew waste such as feces, urine and non-edible part of plants from compartment IVb among others. This compartment is colonized by a consortium of bacteria selected from human fecal material that allows allowed a higher biodegradation efficiency when compared to axenic cultures (Lasseur et al. 2010). Biodegradation is performed under anaerobic thermophilic (55 °C) conditions at acidic pH (5.5) in order to avoid methanogenesis (Poughon et al. 2013).

The outputs of this compartment are volatile fatty acids (VFAs) e.g. acetic, butyric acid and propionic acid as well as ammonium and CO_2 (Gòdia et al. 2004). To achieve separation of liquid fermentation products from undigested solid waste an ultrafiltration membrane is used. The use of this membrane allows the recovery of 95% of VFA and ammonia-N produced in CI (Poughon et al. 2013)

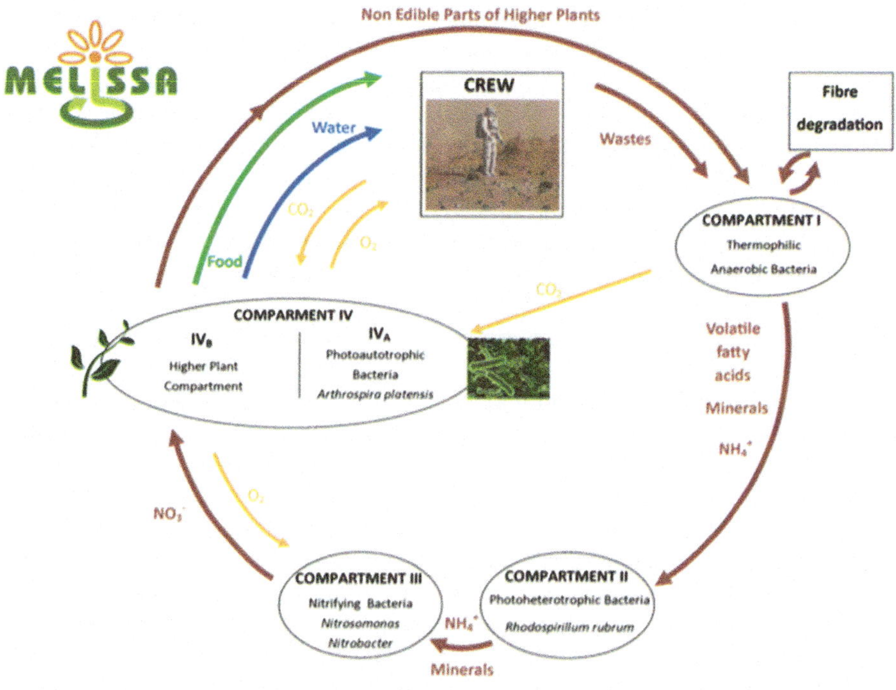

Fig. 16.1 The Micro-Ecological Life support system Alternative (MELiSSA). (Credits: European Space Agency)

Previous studies of compartment CI at pilot scale have shown that it is possible to reach a 90% of degradation efficiency by applying complementary technologies that helps in liquefaction of fiber (using *Fibrobacter succinogenes*), liquefaction and sanitation of recalcitrant organic matter (by high-pressure and temperature unit) and raw waste using hyper-thermophilic organisms (Lasseur et al. 2010).

Compartment II (CII) – The photoheterotrophic compartment: This compartment is colonized by *Rhodospirillum rubrum* S1H (ATCC25903), a spontaneous mutant from *R. rubrum* S1 (ATCC11170, parent strain). Unlike S1 strain S1H contains an increased amount of L-threonine deaminase (15–20-fold) when grown in medium containing malate and ammonium salt (Ning and Gest 1966). This characteristic could be beneficial since it can help to overcome growth inhibition in case of increased amino acids concentration in the loop.

R. rubrum is cultivated under anaerobic photoheterothrophic conditions. This bacterium has been studied at bench scale in a bioreactor of 2.5 l illuminated with halogen lamps. In parallel, pH, temperature and oxygen are measured whereas mixing is done using an internal helix. Design and cultivation conditions have been tested also at pilot scale using carbon sources such as acetate, propionate and butyrate separately. However after long cultivation periods *R. rubrum* cells forms aggregates that adhere to the photobioreactor wall, when acetic acid is used as carbon source (Cabello 2007).

Compartment III (CIII) – The nitrifying compartment: this compartment is colonized by a co-culture *Nitrobacter winogradskyi* ATCC14123 and *Nitrosomonas europaea* ATCC19178. The aim of this compartment is to convert ammonium into nitrate, needed in compartment IVa (Gòdia et al. 2002). *Ns. europaea* oxidizes ammonium (NH_4^+) to nitrite (NO_2^-) whereas *Nb. winogradskyi* converts nitrite to nitrate (NO_3^-). The bioreactor consists of immobilized cells packed-bed column. Cells are fixed as a biofilm on the surface of 4 mm polystyrene beads. Parameters such as pH, dissolved oxygen and temperature are measured and mechanical mixing is done using a magnetic stirrer, gas flow and liquid recirculation (Gòdia et al. 2004).

Compartment IVa and b – Photoautotrophic compartment: Compartment IV has been divided in IVa and IVb. The aim of this compartment is to produce oxygen, food and remove CO_2 (Hendrickx et al. 2006). Compartment IVa is colonized by *Arthrospira* sp. PCC8005, a filamentous cyanobacterium that photosynthetically fixes CO_2 and produces O_2. *Arthrospira* is a natural crucial source of food to many large aquatic organisms, such as fishes. Many *Arthrospira* species are also edible for man and are used as nutritious and health promoting food supplements. A gas-lift photobioreactor of 77 dm^3 of total volume was used to grow *Arthrospira* sp., which had provisions to regulate temperature, carbon dioxide (gas phase), dissolved oxygen, light and biomass. A stable functioning has been obtained in a batch model as well as in the MELiSSA pilot plant (MPP) where even a mathematical model that allows the estimation of mass productivity during long cultivation periods has been tested (Gòdia et al. 2004).

In compartment IVb higher plants e.g. lettuce, red beet and wheat were included in order to provide a healthy diet and improve the recycling system. Water produced during evapotranspiration of higher plants will be used for human consumption to some extent and recycled for the loop. It is foreseen that plants should be cultivated in isolated chambers in order to control environmental changes e.g. temperature, humidity, pressure (Lasseur et al. 2010).

Compartment V - The Crew: All the previous compartments should be able to provide the needs of the crew such as water, oxygen and food using as starting material the crew organic (urine, feces, non-edible part of plants, etc.) and inorganic (CO_2) waste.

16.1.2 Integration of the MEliSSA Loop

Currently active research is going on in order to fully integrate the MELiSSA loop. Besides these efforts only partial integration has been assessed at bench or pilot scale. At pilot scale interconnection of CIII and CIVa has been tested at the liquid phase (Gòdia et al. 2002). A subsequent bench scale integration of CII (2.5 l), CIII (600 ml) and CIVa (2.5 l) compartments has been tested using as initial medium a mixture of VFA (acetate, propionate and butyrate). During this integration, analysis of substrate consumption have shown that some VFAs are not completely used in CII however they can be consumed in the following compartments. Concerning CIII it was observed that the nitrite that was not converted into nitrate passes to CIVa and there it can be used by *Arthrospira* sp. as long as other nitrogen sources are not present (Gòdia et al. 2004). Further integration of CI, CIVb and CV compartments is foreseen to be incorporated at MPP. In the context of the BIORAT project preliminary studies using mice in a closed environment allowed to study the gas exchange between the compartments (Poughon et al. 2009).

16.2 Quorum Sensing in the MELiSSA Loop

So far different QS systems have been described for bacteria, Gram-negative bacteria uses molecules like AHL, AI2 among others for cell-to-cell communication. These molecules also named autoinducers are synthesized by synthases such as LuxI, LuxS, LuxM (Papenfort and Bassler 2016). Communication molecules for Gram-positive bacteria are via peptides in a two component system (TCS). A membrane receptor (sensor kinase) and a cytoplasmic transcription factor are needed to allow the expression of QS genes in a TCS (Ng and Bassler 2009). In the MELiSSA loop Gram-negative and Gram-positive species are used in compartments CI whereas in CII and CIII only Gram-negatives are present. In the MELiSSA loop QS activity has been reported for *R. rubrum* S1H under anaerobic photoheterotrophic and aerobic heterotrophic modes of growth in cultures at batch and bioreactor scale (Cabello 2007; Pycke 2009) as well as under simulated microgravity conditions (Mastroleo et al. 2013). The study of QS in a close life support system such MELiSSA is critical since the presence of active QS signals could act as natural micropollutants that could compromise the other compartments (Pycke 2009). For instance possible transfer of AHLs from CII to CIII could disrupt *Ns europeae* and *Nb winogradskyi* biofilm since both strains display a QS system based on AHLs.

Another possible effect of AHLs on the loop is at the level of CV, the crew. For instance it was reported that 3-oxo-C12-HSL exerts upregulation of pro-inflammatory mediators and induction of apoptosis at higher concentrations (Shiner et al. 2006). Another example of the interaction between the eukaryote and prokaryote world was reported in *Rhodopseudomonas palustris*. This bacterium employs an acyl-HSL synthase for producing *p*-coumaroyl-HSL, by using precursors such as *p*-coumarate, a plant metabolite (Schaefer et al. 2008).

16.2.1 Compartment I

As mentioned above compartment I is colonized by a bacterial consortium, the species conforming this group is not fully characterized yet. However a first study reported two species, *Ruminococcus bromii* and *Petrotoga mobilis* identified by PCR (Cabello 2007). A second study reported *Bacteroides* sp. and *Clostridium* sp. as representatives of Gram-negative and Gram-positive respectively. Both genera were found the most prevalent within the consortium when analyzed by denaturing gradient gel electrophoresis (Poughon et al. 2013).

R. bromii a Gram-positive bacterium abundant in the large human bowel (Abell et al. 2008) has been appointed to play a key role in the degradation of resistant starch in the human colon (Ze et al. 2012). Currently there are no reports of active QS system in *R. bromii* however in close related species *Ruminococcus flavefaciens* and *Ruminococcus albus* found in the intestine, AI-2 has been detected in culture supernatants (Mitsumori et al. 2003; Lukás et al. 2008). Regarding *P. mobilis* there are no reports on QS in this bacterium or in a close relative.

Autoinducer AI-2 is used by for interspecies and intra-species communication (Schauder et al. 2001). Therefore it could be possible a cross-communication among CI colonizing species affecting for instance their biodegradation capacity. An example of interspecies communication is the one observed between *Ruminococcus obeum* and *Vibrio chloerae* during the recovery phase after an acute diarrhea infection. It was observed that AI-2 produced by *R. obeum* repressed several *V. cholerae* colonization factors (Hsiao et al. 2014).

As mentioned above CI works under acidic thermophilic conditions, to our knowledge acidic conditions favors the presence of AHL however high temperatures could disrupt their chemical structure (Yates et al. 2002).

16.2.2 Compartment II

As mentioned above CII is colonized by *R. rubrum* S1H and it should metabolize substrates coming from CI such as VFAs. In this context basic research at bench and batch scale have been performed in order to generate knowledge about VFAs metabolism specifically using acetate, propionate and butyrate (Cabello 2007; Leroy et al. 2015).

Identification of an active QS system was assessed by detecting AHLs in *R. rubrum* cultures grown under photoheterotrophic conditions using succinate as carbon source. LC-MS/MS profiles showed as the most abundant 3-OH-C8-HSL (Pycke 2009). At the genomic level *R. rubrum* displayed a gene (Rru_A3396) encoding a putative AHL synthase (a LuxI-type) and six LuxR-type regulators, one located upstream the synthase and the rest of them distributed in the genome. Moreover homologues to AHL-degrading enzymes such as lactonase AiiA and acylase PvdQ were also identified indicating that *R. rubrum* could degrade AHLs (Carius et al. 2013).

Confirmation that gene Rru_A3396 encoded a AHL synthase was assessed by constructing a knockout mutant of gene Rru_A3396 by double homologous recombination, this QS-mutant strain was named M68. This study revealed that the main AHL synthase is *rruI* and that probably there is a second AHL-synthase since AHL were detected in extracts from M68 cultures. Genome comparison analysis suggests that the second synthase could be a HdtS type. Moreover 3-OH-C8-HSL was also found as the main AHL under relevant MELiSSA culture conditions (Condori et al. 2016) as reported elsewhere (Pycke 2009; Carius et al. 2013; Mastroleo et al. 2013).

Continuous culturing of *R. rubrum* at bioreactor scale under photoheterotrophic (using acetate as the sole carbon source) conditions leads to the formation of cells aggregates and subsequently a thick biofilm over the bioreactor wall (Cabello 2007). Contrary to CIII, biofilm formation in CII is not expected since it blocks completely the external light source therefore the bioreactor has to be stopped.

The fact that QS is active in *R. rubrum* raised the question that it could influence the biofilm formation observed in CII bioreactor. The relation of biofilm formation and QS has been reported for several Gram-negative species such as *Pseudomonas aeruginosa* where QS influence biofilm maturation (Bassler and Losick 2006). Likewise in *R. rubrum*, QS influence biofilm formation this was observed when *R. rubrum* WT and the QS mutant M68 were grown in a flow cell system where M68 did not form biofilm as the WT (Condori 2016). The influence of QS was also studied at the transcriptomic and proteomic level under MELiSSA conditions at batch scale. Results showed that almost the 8% of the genome was differentially expressed, genes encoding proteins related to flagellar structures were downregulated in M68 (Condori 2016). Evidence of positive QS regulation of motility has been reported for *Vibrio harveyi*, probably the same regulation occur in *R. rubrum* since M68 showed lower motility when compared to WT. However whether is a positive or negative regulation remains to be investigated (Yang and Defoirdt 2015).

It is well known that Spaceflight conditions influence bacterial behavior, for instance microgravity can enhance virulence in *Escherichia coli* and *Salmonella typhimurium* (Rosenzweig et al. 2010) or promote biofilm formation as observed in *Pseudomonas aeruginosa* (Kim et al. 2013). Interestingly simulated microgravity conditions influence QS in *R. rubrum* by increasing production of three AHLs (C10-HSL, C12-HSL and 3-OH-C14-HSL) independently of cell density (Mastroleo et al. 2013). Currently there are no reports on the genes regulated by each AHLs in *R. rubrum*, further investigation on the effect of each AHL will unravel the specific response in Space. *R. rubrum* was sent twice to the ISS in order to study the effects of Spaceflight conditions, samples were analyzed post-flight at the transcriptomic and proteomic level. Moreover ground-based studies using ionizing radiation showed that low dose radiation (2mGy) induce stress only at the transcriptomic level (Mastroleo et al. 2009).

16.2.3 Compartment III

CIII colonizers grow in biofilm, this configuration was selected due to the very low bacterial growth rate and because these cells will not be used as food therefore its production is not needed (Gòdia et al. 2002). Currently there are no reports of QS activity in this compartment. However other studies reported AHL production e.g. C6-, C8- and C10-HSL by *Ns. europaea* strain Schmidt. AHLs synthesis is possibly accomplished by NE1184 (Burton et al. 2005) an homolog of the HdtS synthase (Laue et al. 2000). Regarding *Nb winogradsky,* it display an active AHL-based QS that is similar to the LuxIR system, it contains an AHL synthase *nwiI* and its cognate regulator *nwiR*. A first study on the identification of AHL reported the production of C10-HSL and C10:1HSL (Mellbye et al. 2015). Further studies increased the list to C7-HSL, C8-HSL, C9-HSL, C10-HSL and C10:1-HSL, it also reports a new AHL

7,8-*trans*-*N*-(decanoyl) homoserine lactone (Shen et al. 2016). Both studies suggest that QS has an effect on the nitrification process by influencing the production and consumption of environmentally important nitrogen gases (NO, NO_2 and N_2O) (Mellbye et al. 2016).

16.2.4 Compartment IV

At the present time there are no studies reporting the communication system used by *Arthrospira* PCC8005. However AHL-based communication does not restrict to the bacterial world for instance the cyanobacterium *Gloeothece* PCC6909 base its communication system on AHL, specifically in C8-HSL. However QS function on biofilm formation remains unclear (Sharif et al. 2008). Regarding the role that QS could play on cyanobacteria, proteomic studies in *Spirulina platensis* suggest a potential role under low temperature stress condition since S-adenosyl homocysteine (SAH) hydrolase was found upregulated during cold-induced stress studies (Hongsthong et al. 2008). Interestingly cyanobacteria is also equipped with quorum quenching genes for instance *Anabaena* sp. PCC7120 possesses AiiC an AHL-acylase (Romero et al. 2008). In nature, the presence of this enzyme could serve as a defense mechanism against foreign AHL since these molecules inhibit growth and can have a toxic effect in *Anabaena* (Romero et al. 2011).

16.3 Possible Inhibition of QS in the MELiSSA Loop

Currently only in CII unwanted biofilm was formed endangering the functioning of the MELiSSA loop. Biofilm formation in *R. rubrum* is related to QS. Among the possible solutions to avoid biofilm formation are: replacing the WT by strain M68 or the addition of QS inhibitors. As mentioned above the mutant strain M68 do not form biofilm and exhibit a similar growth rate than WT (at batch scale) when growing under photoheterotrophic conditions using acetate as the sole carbon source (Condori et al. 2016). However genetic modified organisms are not allowed to be part of the MELiSSA loop. Therefore QS inhibitors could be a solution to avoid the biofilm observed in the bioreactor. Since *R. rubrum* produces AHLs, QS inhibition could be achieved by different methods i.e. using AHL analogues, decreasing AHLs production or inhibition of AHL synthesis (Kalia 2013). Addition of QS inhibitors in CII requires a complete understanding of the QS system in *R. rubrum* at the level of gene regulation since it is unknown which AHL binds to which regulators and subsequently which of these form complexes. By adding analogues of AHL the LuxR-type regulators could be inhibited however as mentioned above *R. rubrum* display six regulators and the genes under the control of each one remains unknown. Identification of such genes would allow to define the choice of QS inhibitors. Regarding the source of the QS inhibitors; the best choice would be from a prokaryotes organism than eukaryote source since some of the latter require calcium ion for activity (Kalia 2013).

As mentioned for AHL, the use of analogues imply a possible effect over the complete loop since they could pass from CII into the other compartments. The same potential risk is latent if external AHL-degrading enzymes, antibodies or any molecule is added to the loop. Perhaps a better solution could be to take advantage of the genes encoding putative AHL-degrading enzymes lactonase AiiA and acylase PvdQ present in *R. rubrum*. As reported previously the best choice would be to take advantage of an lactonase instead of an acylase since the latter depends on acyl chain length.

16.4 Opinion

As presented in this chapter the biotechnological solution for crew survival to long space manned mission is the life support system that includes the use of microorganisms. Several projects have been studied over the past 50 years. However quorum sensing which represents one of the factors that influence bacterial behavior, was not intensively studied in life support systems so far. Recently the relation of QS and biofilm formation in the compartment II of the European project MELiSSA showed the importance of studying cell-to-cell communication.

References

Abell GC, Cooke CM, Bennett CN, Conlon MA, McOrist AL (2008) Phylotypes related to *Ruminococcus bromii* are abundant in the large bowel of humans and increase in response to a diet high in resistant starch. FEMS Microbiol Ecol 66:505–515. https://doi.org/10.1111/j.1574-6941.2008.00527.x

Bartsev SI, Mezhevikin VV, Okhonin VA (1996) BIOS-4 as an embodiment of CELSS development conception. Adv Space Res 18(1/2):201–204. https://doi.org/10.1016/0273-1177(95)00809-S

Bassler BL, Losick R (2006) Bacterially speaking. Cell 125:237–246. https://doi.org/10.1016/j.cell.2006.04.001

Bluem V, Paris F (2003) Possible applications of aquatic bioregenerative life support modules for food production in a Martian base. Adv Space Res 31:77–86. https://doi.org/10.1016/S0273-1177(02)00659-2

Burton EO, Read HW, Pellitteri MC, Hickey WJ (2005) Identification of acyl-homoserine lactone signal molecules produced by *Nitrosomonas europaea* strain Schmidt. Appl Environ Microbiol 71:4906–4909. https://doi.org/10.1128/AEM.71.8.4906-4909.2005

Cabello F (2007) Cultivo en bioreactores de *Rhodospirillum rubrum* en condiciones fotoheterotró-ficas. (Doctor), Universitat Autònoma de Barcelona, Barcelona

Carius L, Carius AB, McIntosh M, Grammel H (2013) Quorum sensing influences growth and photosynthetic membrane production in high-cell-density cultivations of *Rhodospirillum rubrum*. BMC Microbiol 13:189. https://doi.org/10.1186/1471-2180-13-189

Condori S (2016) Unraveling the quorum sensing system of the photosynthetic bacterium *Rhodospirillum rubrum* S1H under light anaerobic conditions. (Doctor), University of Mons, Mons

Condori S, Atkinson S, Leys N, Wattiez R, Mastroleo F (2016) Construction and phenotypic characterization of M68, an RruI quorum sensing knockout mutant of the photosynthetic alphaproteobacterium *Rhodospirillum rubrum*. Res Microbiol 167:380–392. https://doi.org/10.1016/j.resmic.2016.02.006

Erickson JD, Eckelkamp RE, Barta DJ, Dragg J (1996) Mission simulation as an approach to develop requirements for automation in advanced life support systems. Adv Space Res 18:191–196

Farges B, Poughon L, Creuly C, Cornet J-F, Dussap CG, Lasseur C (2008) Dynamic aspects and controllability of the MELiSSA project: a bioregenerative system to provide life support in space. Appl Biochem Biotechnol 151:686. https://doi.org/10.1007/s12010-008-8292-2

Gòdia F, Albiol J, Montesinos JL, Perez J, Creus N, Cabello F, Mengual X, Montras A, Lasseur C (2002) MELISSA: a loop of interconnected bioreactors to develop life support in space. J Biotechnol 99:319–330. https://doi.org/10.1016/S0168-1656(02)00222-5

Gòdia F, Albiol J, Pérez J, Creus N, Cabello F, Montràs A, Masot A, Lasseur C (2004) The MELISSA pilot plant facility as an integration test-bed for advanced life support systems. Adv Space Res 34:1483–1493. https://doi.org/10.1016/j.asr.2003.08.038

Hendrickx L, De Wever H, Hermans V, Mastroleo F, Morin N, Wilmotte A, Janssen P, Mergeay M (2006) Microbial ecology of the closed artificial ecosystem MELiSSA (Micro-Ecological Life Support System Alternative): reinventing and compartmentalizing the Earth's food and oxygen regeneration system for long-haul space exploration missions. Res Microbiol 157:77–86. https://doi.org/10.1016/j.resmic.2005.06.014

Hongsthong A, Sirijuntarut M, Prommeenate P, Lertladaluck K, Porkaew K, Cheevadhanarak S, Tanticharoen M (2008) Proteome analysis at the subcellular level of the cyanobacterium *Spirulina platensis* in response to low-temperature stress conditions. FEMS Microbiol Lett 288:92–101. https://doi.org/10.1111/j.1574-6968.2008.01330.x

Hsiao A, Ahmed AMS, Subramanian S, Griffin NW, Drewry LL, Petri WA, Haque R, Ahmed T, Gordon JI (2014) Members of the human gut microbiota involved in recovery from *Vibrio cholerae* infection. Nature 515:423–426. https://doi.org/10.1038/nature13738

Kalia VC (2013) Quorum sensing inhibitors: an overview. Biotechnol Adv 31:224–245. https://doi.org/10.1016/j.biotechadv.2012.10.004

Kim W, Tengra FK, Young Z, Shong J, Marchand N, Chan HK, Pangule RC, Parra M, Dordick JS, Plawsy JL, Collins CH (2013) Spaceflight promotes biofilm formation by *Pseudomonas aeruginosa*. PLoS One 8:e62437. https://doi.org/10.1371/journal.pone.0062437

Lasseur C, Brunet J, de Weever H, Dixon M, Dussap G, Godia F, Leys N, Mergeay M, Van Der Straeten D (2010) MELiSSA: the European project of closed life support system. Gravitational Space Biol 23:12

Laue BE, Jiang Y, Chhabra SR, Jacob S, Stewart GS, Hardman A, Downie JA, O'Gara F, Williams P (2000) The biocontrol strain *Pseudomonas fluorescens* F113 produces the Rhizobium small bacteriocin, N-(3-hydroxy-7-cis-tetradecenoyl) homoserine lactone, via HdtS, a putative novel N-acylhomoserine lactone synthase. Microbiology 146:2469–2480. https://doi.org/10.1099/00221287-146-10-2469

Leroy B, De Meur Q, Moulin C, Wegria G, Wattiez R (2015) New insight into the photoheterotrophic growth of the isocytrate lyase-lacking purple bacterium *Rhodospirillum rubrum* on acetate. Microbiology 161:1061–1072. https://doi.org/10.1099/mic.0.000067

Lukás F, Gorenc G, Kopečný J (2008) Detection of possible AI-2-mediated quorum sensing system in commensal intestinal bacteria. Folia Microbiol 53:221–224. https://doi.org/10.1007/s12223-008-0030-1

Mastroleo F, Van Houdt R, Leroy B, Benotmane MA, Janssen A, Mergeay M, Vanhavere F, Hendrickx L, Wattiez R, Leys N (2009) Experimental design and environmental parameters affect *Rhodospirillum rubrum* S1H response to space flight. ISME J 3:1402–1419. https://doi.org/10.1038/ismej.2009.74

Mastroleo F, Van Houdt R, Atkinson S, Mergeay M, Hendrickx L, Wattiez R, Leys N (2013) Modelled microgravity cultivation modulates N-acylhomoserine lactone production in *Rhodospirillum rubrum* S1H independently of cell density. Microbiology 159:2456–2466. https://doi.org/10.1099/mic.0.066415-0

Mellbye BL, Bottomley PJ, Sayavedra-Soto LA (2015) Nitrite-oxidizing bacterium *Nitrobacter winogradskyi* produces N-acyl-homoserine lactone autoinducers. Appl Environ Microbiol 81:5917–5926. https://doi.org/10.1128/AEM.01103-15

Mellbye BL, Giguere AT, Bottomley PJ, Sayavedra-Soto LA (2016) Quorum quenching of *Nitrobacter winogradskyi* suggests that quorum sensing regulates fluxes of nitrogen oxide(s) during nitrification. MBio 7(5):e01753–e01716. https://doi.org/10.1128/mBio.01753-16

Mergeay M, Verstraete W, Dubertret G, Lefort-Tran M, Chipaux C, Binot R (1988) 'MELiSSA' – a microorganisms-based model for 'CELSS' development. Paper presented at the 3rd European symposium on space thermal control and life support systems, Noordwijk, The Netherlands

Mitsumori M, Xu L, Kajikawa H, Kurihara M, Tajima K, Hai J, Takenaka A (2003) Possible quorum sensing in the rumen microbial community: detection of quorum-sensing signal molecules from rumen bacteria. FEMS Microbiol Lett 219:47–52. https://doi.org/10.1016/S0378-1097(02)01192-8

Nelson M, Pechurkin NS, Allen JP, Somova LA, Gitelson JI (2009) In: L.K.W. et al (eds) Environmental biotechnology, vol 10. Humana Press, New York

Ng WL, Bassler BL (2009) Bacterial quorum-sensing network architectures. Annu Rev Genet 43:197–222. https://doi.org/10.1146/annurev-genet-102108-134304

Ning C, Gest H (1966) Regulation of L-isoleucine biosynthesis in the photosynthetic bacterium *Rhodospirillum rubrum*. Proc Natl Acad Sci U S A 56:1823–1827

Papenfort K, Bassler BL (2016) Quorum sensing signal-response systems in Gram-negative bacteria. Nat Rev Microbiol 14:576–588. https://doi.org/10.1038/nrmicro.2016.89

Poughon L, Farges B, Dussap CG, Godia F, Lasseur C (2009) Simulation of the MELiSSA closed loop system as a tool to define its integration strategy. Adv Space Res 44:1392–1403. https://doi.org/10.1016/j.asr.2009.07.021

Poughon L, Creuly C, Farges B, Dussap CG, Schiettecatte W, Jovetic S, De Wever H (2013) Test of an anaerobic prototype reactor coupled with a filtration unit for production of VFAs. Bioresour Technol 145:240–247. https://doi.org/10.1016/j.biortech.2012.12.052

Pycke B (2009) The fate and effects of micropollutants in a biological life support system. (Doctor), Ghent University, Belgium

Romero M, Diggle SP, Heeb S, Cámara M, Otero A (2008) Quorum quenching activity in *Anabaena* sp. PCC 7120: identification of AiiC, a novel AHL-acylase. FEMS Microbiol Lett 280:73–80. https://doi.org/10.1111/j.1574-6968.2007.01046.x

Romero M, Muro-Pastor AM, Otero A (2011) Quorum sensing N-acylhomoserine lactone signals affect nitrogen fixation in the cyanobacterium *Anabaena* sp. PCC7120. FEMS Microbiol Lett 315:101–108. https://doi.org/10.1111/j.1574-6968.2010.02175.x

Rosenzweig JA, Abogunde O, Thomas K, Lawal A, Nguyen YU, Sodipe A, Jejelowo O (2010) Spaceflight and modeled microgravity effects on microbial growth and virulence. Appl Microbiol Biotechnol 85:885–891. https://doi.org/10.1007/s00253-009-2237-8

Schaefer AL, Greenberg EP, Oliver CM, Oda Y, Huang JJ, Bittan-Banin G, Peres CM, Schmidt S, Juhaszova K, Sufrin JR, Harwood CS (2008) A new class of homoserine lactone quorum-sensing signals. Nature 454:595–599. https://doi.org/10.1038/nature07088

Schauder S, Shokat K, Surette MG, Bassler BL (2001) The LuxS family of bacterial autoinducers: biosynthesis of a novel quorum-sensing signal molecule. Mol Microbiol 41:463–476. https://doi.org/10.1046/j.1365-2958.2001.02532.x

Sharif DI, Gallon J, Smith CJ, Dudley E (2008) Quorum sensing in cyanobacteria: N-octanoyl-homoserine lactone release and response, by the epilithic colonial cyanobacterium *Gloeothece* PCC6909. ISME J 2:1171–1182. https://doi.org/10.1038/ismej.2008.68

Shen Q, Gao J, Liu J, Liu S, Liu Z, Wang Y, Guo B, Zhuang X, Zhuang G (2016) A new acyl-homoserine lactone molecule generated by *Nitrobacter winogradskyi*. Sci Rep 6:22903. https://doi.org/10.1038/srep22903

Shiner EK, Terentyev D, Bryan A, Sennoune S, Martinez-Zaguilan R, Li G, Gyorke S, Williams SC, Rumbaugh KP (2006) Pseudomonas aeruginosa autoinducer modulates host cell responses through calcium signalling. Cell Microbiol 8:1601–1610. https://doi.org/10.1111/j.1462-5822.2006.00734.x

Tako Y, Arai R, Tsuga S, Komatsubara O, Masuda T, Nozoe S, Nitta K (2010) CEEF: closed ecology experiment facilities. Gravitational Space Biol 23

Yang Q, Defoirdt T (2015) Quorum sensing positively regulates flagellar motility in pathogenic *Vibrio harveyi*. Environ Microbiol 17:960–968. https://doi.org/10.1111/1462-2920.12420

Yates EA, Philipp B, Buckley C, Atkinson S, Chhabra SR, Sockett RE, Goldner M, Dessaux Y, Camara M, Smith H, Williams P (2002) N-Acylhomoserine lactones undergo lactonolysis in a pH-, temperature-, and acyl chain length-dependent manner during growth of *Yersinia pseudotuberculosis* and *Pseudomonas aeruginosa*. Infect Immun 70:5635–5646. https://doi.org/10.1128/iai.70.10.5635-5646.2002

Ze X, Duncan SH, Louis P, Flint HJ (2012) Ruminococcus bromii is a keystone species for the degradation of resistant starch in the human colon. ISME J 6:1535–1543. https://doi.org/10.1038/ismej.2012.4

17 Novel Super-Regulators of Quorum Sensing in *Pseudomonas aeruginosa*

Chuanmin Zhou and Min Wu

Abstract

Quorum sensing (QS) systems are a vital network in *Pseudomonas aeruginosa* for regulating cell to cell communication. Many of this bacterial virulence factors are controlled by or associated with QS. As QS system may directly regulate up to 10% genes of *P. aeruginosa*, this system is of fundamental importance in bacterial physiology and pathogenesis. However, it is still unclear how the QS genes regulate their targets and how QS circuits are modulated by other regulators. Here, we review how a series of recently identified critical regulators, named "super-regulators" in *P. aeruginosa,* participate in QS signaling to modulate the expression of its effectors.

Keywords

Quorum sensing · *Pseudomonas aeruginosa* · Two-component systems;

Abbreviation

Bfm	Biofilm maturation
CdpR	ClpAP-degradation and pathogenicity Regulatorregulator
QscR	Quorum-sensing control repressor

C. Zhou
Department of Biomedical Sciences, University of North Dakota, Grand Forks, ND, USA

Wuhan University School of Health Sciences, Wuhan University,
Wuhan, Wuhan Province, People's Republic of China

M. Wu (✉)
Department of Biomedical Sciences, University of North Dakota, Grand Forks, ND, USA
e-mail: min.wu@med.und.edu

V. C. Kalia (ed.), *Quorum Sensing and its Biotechnological Applications*,
https://doi.org/10.1007/978-981-13-0848-2_17

QsrO	QS-repressing ORF
RpoN	RNA polymerase sigma-54 factor
RpoS	RNA polymerase, sigma S
VqsM	Virulence and QS modulator
VqsR	Virulence and quorum-sensing regulator

17.1 Introduction

P. aeruginosa is a Gram-negative opportunistic bacterial pathogen that can cause acute and chronic pulmonary infection in immunocompromised individuals, such as patients with Cystic fibrosis (CF), Chronic Obstructive Pulmonary Disease (COPD), severe burns or cancer (Chugani and Greenberg 2007; Lyczak et al. 2000, 2002; Sousa and Pereira 2014). A number of critical genes in this bacterium govern the pathogenesis and physiology to coordinate a massive invasion to its mammalian hosts. For example, at least three different quorum sensing (QS) systems are found in *P. aeruginosa*: two LuxI/LuxR-type QS circuits LasI/LasR, RhlI/RhlR, and non-LuxI/LuxR-type QS circuits *Pseudomonas* quinolone signal (PQs) system. The QS system is a global regulatory mechanism controlling hundreds of genes including a large cluster of virulence factors. Up to 10% of the predicted 5570 genes are thought to be regulated by QS systems (Schuster and Greenberg 2006). Correspondingly, there are quite a few super-regulators participating in QS circuit regulation though the underlying mechanisms remain to be further elaborated.

Two-component systems (TCS), consisting of a sensor histidine protein kinase (HK), and a response regulator protein (RR), play a key role in sensing environmental challenges (Stock et al. 2000). Normally, HK is a membrane protein and RR is a cytoplasmic partner. HK is stimulated by extracellular signal and autophosphorylates itself. Then the activated HK phosphotransfers to the RR, and in turn the phosphorylated RR regulates the downstream effectors in response to the stimuli (Stock et al. 2000). More than 100 TCS genes have been found in *P. aeruginosa* (Rodrigue et al. 2000; Stover et al. 2000). However, the interactions between TCSs and QS are poorly understood (Rodrigue et al. 2000; Wang et al. 2013; Cao et al. 2014).

In recent years, we have witnessed a huge amount of fundamental discoveries in QS functions and novel regulators that impact QS expression and action. Herein, we review the progress in super-regulators of QS and TCS, dealing with the latest understanding of their functions and underlying molecular mechanisms.

17.2 RsaL

RsaL is a global regulator in *P. aeruginosa* (controlling more than 340 genes), which is involved in inhibiting secreted virulence factors, decreasing twitching and swarming motility, and promoting biofilm formation (Rampioni et al. 2009). The *rsaL* gene is located at the intergenic region between *lasR* and *lasI* (so named due to its location nearby the Las genes), encoding an 11 kDa protein (Gambello and Iglewski 1991; de Kievit et al. 1999), which can be activated by N-(3-oxododecanoyl)-L-HSL (3OC12-HSL). Previous research found that *rsaL* mutation strain produced elevated levels 3OC12-HSL, indicating RsaL repressing *lasI* transcription (Rampioni et al. 2006). Researchers reported that RsaL and LasR shared similar binding sites in *lasI* promoter (Rampioni et al. 2006). Recently, the Liang Lab performed ChIP-seq assay and found that RsaL promotes expression of *pqsH* and *cdpR* by binding to the intergenic region between *pqsH* and *cdpR* (Kang et al. 2017). Thus, RsaL is regulated by QS, and in turn, RsaL may participate in regulation of other genes by the QS pathway, forming an RsaL-QS circuit.

17.3 CdpR

A new QS regulator PA2588 was found in *P. aeruginosa* and named as CdpR, regulating expression of virulence factors and pathogenicity (Zhao et al. 2016). CdpR is indicated as an AraC-family regulator and functions in the downstream of PQs effector gene *pqsH*. Previously, CdpR was reported to be regulated by LasI and VqsM by directly targeting the promoter (Liang et al. 2014; Zhao et al. 2016). CdpR shares the promoter region with *pqsH* and hence binds to the promoter motif of *pqsH*, which regulates the transcription of *pqsH* and activates PqsH to impact QS systems (Zhao et al. 2016).

17.4 VqsM

VqsM is a global QS modulator that was first reported in 2005 (Dong et al. 2005). Approximately 100 genes that are promoted by VqsM are also shown to be regulated by QS systems, including *rhlR, rsaL, vqsR, lasI* and *rhlI* (Dong et al. 2005). As a QS regulator, VqsM promotes the production of N-acyl homoserine lactones (AHLs) through VqsR, RpoS and PprB (Dong et al. 2005). The expression of VqsM may be influenced by a negative QS regulator QsrO (Köhler et al. 2014). Recent research reveals that VqsM directly binds to the promoter of *lasI* and correspondingly controls the transcription of *lasI*, while indirectly regulates the Rhl system (Liang et al. 2014). Despite the recent insight into the role of VqsM in QS signaling, the underlying regulatory mechanisms and the interactive network remain to be further characterized.

17.5 RpoN

RpoN participates in regulating the production of many virulence factors, including alginate, rhamnolipid, and lipase (Studholme and Buck 2000; Hendrickson et al. 2001). Many virulence activities controlled by RpoN are also regulated by Rhl. Thompson and colleagues find that transcription of *rhlI* is reduced in the *rpoN* deletion mutant strain in an M9 minimal medium, while equivalent results are not observed in the complex medium (Thompson et al. 2003). RpoN positively regulates RhlI by binding its promoter, which is consistent with previous reports (Studholme and Buck 2000; Thompson et al. 2003). However, the detailed mechanisms of how RpoN regulates virulence in *P. aeruginosa* are elusive. Recently, two scientific papers demonstrate that RpoN regulates QS signaling by modulating PQs molecules *pqsR* and *pqsE* (Cai et al. 2015; Viducic et al. 2016). Transcription of *pqsA* is significantly lower in mutant strain than the wild-type (Cai et al. 2015). RpoN regulates PqsR by binding with the *pqsR* sequence (Cai et al. 2015; Schulz et al. 2015). However, another report shows opposite results in which *rpoN* mutant promotes the expression of *pqsA, pqsH,* and *pqsR* versus their corresponding controls (Viducic et al. 2016).

17.6 QsrO

QsrO (PA2226) is a novel regulator of QS found in 2014 and is located at the upstream of *vqsM* (Kohler et al. 2014). QsrO is able to inhibit three different QS systems in *P. aeruginosa* (Kohler et al. 2014). Deletion of QsrO results in a subdued QS phenotype with both PAO1 and PA14 strains by reducing activities of VqsM, etc. Co-expression of QsrO and PA2225 can block the activation of the type III secretion system, suggesting that QsrO is a negative-regulator for the QS intracellular circuits in *P. aeruginosa.*

17.7 QscR

QscR is an orphan LuxR-type homolog regulator that binds to 3OC12HSL (Lee et al. 2006; Oinuma and Greenberg 2011). QscR can also form heterodimers with LasR and RhlR, thereby inhibiting QS signaling (Ledgham et al. 2003a).

17.8 VqsR

VqsR is also a global regulator in *P. aeruginosa* (controlling approximately 200 genes). VqsR promotes QS signaling and production of virulence factors confirmed by *vqsR* deletion mutation (Juhas et al. 2004, 2005). VqsR is activated by *las* system via binding the las box (Hentzer et al. 2003; Li et al. 2007). VqsR expression is also inhibited in *pqsA*, *pqsR* and *pqsE* mutant strains (Viducic et al. 2017). Furthermore, RpoS is also reported to participate in the interaction between PQs and VqsR (Viducic et al. 2017), and in turn, VqsR promotes QS intercellular communication by binding to an inverted repeat sequence in *qscR* (Liang et al. 2012).

17.9 BfmR/S

The BfmR/S TCS is critical regulators of biofilm formation and their expression is highly induced in cystic fibrosis patients (Son et al. 2007). BfmR is also negatively regulated by BfmS (Cao et al. 2014). BfmR, like LasR and Vfr, directly binds to the *rhlR* promoter, reducing the production of N-butanoyl-L-homoserine lactone (C4-HSL) (Croda-Garcia et al. 2011; Balasubramanian et al. 2013; Cao et al. 2014). It is possible that BfmR/S TCS takes part in the regulation of *P. aeruginosa* virulence in cystic fibrosis disease.

17.10 Conclusion

The QS signaling involves a global regulatory system. Transcription of hundreds of genes are controlled by the QS system (Schuster and Greenberg 2006). Recent progress suggests that super-regulators critically regulate QS circuits. A number of researchers continuously discover new regulators and their action mechanisms for QS regulatory network to broaden our horizons to better understand the complex signaling circuits (Table 17.1 and Fig. 17.1).

In brief, four different patterns are currently described in super-regulators for regulating QS systems. The first mechanism is directly promoting or inhibiting the transcription of QS genes by targeting the related promoters (Albus et al. 1997; Siehnel et al. 2010). The second is binding to the LasR or RhlR and form dimers for inhibiting Lux-R, such as QscR (Ledgham et al. 2003a, b; Lee et al. 2006; Oinuma and Greenberg 2011). The third mechanism is that super-regulators may control the first group of the super-regulators and subsequently regulate the expression of QS genes (Zhao et al. 2016). Finally, sRNAs are also involved in regulating the expression of QS molecules, such as RsmY and RsmZ (Kay et al. 2006; O'Callaghan et al. 2011).

P. aeruginosa is the most common hospital opportunistic pathogen, causing great threat to nosocomial infection for immunocompromised patients (Lyczak et al. 2000, 2002; Chugani and Greenberg 2007; Sousa and Pereira 2014). QS signaling is an important mechanism for *P. aeruginosa* to produce a series of virulence factors and thereby enhancing its pathogenicity. *P. aeruginosa* QS may serve as a therapeutic target for curing cystic fibrosis patients by downregulating its virulence. The master QS factor LasR is located at the top of QS circuits (Rutherford and Bassler 2012). Virulence factor production and initial infection are known to be LasI/LasR dependent (Rutherford and Bassler 2012). It is highly likely that LasR inhibitors can be used as an effective treatment method. Some studies on inhibitor development have revealed that LasR antagonists or autoinducer (AI) analogues with the conserved ligand binding to LasR may inhibit the function of LasR (Kim et al. 2008; Mattmann and Blackwell 2010; McInnis and Blackwell 2011).

Table 17.1 QS super-regulators in *P. aeruginosa*

Regulators	Quorum sensing targeting	References
AlgR2	Down-regulate transcription of *lasR* and *rhlR (not discussed in the chapter)*	Ledgham et al. (2003b)
BfmR	Down-regulate transcription of *rhlR*	Cao et al. (2014)
CdpR	Down-regulate transcription of *pqsH*	Zhao et al. (2016)
DksA	Down-regulate transcription of *rhlI (not discussed in the chapter)*	Branny et al. (2001) and Jude et al. (2003)
GacA/S	Up-regulate transcription of *lasR and rhlR (not discussed in the chapter)*	Parkins et al. (2001)
QslA	Negative regulator (anti-activator) of LasR and PqsR proteins *(not discussed in the chapter)*	Seet and Zhang (2011) and Fan et al. (2013)
QteE	Down-regulate transcription of *lasR* and *rhlR (not discussed in the chapter)*	Siehnel et al. (2010)
QscO	Down-regulate transcription of QS	Kohler et al. (2014)
QscR	Negative regulator (anti-activator) of LasR protein	Ledgham et al. (2003a, b); Lee et al. (2006) and Oinuma and Greenberg (2011)
PprB	Up-regulate transcription of *lasI*, *rhlI* and *rhlR (not discussed in the chapter)*	Dong et al. (2005) and Pugsley (2008)
RpoN	Down-regulate transcription of *lasR* and *rhlR*	Heurlier et al. (2003) and Thompson et al. (2003)
RpoS	Down-regulate transcription of *rhlI*	Schuster et al. (2004)
RsaL	Down-regulate transcription of *lasI* and *pqsH*	Rampioni et al. (2006) and Kang et al. (2017)
RsmA	Down-regulate transcription of *lasI (not discussed in the paper)*	Pessi et al. (2001)
Vfr	Up-regulate transcription of *lasR* and *rhlR (not discussed in the chapter)*	Albus et al. (1997)
VqsM	Up-regulate transcription of *lasI*	Dong et al. (2005) and Liang et al. (2014)
VqsR	Up-regulate transcription of *pqsR*	Liang et al. (2012)

New super-QS-regulators, such as Vfr, VqsM, and VqsR, are also global regulators in *P. aeruginosa*, responsible for bacterial virulence factor production, biofilm formation, etc. (Albus et al. 1997; Dong et al. 2005; Liang et al. 2012, 2014). These new regulators provide broader targets for switching on or off the QS systems, indicating potential for controlling bacterial infection.

17.11 Opinion

P. aeruginosa is an important, common opportunistic pathogen, causing severe infection in immunocompromised patients, with threatening mortality. Some known inhibitors for LuxI/LuxR may provide a therapeutic method for controlling a *P. aeruginosa* in vivo by reducing its virulence. We posit a presumption that super-regulators of QS can also be new targets to design improved treatments for acute or chronic respiratory infection.

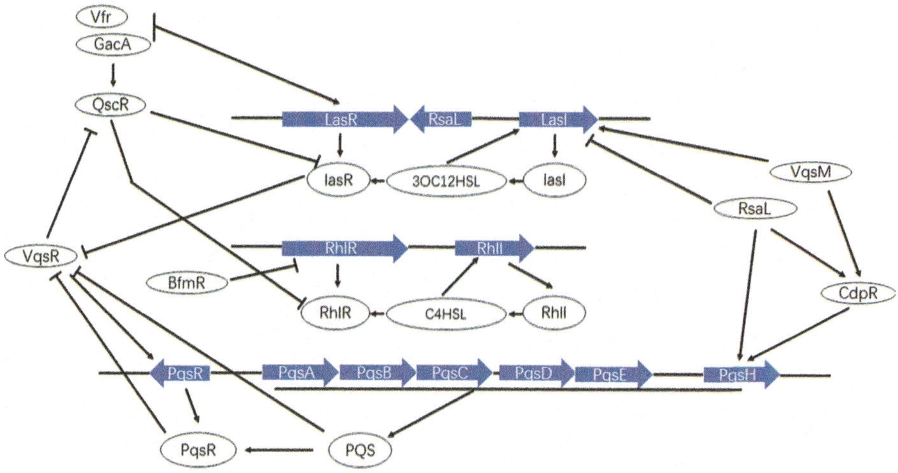

Fig. 17.1 Schematic overview of the interconnection between QS super-regulators and the QS system in *Pseudomonas aeruginosa*

References

Albus AM, Pesci EC, Runyen-Janecky LJ, West SE, Iglewski BH (1997) Vfr controls quorum sensing in *Pseudomonas aeruginosa*. J Bacteriol 179:3928–3935

Balasubramanian D, Schneper L, Kumari H, Mathee K (2013) A dynamic and intricate regulatory network determines *Pseudomonas aeruginosa* virulence. Nucleic Acids Res 41:1–20. https://doi.org/10.1093/nar/gks1039

Branny P, Pearson JP, Pesci EC, Kohler T, Iglewski BH, Van Delden C (2001) Inhibition of quorum sensing by a *Pseudomonas aeruginosa* dksA homologue. J Bacteriol 183:1531–1539. https://doi.org/10.1128/JB.183.5.1531-1539.2001

Cai Z, Liu Y, Chen Y, Yam JK, Chew SC, Chua SL, Wang K, Givskov M, Yang L (2015) RpoN regulates virulence factors of *Pseudomonas aeruginosa* via modulating the PqsR quorum sensing regulator. Int J Mol Sci 16:28311–28319. https://doi.org/10.3390/ijms161226103

Cao Q, Wang Y, Chen F, Xia Y, Lou J, Zhang X, Yang N, Sun X, Zhang Q, Zhuo C, Huang X, Deng X, Yang CG, Ye Y, Zhao J, Wu M, Lan L (2014) A novel signal transduction pathway that modulates rhl quorum sensing and bacterial virulence in *Pseudomonas aeruginosa*. PLoS Pathog 10:e1004340. https://doi.org/10.1371/journal.ppat.1004340

Chugani S, Greenberg EP (2007) The influence of human respiratory epithelia on *Pseudomonas aeruginosa* gene expression. Microb Pathog 42:29–35. https://doi.org/10.1016/j.micpath.2006.10.004

Croda-Garcia G, Grosso-Becerra V, Gonzalez-Valdez A, Servin-Gonzalez L, Soberon-Chavez G (2011) Transcriptional regulation of *Pseudomonas aeruginosa* rhlR: role of the CRP orthologue Vfr (virulence factor regulator) and quorum-sensing regulators LasR and RhlR. Microbiology 157:2545–2555. https://doi.org/10.1099/mic.0.050161-0

de Kievit T, Seed PC, Nezezon J, Passador L, Iglewski BH (1999) RsaL, a novel repressor of virulence gene expression in *Pseudomonas aeruginosa*. J Bacteriol 181:2175–2184

Dong YH, Zhang XF, Xu JL, Tan AT, Zhang LH (2005) VqsM, a novel AraC-type global regulator of quorum-sensing signalling and virulence in *Pseudomonas aeruginosa*. Mol Microbiol 58:552–564. https://doi.org/10.1111/j.1365-2958.2005.04851.x

Fan H, Dong Y, Wu D, Bowler MW, Zhang L, Song H (2013) QsIA disrupts LasR dimerization in antiactivation of bacterial quorum sensing. Proc Natl Acad Sci U S A 110:20765–20770. https://doi.org/10.1073/pnas.1314415110

Gambello MJ, Iglewski BH (1991) Cloning and characterization of the *Pseudomonas aeruginosa* lasR gene, a transcriptional activator of elastase expression. J Bacteriol 173:3000–3009

Hendrickson EL, Plotnikova J, Mahajan-Miklos S, Rahme LG, Ausubel FM (2001) Differential roles of the *Pseudomonas aeruginosa* PA14 rpoN gene in pathogenicity in plants, nematodes, insects, and mice. J Bacteriol 183:7126–7134. https://doi.org/10.1128/JB.183.24.7126-7134.2001

Hentzer M, Wu H, Andersen JB, Riedel K, Rasmussen TB, Bagge N, Kumar N, Schembri MA, Song Z, Kristoffersen P, Manefield M, Costerton JW, Molin S, Eberl L, Steinberg P, Kjelleberg S, Hoiby N, Givskov M (2003) Attenuation of *Pseudomonas aeruginosa* virulence by quorum sensing inhibitors. EMBO J 22:3803–3815. https://doi.org/10.1093/emboj/cdg366

Heurlier K, Denervaud V, Pessi G, Reimmann C, Haas D (2003) Negative control of quorum sensing by RpoN (sigma54) in *Pseudomonas aeruginosa* PAO1. J Bacteriol 185:2227–2235. https://doi.org/10.1128/JB.185.7.2227-2235.2003

Jude F, Kohler T, Branny P, Perron K, Mayer MP, Comte R, van Delden C (2003) Posttranscriptional control of quorum-sensing-dependent virulence genes by DksA in *Pseudomonas aeruginosa*. J Bacteriol 185:3558–3566. https://doi.org/10.1128/JB.185.12.3558-3566.2003

Juhas M, Wiehlmann L, Huber B, Jordan D, Lauber J, Salunkhe P, Limpert AS, von Gotz F, Steinmetz I, Eberl L, Tummler B (2004) Global regulation of quorum sensing and virulence by VqsR in *Pseudomonas aeruginosa*. Microbiology 150:831–841. https://doi.org/10.1099/mic.0.26906-0

Juhas M, Wiehlmann L, Salunkhe P, Lauber J, Buer J, Tummler B (2005) GeneChip expression analysis of the VqsR regulon of *Pseudomonas aeruginosa* TB. FEMS Microbiol Lett 242:287–295. https://doi.org/10.1016/j.femsle.2004.11.020

Kang H, Gan J, Zhao J, Kong W, Zhang J, Zhu M, Li F, Song Y, Qin J, Liang H (2017) Crystal structure of *Pseudomonas aeruginosa* RsaL bound to promoter DNA reaffirms its role as a global regulator involved in quorum-sensing. Nucleic Acids Res 45:699–710. https://doi.org/10.1093/nar/gkw954

Kay E, Humair B, Denervaud V, Riedel K, Spahr S, Eberl L, Valverde C, Haas D (2006) Two GacA-dependent small RNAs modulate the quorum-sensing response in *Pseudomonas aeruginosa*. J Bacteriol 188:6026–6033. https://doi.org/10.1128/JB.00409-06

Kim C, Kim J, Park HY, Park HJ, Lee JH, Kim CK, Yoon J (2008) Furanone derivatives as quorum-sensing antagonists of *Pseudomonas aeruginosa*. Appl Microbiol Biotechnol 80:37–47. https://doi.org/10.1007/s00253-008-1474-6

Kohler T, Ouertatani-Sakouhi H, Cosson P, van Delden C (2014) QsrO a novel regulator of quorum-sensing and virulence in *Pseudomonas aeruginosa*. PLoS One 9:e87814. https://doi.org/10.1371/journal.pone.0087814

Ledgham F, Soscia C, Chakrabarty A, Lazdunski A, Foglino M (2003a) Global regulation in *Pseudomonas aeruginosa*: the regulatory protein AlgR2 (AlgQ) acts as a modulator of quorum sensing. Res Microbiol 154:207–213. https://doi.org/10.1016/S0923-2508(03)00024-X

Ledgham F, Ventre I, Soscia C, Foglino M, Sturgis JN, Lazdunski A (2003b) Interactions of the quorum sensing regulator QscR: interaction with itself and the other regulators of *Pseudomonas aeruginosa* LasR and RhlR. Mol Microbiol 48:199–210. https://doi.org/10.1046/j.1365-2958.2003.03423.x

Lee JH, Lequette Y, Greenberg EP (2006) Activity of purified QscR, a *Pseudomonas aeruginosa* orphan quorum-sensing transcription factor. Mol Microbiol 59:602–609. https://doi.org/10.1111/j.1365-2958.2005.04960.x

Li LL, Malone JE, Iglewski BH (2007) Regulation of the *Pseudomonas aeruginosa* quorum-sensing regulator VqsR. J Bacteriol 189:4367–4374. https://doi.org/10.1128/JB.00007-07

Liang H, Deng X, Ji Q, Sun F, Shen T, He C (2012) The *Pseudomonas aeruginosa* global regulator VqsR directly inhibits QscR to control quorum-sensing and virulence gene expression. J Bacteriol 194:3098–3108. https://doi.org/10.1128/JB.06679-11

Liang H, Deng X, Li X, Ye Y, Wu M (2014) Molecular mechanisms of master regulator VqsM mediating quorum-sensing and antibiotic resistance in *Pseudomonas aeruginosa*. Nucleic Acids Res 42:10307–10320. https://doi.org/10.1093/nar/gku586

Lyczak JB, Cannon CL, Pier GB (2000) Establishment of *Pseudomonas aeruginosa* infection: lessons from a versatile opportunist. Microbes Infect 2:1051–1060

Lyczak JB, Cannon CL, Pier GB (2002) Lung infections associated with cystic fibrosis. Clin Microbiol Rev 15:194–222. https://doi.org/10.1128/CMR.15.2.194-222.2002

Mattmann ME, Blackwell HE (2010) Small molecules that modulate quorum sensing and control virulence in *Pseudomonas aeruginosa*. J Org Chem 75:6737–6746. https://doi.org/10.1021/jo101237e

McInnis CE, Blackwell HE (2011) Thiolactone modulators of quorum sensing revealed through library design and screening. Bioorg Med Chem 19:4820–4828. https://doi.org/10.1016/j.bmc.2011.06.071

O'Callaghan J, Reen FJ, Adams C, O'Gara F (2011) Low oxygen induces the type III secretion system in *Pseudomonas aeruginosa* via modulation of the small RNAs rsmZ and rsmY. Microbiology 157:3417–3428. https://doi.org/10.1099/mic.0.052050-0

Oinuma K, Greenberg EP (2011) Acyl-homoserine lactone binding to and stability of the orphan *Pseudomonas aeruginosa* quorum-sensing signal receptor QscR. J Bacteriol 193:421–428. https://doi.org/10.1128/JB.01041-10

Parkins MD, Ceri H, Storey DG (2001) *Pseudomonas aeruginosa* GacA, a factor in multihost virulence, is also essential for biofilm formation. Mol Microbiol 40:1215–1226

Pessi G, Williams F, Hindle Z, Heurlier K, Holden MT, Camara M, Haas D, Williams P (2001) The global posttranscriptional regulator RsmA modulates production of virulence determinants and N-acylhomoserine lactones in *Pseudomonas aeruginosa*. J Bacteriol 183:6676–6683. https://doi.org/10.1128/JB.183.22.6676-6683.2001

Pugsley A (2008) The two-component response regulator PprB modulates quorum-sensing signal production and global gene expression in *Pseudomonas aeruginosa*. Mol Microbiol 69:780. https://doi.org/10.1111/j.1365-2958.2008.06351.x

Rampioni G, Bertani I, Zennaro E, Polticelli F, Venturi V, Leoni L (2006) The quorum-sensing negative regulator RsaL of *Pseudomonas aeruginosa* binds to the lasI promoter. J Bacteriol 188:815–819. https://doi.org/10.1128/JB.188.2.815-819.2006

Rampioni G, Schuster M, Greenberg EP, Zennaro E, Leoni L (2009) Contribution of the RsaL global regulator to *Pseudomonas aeruginosa* virulence and biofilm formation. FEMS Microbiol Lett 301:210–217. https://doi.org/10.1111/j.1574-6968.2009.01817.x

Rodrigue A, Quentin Y, Lazdunski A, Mejean V, Foglino M (2000) Two-component systems in *Pseudomonas aeruginosa*: why so many? Trends Microbiol 8:498–504

Rutherford ST, Bassler BL (2012) Bacterial quorum sensing: its role in virulence and possibilities for its control. Cold Spring Harb Perspect Med 2:a0124727. https://doi.org/10.1101/cshperspect.a012427

Schulz S, Eckweiler D, Bielecka A, Nicolai T, Franke R, Dotsch A, Hornischer K, Bruchmann S, Duvel J, Haussler S (2015) Elucidation of sigma factor-associated networks in *Pseudomonas aeruginosa* reveals a modular architecture with limited and function-specific crosstalk. PLoS Pathog 11:e1004744. https://doi.org/10.1371/journal.ppat.1004744

Schuster M, Greenberg EP (2006) A network of networks: quorum-sensing gene regulation in *Pseudomonas aeruginosa*. Int J Med Microbiol 296:73–81. https://doi.org/10.1016/j.ijmm.2006.01.036

Schuster M, Hawkins AC, Harwood CS, Greenberg EP (2004) The *Pseudomonas aeruginosa* RpoS regulon and its relationship to quorum sensing. Mol Microbiol 51:973–985. https://doi.org/10.1046/j.1365-2958.2003.03886.x

Seet Q, Zhang LH (2011) Anti-activator QslA defines the quorum sensing threshold and response in *Pseudomonas aeruginosa*. Mol Microbiol 80:951–965. https://doi.org/10.1111/j.1365-2958.2011.07622.x

Siehnel R, Traxler B, An DD, Parsek MR, Schaefer AL, Singh PK (2010) A unique regulator controls the activation threshold of quorum-regulated genes in *Pseudomonas aeruginosa*. Proc Natl Acad Sci U S A 107:7916–7921. https://doi.org/10.1073/pnas.0908511107

Son MS, Matthews WJ Jr, Kang Y, Nguyen DT, Hoang TT (2007) In vivo evidence of *Pseudomonas aeruginosa* nutrient acquisition and pathogenesis in the lungs of cystic fibrosis patients. Infect Immun 75:5313–5324. https://doi.org/10.1128/IAI.01807-06

Sousa AM, Pereira MO (2014) *Pseudomonas aeruginosa* diversification during infection development in cystic fibrosis lungs – a review. Pathogens 3:680–703. https://doi.org/10.3390/pathogens3030680

Stock AM, Robinson VL, Goudreau PN (2000) Two-component signal transduction. Annu Rev Biochem 69:183–215. https://doi.org/10.1146/annurev.biochem.69.1.183

Stover CK, Pham XQ, Erwin AL, Mizoguchi SD, Warrener P, Hickey MJ, Brinkman FS, Hufnagle WO, Kowalik DJ, Lagrou M, Garber RL, Goltry L, Tolentino E, Westbrock-Wadman S, Yuan Y, Brody LL, Coulter SN, Folger KR, Kas A, Larbig K, Lim R, Smith K, Spencer D, Wong GK, Wu Z, Paulsen IT, Reizer J, Saier MH, Hancock RE, Lory S, Olson MV (2000) Complete genome sequence of *Pseudomonas aeruginosa* PAO1, an opportunistic pathogen. Nature 406:959–964. https://doi.org/10.1038/35023079

Studholme DJ, Buck M (2000) The biology of enhancer-dependent transcriptional regulation in bacteria: insights from genome sequences. FEMS Microbiol Lett 186:1–9

Thompson LS, Webb JS, Rice SA, Kjelleberg S (2003) The alternative sigma factor RpoN regulates the quorum sensing gene rhlI in *Pseudomonas aeruginosa*. FEMS Microbiol Lett 220:187–195. https://doi.org/10.1016/S0378-1097(03)00097-1

Viducic D, Murakami K, Amoh T, Ono T, Miyake Y (2016) RpoN modulates carbapenem tolerance in *Pseudomonas aeruginosa* through *Pseudomonas* quinolone signal and PqsE. Antimicrob Agents Chemother 60:5752–5764. https://doi.org/10.1128/AAC.00260-16

Viducic D, Murakami K, Amoh T, Ono T, Miyake Y (2017) Role of the interplay between quorum sensing regulator VqsR and the *Pseudomonas* quinolone signal in mediating carbapenem tolerance in *Pseudomonas aeruginosa*. Res Microbiol 168:450–460. https://doi.org/10.1016/j.resmic.2017.02.007

Wang D, Seeve C, Pierson LS 3rd, Pierson EA (2013) Transcriptome profiling reveals links between ParS/ParR, MexEF-OprN, and quorum sensing in the regulation of adaptation and virulence in *Pseudomonas aeruginosa*. BMC Genomics 14:618. https://doi.org/10.1186/1471-2164-14-618

Zhao J, Yu X, Zhu M, Kang H, Ma J, Wu M, Gan J, Deng X, Liang H (2016) Structural and molecular mechanism of CdpR involved in quorum-sensing and bacterial virulence in *Pseudomonas aeruginosa*. PLoS Biol 14:e1002449. https://doi.org/10.1371/journal.pbio.1002449

Applications of Serine/Threonine Protein Kinases (STPK): A Bus for Dormancy Exit

18

Bhagwan Rekadwad

Abstract

As the response to unfavorable growth conditions, bacteria transform into the dormant state with the concomitant formation of the specialized dormant forms/structure characterized by low metabolic activity and resistance to hostile conditions. Such dormant cells can be reactivated under the influence of several factors including proteins of such as muropeptides, Resuscitation promoting factor (Rpf) and STPKs family, which possess peptidoglycan hydrolase activity were considered to belong to the group of the autocrine growth factors of the bacteria. Remarkable interest toward Rpf-STPKs family is determined by its participation in resuscitation of the dormant forms of various bacteria and their genes, what in turn into its application in microbial processes and in biotechnology such as breaking bacterial/endospore dormancy, in host pathogen interaction, in depression of neurons, in cell shape control and cell division etc.

18.1 Introduction

The bacterial spores possesses an outer multilayered shell called as coat that protects the bacterial dormant genome during stressful conditions such acid and alkaline environment, low water activity, high or low temperature, in the presence of harmful ions etc. (McKenney et al. 2013). The bacterial endospore composed of dozens of proteins helps them during the period of dormancy and in breaking the dormancy (Boone and Driks 2016; Korza et al. 2016).

B. Rekadwad (✉)
National Centre for Microbial Resource, National Centre for Cell Science, Pune, India

V. C. Kalia (ed.), *Quorum Sensing and its Biotechnological Applications*,
https://doi.org/10.1007/978-981-13-0848-2_18

It is an exceptionally fascinating practical component of bacterial endospore is that spores are metabolically inert/lethargic however have high imperviousness to the unforgiving condition can break torpidity with extraordinary move inside as moment of detecting fitting signs through empowering little particles which might be protein, chemical, sugar or amino acid and change over theirs into to utilitarian and functional vegetative cell with intrinsic characters.

Number of molecules either inherent or exists in surrounding environments such as muropeptides, STPKs, lipooxygenases, siderophores through signalling pathways break the dormancy of spores and allow them to develop into dividing cell with full native characters (Fischer et al. 2017; Grandchamp et al. 2017; Donato et al. 2017). The microbial communities involved signalling pathways/quorum sensing has ability to complete complicated tasks and communicate among and between the members of communities regardless of their active or inactive forms (Kalia et al. 2015; Koul and Kalia 2017). The present chapter describes the applications of such small molecules having applications in microbial processes and in biotechnology.

18.2 Applications of Bacterial Serine/Threonine Protein Kinases (STPK): Sending Wake Up Signal to Break Bacterial Endospore Dormancy

Germination of bacterial endospores (e.g. in *Bacillus*) triggered by the impacted by nutrients when passes through the inner side of spore membrane and by sending membrane binding protein kinases. Muropeptides are found in environment as well as in the bacterial cell wall (Dworkin and Shah 2010; Boudreau et al. 2012). In bacteria, muropeptides are normally released during growth of bacteria i.e. during cell division or doubling. It is required in less than pictogram per millilitres quantity. Normally, in the environment, muropeptides found in environment at high concentration. This will trigger the spore germination. Therefore, Gram-positive bacteria spore are germinated in the environment through the signal produced by the muropeptide may be initiated by unique phenomenon quorum sensing (Fatima et al. 2010; Vidal et al. 2011; Shukla et al. 2014; Kalia and Kumar 2015; Siddiqui et al. 2015; Scott and Hasty 2016; Rekadwad and Khobragade 2017a). Not only muropeptide, amino acid such as *meso*-diaminopimelate (Dpm) ineffectively elicits spore germination in the *Bacillus subtilis* (Setlow 2006, 2008; Shah et al. 2008). Like muropeptides, bryostatin of the PrkC family is a well activator of STPK. There are many clues given by the scientist worldwide on germination of bacterial endospores. Germination of endospores added a new dimension and more study needs to be carried out in this context i.e. pathways exist for germination of spores (Fig. 18.1) by muropeptides and SPTK (Pompeo et al. 2016; Nikitushkin et al. 2016; Arora et al. 2017).

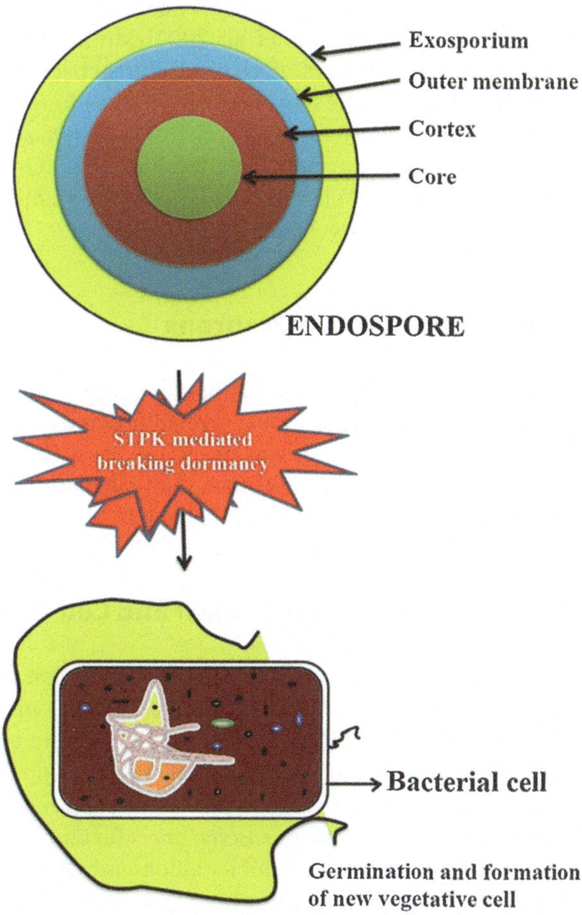

Fig. 18.1 SPTK mediated breaking of bacterial endospore dormancy

18.3 Role of STPK in Host Pathogen Interaction

In host pathogen interaction bacterial adaptation to the host environment and disruption in host immune system and its responses are crucial. This type of host pathogen interactions mediated by quorum sensing i.e. through the sensor or signalling molecules produced protein molecules belonging to STPK family. It seems to be a molecular quorum sensing (Kobir et al. 2011; Cousin et al. 2013; Bazire and Dufour 2014; Kalia 2014, 2015; Kaur et al. 2015; Pooja et al. 2015; Koul et al. 2016; Koul and Kalia 2017; Kalia et al. 2017; Ray and Kalia 2017; Rekadwad and Khobragade 2017a, b). These protein/enzymes produced have dual roles viz. it sense the environment and to weaken the specific host immune system/processes involved in defence mechanism of hosts immune system (Echenique et al. 2004; Hussain et al. 2006; Alber 2009; Kristich et al. 2007; Molle and Kremer 2010; Canova and Molle 2014). STPK protein produced by different bacterial species having eukaryotic STPK like characteristic exist in genus such as *Enterococcus faecalis, Mycobacterium, Listeria monocytogenes, Pseudomonas aeruginosa, Staphylococcus aureus, Streptococcus* spp. and *Yersinia* spp. (Greenstein et al. 2005; Truong-Bolduc et al. 2008; Wehenkel et al. 2008; Beltramini et al. 2009; Truong-Bolduc and Hooper 2010; Ulrych et al. 2016)

18.4 Role STPK in Depression of Neurons

The adenylate cyclase inhibitor (ACI) is inhibitor of large group of kinases (Makhnovskii et al. 2011). In common snail, the reversible depression is linked with activation of variety of STPK enzymes. It was observed that it follows the mathematical model confirmed from experimental data and calculation from obtained results (Makhnovskii et al. 2013; Pivovarov et al. 2014a, b).

18.5 Role of STPKs in Cell Shape Control and Cell Division

Intensive genome examination of *Corynebacterium glutamicum* uncovered the nearness of four putative qualities (PknA, PknB, PknG, and PknL) encoding STPKs, all of which have a place with the PKN2 group of prokaryotic protein kinases (PKs) that are most firmly identified with the eukaryotic STPKs (Goldová et al. 2011). These have role in control of cell shape and synthesis of peptidoglycan amid cell division. Above four genes have different functions; one's functions rely on other. It was proved that PknA, PknB and PknL exhibits autokinase activity while PknG assumes vital part in falls of phosphorylation whose systems completely depend on PknA action. The confinement of phosphor acceptors differs from one kinases to the next. If there should be an occurrence of *Corynebacterium glutamicum* impedance either in PknG or PknL brings about creation of reasonable mutants having a model morphotypes and shifted development rate. Some restrictive mutants contain halfway exhaustion of PknA and PknB qualities. As consequences of quality expression cells were lengthened i.e. there was deformity in cell division (Wehenkel et al. 2006; Fiuza et al. 2008; Margolin 2009; Molle and Kremer 2010). Be that as it may, over articulation of said PknA and PknB qualities stops apical development on bar formed bacterium which mightily empower cell to take a coccoid-like morphology (Donovan and Bramkamp 2014; Monteiro et al. 2015; Zhou et al. 2016; Kysela et al. 2016; Errington 2017). More work should be completed to see how the predetermined number of kinases perceives a vital number of substrates and how they take an interest in numerous mind boggling flagging pathways.

References

Alber T (2009) Signaling mechanisms of the *Mycobacterium tuberculosis* receptor Ser/Thr protein kinases. Curr Opin Struct Biol 19:650–657. https://doi.org/10.1016/j.sbi.2009.10.017

Arora G, Sajid A, Virmani R, Singhal A, Santosh Kumar CM, Dhasmana N, Khanna T, Maji A, Misra R, Molle V, Becher D, Gerth U, Mande SC, Singh Y (2017) Ser/Thr protein kinase PrkC-mediated regulation of GroEL is critical for biofilm formation in *Bacillus anthracis*. Biofilms Microbiomes 3:7. https://doi.org/10.1038/s41522-017-0015-4

Bazire A, Dufour A (2014) The *Pseudomonas aeruginosa* rhlG and rhlAB genes are inversely regulated and RhlG is not required for rhamnolipid synthesis. BMC Microbiol 14:160. https://doi.org/10.1186/1471-2180-14-160

Beltramini AM, Mukhopadhyay CD, Pancholi V (2009) Modulation of cell wall structure and antimicrobial susceptibility by a *Staphylococcus aureus* eukaryote-like serine/threonine kinase and phosphatase. Infect Immun 77:1406–1416. https://doi.org/10.1128/IAI.01499-08

Boone T, Driks A (2016) Protein synthesis during germination: shedding new light on a classical question. J Bacteriol 198:3251–3253. https://doi.org/10.1128/JB.00721-16

Boudreau MA, Fisher JF, Mobashery S (2012) Messenger functions of the bacterial cell wall-derived muropeptides. Biochemistry 51:2974–2990. https://doi.org/10.1021/bi300174x

Canova MJ, Molle V (2014) Bacterial serine/threonine protein kinases in host-pathogen interactions. J Biol Chem 289:9473–9479. https://doi.org/10.1074/jbc.R113.529917

Cousin C, Derouiche A, Shi L, Pagot Y, Poncet S, Mijakovic I (2013) Protein-serine/threonine/tyrosine kinases in bacterial signalling and regulation. FEMS Microbiol Lett 346:11–19. https://doi.org/10.1111/1574-6968.12189

Donato V, Ayala FR, Cogliati S, Bauman C, Costa GB, Leñini C, Grau R (2017) *Bacillus subtilis* biofilm extends *Caenorhabditis elegans* longevity through down regulation of the insulin-like signalling pathway. Nat Comm 8:14332. https://doi.org/10.1038/ncomms14332

Donovan C, Bramkamp M (2014) Cell division in Corynebacterineae. Front Microbiol 5:132. https://doi.org/10.3389/fmicb.2014.00132

Dworkin J, Shah IM (2010) Exit from dormancy in microbial organisms. Nat Rev Microbiol 8:890–896. https://doi.org/10.1038/nrmicro2453

Echenique J, Kadioglu A, Romao S, Andrew PW, Trombe MC (2004) Protein serine/threonine kinase StkP positively controls virulence and competence in *Streptococcus pneumoniae*. Infect Immun 72:2434–2437. https://doi.org/10.1128/IAI.72.4.2434-2437.2004

Errington J (2017) Cell wall-deficient, L-form bacteria in the 21st century: a personal perspective. Biochem Soc Trans 45:287–295. https://doi.org/10.1042/BST20160435

Fatima Q, Zahin M, Khan MSA, Ahemad I (2010) Modulation of quorum sensing controlled behaviour of bacteria by growing seedling, seed and seedling extracts of leguminous plants. Indian J Microbiol 50:238–242. https://doi.org/10.1007/s12088-010-0025-x

Fischer GJ, Bacon W, Yang J, Palmer JM, Dagenais T, Hammock BD, Keller NP (2017) Lipoxygenase activity accelerates programmed spore germination in *Aspergillus fumigates*. Front Microbiol 8:831. https://doi.org/10.3389/fmicb.2017.00831

Fiuza M, Canova MJ, Zanella-Cle'on I, Becchi M, Cozzone J, Mateos LM, Kremer L, Gil JA, Molle V (2008) From the characterization of the four serine/threonine protein kinases (PknA/B/G/L) of *Corynebacterium glutamicum* toward the role of PknA and PknB in cell division. J Biol Chem 283:18099–18112. https://doi.org/10.1074/jbc.M802615200

Goldová J, Ulrych A, Hercík K, Branny P (2011) A eukaryotic-type signalling system of Pseudomonas aeruginosa contributes to oxidative stress resistance, intracellular survival and virulence. BMC Genomics 12:437. https://doi.org/10.1186/1471-2164-12-437

Grandchamp GM, Caro L, Shank EA (2017) Pirated siderophores promote sporulation in *Bacillus subtilis*. Appl Environ Microbiol 83:e03293–e03216. https://doi.org/10.1128/AEM.03293-16

Greenstein AE, Grundner C, Echols N, Gay LM, Lombana TN, Miecskowski CA, Pullen KE, Sung PY, Alber T (2005) Structure/function studies of Ser/Thr and Tyr protein phosphorylation in *Mycobacterium tuberculosis*. J Mol Microbiol Biotechnol 9:167–181. https://doi.org/10.1159/000089645

Hussain H, Branny P, Allan E (2006) A eukaryotic-type serine/threonine protein kinase is required for biofilm formation, genetic competence, and acid resistance in *Streptococcus mutans*. J Bacteriol 188:1628–1632. https://doi.org/10.1128/JB.188.4.1628-1632.2006

Kalia VC (2014) Microbes, antimicrobials and resistance: the battle goes on. Indian J Microbiol 54:1–2. https://doi.org/10.1007/s12088-013-0443-7

Kalia VC (2015) Microbes: the most friendly beings? In: Kalia VC (ed) Quorum sensing vs quorum quenching: a battle with no end in sight. Springer, India, pp 1–5. https://doi.org/10.1007/978-81-322-1982-8_1

Kalia VC, Kumar P (2015) Genome wide search for biomarkers to diagnose *Yersinia* infections. Indian J Microbiol 55:366. https://doi.org/10.1007/s12088-015-0552-6

Kalia VC, Kumar P, Pandian STK, Sharma P (2015) Biofouling control by quorum quenching. Hb_25. In: Kim SK (ed) Springer handbook of marine biotechnology, vol 15. Springer, Berlin/Heidelberg, pp 431–440

Kalia VC, Prakash J, Koul S, Ray S (2017) Simple and rapid method for detecting biofilm forming bacteria. Indian J Microbiol 57:109–111. https://doi.org/10.1007/s12088-016-0616-2

Kaur G, Rajesh S, Princy SA (2015) Plausible drug targets in the *Streptococcus* mutans quorum sensing pathways to combat dental biofilms and associated risks. Indian J Microbiol 55:349. https://doi.org/10.1007/s12088-015-0534-8

Kobir A, Shi L, Boskovic A, Grangeasse C, Franjevic D, Mijakovic I (2011) Protein phosphorylation in bacterial signal transduction. Biochim Biophys Acta 18:989–994. https://doi.org/10.1016/j.bbagen.2011.01.006

Korza G, Setlow B, Rao L, Li Q, Setlow P (2016) Changes in Bacillus spore small molecules, rRNA, germination, and outgrowth after extended sublethal exposure to various temperatures: evidence that protein synthesis is not essential for spore germination. J Bacteriol 198:3254–3264. https://doi.org/10.1128/JB.00583-16

Koul S, Prakash J, Mishra A, Kalia VC (2016) Potential emergence of multi-quorum sensing inhibitor resistant (MQSIR) bacteria. Indian J Microbiol 56:1–18. https://doi.org/10.1007/s12088-015-0558-0

Koul S, Kalia VC (2017) Multiplicity of quorum quenching enzymes: a potential mechanism to limit quorum sensing bacterial population. Indian J Microbiol 57:100–108. https://doi.org/10.1007/s12088-016-0633-1

Kristich CJ, Wells CL, Dunny GM (2007) A eukaryotic-type Ser/Thr kinase in *Enterococcus faecalis* mediates antimicrobial resistance and intestinal persistence. Proc Natl Acad Sci U S A 104:3508–3513. https://doi.org/10.1073/pnas.0608742104

Kysela DT, Randich AM, Caccamo PD, Brun YV (2016) Diversity takes shape: understanding the mechanistic and adaptive basis of bacterial morphology. PLoS Biol 14:e1002565. https://doi.org/10.1371/journal.pbio.1002565

Makhnovskii DA, Tretyakova MS, Murzina GB, Pivovarov AS (2011) Endocytosis of cholinoreceptors in the mechanism of depression of the cholinosensitivity of common snail neurons in a cellular model of habituation. Neurosci Behav Physiol 41:617. https://doi.org/10.1007/s11055-011-9464-z

Makhnovskii DA, Murzina GB, Tretyakova MS, Pivovarov AS (2013) The role of serine/threonine and tyrosine protein kinases in depression of the cholinosensitivity of neurons in the common snail in a cellular analog of habituation. Neurosci Behav Physiol 43:22. https://doi.org/10.1007/s11055-012-9686-8

Margolin W (2009) Sculpting the bacterial cell. Curr Biol 19:R812–R822. https://doi.org/10.1016/j.cub.2009.06.033

McKenney PT, Driks A, Eichenberger P (2013) The *Bacillus subtilis* endospore: assembly and functions of the multilayered coat. Nat Rev Microbiol 11:33–44. https://doi.org/10.1038/nrmicro2921

Molle V, Kremer L (2010) Division and cell envelope regulation by Ser/Thr phosphorylation: *Mycobacterium* shows the way. Mol Microbiol 75:1064–1077. https://doi.org/10.1111/j.1365-2958.2009.07041.x

Monteiro JM, Fernandes PB, Vaz F, Pereira A, Tavares AC, Ferreira MT, Pereira PM, Veiga H, Kuru E, VanNieuwenhze MS, Brun YV, Filipe SR, Pinhoa MG (2015) Cell shape dynamics during the staphylococcal cell cycle. Nat Commun 6:8055. https://doi.org/10.1038/ncomms9055

Nikitushkin VD, Demina GR, Kaprelyants AS (2016) Rpf Proteins are the factors of reactivation of the dormant forms of actinobacteria. Biochemist 81:1719–1734. https://doi.org/10.1134/S0006297916130095

Pivovarov AS, Murzina GB, Tret'yakova MS, Makhnovskii DA (2014a) The role of serine/threonine and tyrosine protein phosphatases in common snail command neurons in a cellular analog of habituation. Neurosci Behav Physiol 44:640. https://doi.org/10.1007/s11055-014-9964-8

Pivovarov AS, Murzina GB, Makhnovsky DA, Vasil'eva NA, Tret'yakova MS (2014b) The role of myosins in depression of neuron sensitivity to acetylcholine in a cellular analog of habituation in the common snail. Neurosci Behav Physiol 44:1039. https://doi.org/10.1007/s11055-014-0021-4

Pompeo F, Foulquier E, Galinier A (2016) Impact of serine/threonine protein kinases on the regulation of sporulation in *Bacillus subtilis*. Front Microbiol 7:568. https://doi.org/10.3389/fmicb.2016.00568

Pooja S, Pushpananthan M, Jayashree S, Gunasekaran P, Rajendhran J (2015) Identification of periplasmic a-amlyase from cow dung metagenome by product induced gene expression profiling (Pigex). Indian J Microbiol 55:57–65. https://doi.org/10.1007/s12088-014-0487-3

Ray S, Kalia VC (2017) Biomedical applications of polyhydroxyalkanaotes. Indian J Microbiol 57:39–47. https://doi.org/10.1007/s12088-016-0651-7

Rekadwad BN, Khobragade CN (2017a) Bacterial quorum sensing (QS) in rhizosphere (paddy soil): understanding soil signaling and N- recycling for increased crop production. In: Kalia VC (ed) Microbial applications vol.2 – biomedicine, agriculture and industry. Springer International Publishing, Switzerland., AG, Springer Nature, pp 119–131. https://doi.org/10.1007/978-3-319-52669-0_6

Rekadwad BN, Khobragade CN (2017b) Microbial biofilm: role in crop productivity. In: KALIA VC (ed) Microbial applications vol.2 – biomedicine, agriculture and industry. Springer International Publishing, Switzerland., AG, Springer Nature, pp 107–118. https://doi.org/10.1007/978-3-319-52669-0_5

Scott SR, Hasty J (2016) Quorum sensing communication modules for microbial consortia. ACS Synth Biol 5:969–977. https://doi.org/10.1021/acssynbio.5b00286

Setlow P (2006) Spores of *Bacillus subtilis*: their resistance to and killing by radiation, heat and chemicals. J Applied Microbiol 101:514–525. https://doi.org/10.1111/j.1365-2672.2005.02736.x

Setlow P (2008) Dormant spores receive an unexpected wake-up call. Cell 135:410–412. https://doi.org/10.1016/j.cell.2008.10.006

Shah IM, Laaberki MH, Popham DL, Dworkin J (2008) A eukaryotic-like Ser/Thr kinase signals bacteria to exit dormancy in response to peptidoglycan fragments. Cell 135:486–496. https://doi.org/10.1016/j.cell.2008.08.039

Shukla SK, Mangwani N, Rao TS, Das S (2014) Biofilm-mediated bioremediation of polycyclic aromatic hydrocarbons. In: Das S (ed) Microbial biodegradation and bioremediation. Elsevier, Oxford, pp 203–232. https://doi.org/10.1016/B978-0-12-800021-2.00008-X

Siddiqui MF, Rzechowicz M, Harvey W, Zularisam AW, Anthony GF (2015) Quorum sensing based membrane biofouling control for water treatment: a review. J Water Proc Eng 30:112–122. https://doi.org/10.1016/j.jwpe.2015.06.003

Truong-Bolduc QC, Hooper DC (2010) Phosphorylation of MgrA and its effect on expression of the NorA and NorB efflux pumps of *Staphylococcus aureus*. J Bacteriol 192:2525–2534. https://doi.org/10.1128/JB.00018-10

Truong-Bolduc QC, Ding Y, Hooper DC (2008) Posttranslational modification influences the effects of MgrA on *norA* expression in *Staphylococcus aureus*. J Bacteriol 190:7375–7381. https://doi.org/10.1128/JB.01068-08

Ulrych A, Holečková N, Goldová J, Doubravová L, Benada O, Kofroňová O, Halada P, Branny P (2016) Characterization of pneumococcal Ser/Thr protein phosphatase phpP mutant and identification of a novel PhpP substrate, putative RNA binding protein Jag. BMC Microbiol 16:247. https://doi.org/10.1186/s12866-016-0865-6

Vidal JE, Ludewick HP, Kunkel RM, Zähner D, Klugman KP (2011) The LuxS-dependent quorum-sensing system regulates early biofilm formation by *Streptococcus pneumoniae* strain D39. Infect Immun 79:4050–4060. https://doi.org/10.1128/IAI.05186-11

Wehenkel A, Fernandez P, Bellinzoni M, Catherinot V, Barilone N, Labesse G, Jackson M, Alzari PM (2006) The structure of PknB in complex with mitoxantrone, an ATP-competitive inhibitor, suggests a mode of protein kinase regulation in mycobacteria. FEBS Lett 580:3018–3022. https://doi.org/10.1016/j.febslet.2006.04.046

Wehenkel A, Bellinzoni M, Graña M, Duran R, Villarino A, Fernandez P, Andre-Leroux G, England P, Takiff H, Cerveñansky C, Cole ST, Alzari PM (2008) Mycobacterial Ser/Thr protein kinases and phosphatases: physiological roles and therapeutic potential. Biochim Biophys Acta 1784:193–202. https://doi.org/10.1016/j.bbapap.2007.08.006

Zhou X, Halladin DK, Theriot JA (2016) Fast mechanically driven daughter cell separation is widespread in actinobacteria. mBio 7:4e00952–4e00916. https://doi.org/10.1128/mBio.00952-16

The manufacturer's authorised representative in the EU is Springer Nature Customer Service Centre GmbH, Europaplatz 3, 69115 Heidelberg, Germany. If you have any concerns regarding our products, please contact ProductSafety@springernature.com

Printed and bound by CPI Group (UK) Ltd, Croydon, CR0 4YY

30/07/2026

02181664-0002